T0291614

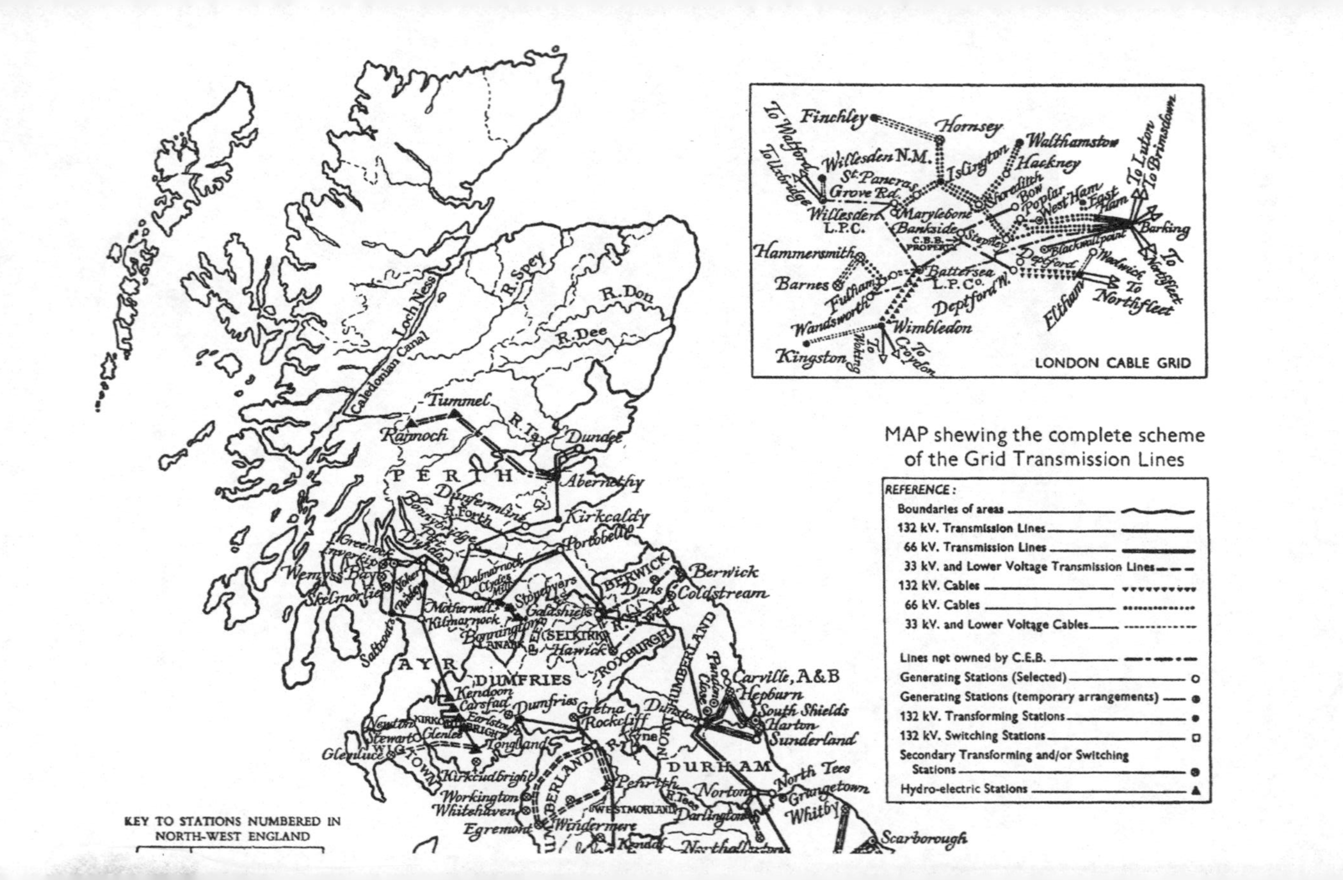
MAP shewing the complete scheme of the Grid Transmission Lines
REFERENCE:
Boundaries of areas
132 kV. Transmission Lines
66 kV. Transmission Lines
33 kV. and Lower Voltage Transmission Lines
132 kV. Cables
66 kV. Cables
33 kV. and Lower Voltage Cables
Lines not owned by C.E.B.
Generating Stations (Selected)
Generating Stations (temporary arrangements)
132 kV. Transforming Stations
132 kV. Switching Stations
Secondary Transforming and/or Switching Stations
Hydro-electric Stations
LONDON CABLE GRID
Finchley
Hornsey
Walthamstow
To Watford
To Uxbridge
Willesden N.M.
St Pancras
Islington
Hackney
Grove Rd.
Willesden L.P.C.
Marylebone
Shoreditch
Bow
Poplar
West Ham
East Ham
To Luton
To Brimsdown
Barking
Bankside
C.E.B. PROPERTY
Stepney
Blackwallpoint
Hammersmith
Deptford
Woolwich
To Northfleet
Barnes
Fulham
Battersea L.P.Co.
Deptford W.
Wandsworth
Wimbledon
Eltham
Kingston
To Woking
To Croydon
Loch Ness
Caledonian Canal
R. Spey
R. Don
R. Dee
Tummel
Rannoch
R. Tay
Dundee
PERTH
Abernethy
Dunfermline
R. Forth
Bonnybridge
Kirkcaldy
Greenock
Inverkip
Port Dundas
Portobello
Wemyss Bay
Dalmarnock
Clyde Mill
Skelmorlie
Yoker
Paisley
Motherwell
Kilmarnock
Stobo
BERWICK
Duns
Berwick
Coldstream
Galashiels
R. Tweed
Saltcoats
Bonnington
(SELKIRK)
Hawick
ROXBURGH
AYR
DUMFRIES
Kendoon
Carsfad
Earlstoun
Dumfries
Gretna
Rockcliff
Dunston
NORTHUMBERLAND
Pandon Close
Carville, A&B
Hepburn
South Shields
Harton
Sunderland
Newton Stewart
Glenlee
Tongland
Glenluce
WIGTOWN
Kirkcudbright
R. Tyne
DURHAM
North Tees
Grangetown
Workington
Whitehaven
Penrith
R. Tees
Norton
Whitby
WESTMORLAND
Darlington
Egremont
Windermere
Kendal
Scarborough
KEY TO STATIONS NUMBERED IN NORTH-WEST ENGLAND

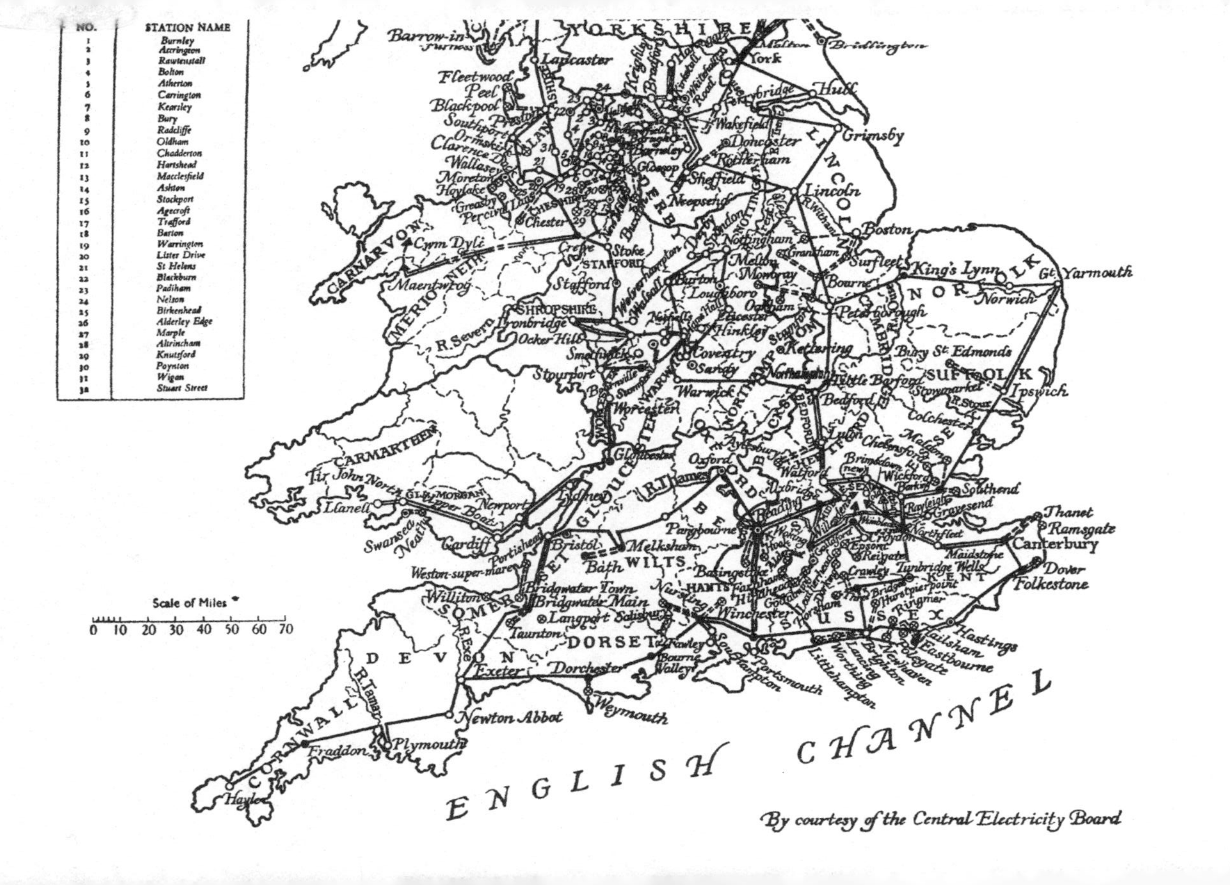

NO.	STATION NAME
1	Burnley
2	Accrington
3	Rawtenstall
4	Bolton
5	Atherton
6	Carrington
7	Kearsley
8	Bury
9	Radcliffe
10	Oldham
11	Chadderton
12	Hartshead
13	Macclesfield
14	Ashton
15	Stockport
16	Agecroft
17	Trafford
18	Barton
19	Warrington
20	Lister Drive
21	St Helens
22	Blackburn
23	Padiham
24	Nelson
25	Birkenhead
26	Alderley Edge
27	Marple
28	Altrincham
29	Knutsford
30	Poynton
31	Wigan
32	Stuart Street

By courtesy of the Central Electricity Board

MAGNETISM AND ELECTRICITY

By courtesy of the Metropolitan-Vickers Electrical Co. Ltd.

A million volt power frequency arc.

MAGNETISM AND ELECTRICITY

by

A. E. E. McKENZIE, M.A.

Trinity College, Cambridge

'The beauty of electricity...is not that the power is mysterious or unexpected... but that it is under law...."

MICHAEL FARADAY

CAMBRIDGE

AT THE UNIVERSITY PRESS

1952

CAMBRIDGE UNIVERSITY PRESS
Cambridge, New York, Melbourne, Madrid, Cape Town,
Singapore, São Paulo, Delhi, Mexico City

Cambridge University Press
The Edinburgh Building, Cambridge CB2 8RU, UK

Published in the United States of America by Cambridge University Press, New York

www.cambridge.org
Information on this title: www.cambridge.org/9781107622678

First edition 1938
First published 1941
Reprinted, with corrections 1941, 1942, 1944, 1945, 1947, 1948, 1952
First paperback edition 2013

A catalogue record for this publication is available from the British Library

ISBN 978-1-107-62267-8 Paperback

CONTENTS

PREFACE

This volume completes my series of school certificate physics text-books.

It is, like the others, a learning rather than a teaching manual —a readable book for the boy. Some historical detail has been included to give an impression of the way science grows, and to stress the cultural rather than the technical aspect of the subject.

Magnetism and electricity lends itself, perhaps more than any other branch of elementary physics, to the exposition of the role of hypothesis and theory in the progress of science: Faraday's theory of lines of force, the molecular theory of magnetism, the two-fluid, one-fluid, and electron theories of electricity, the ionic theory, theories of terrestrial magnetism, the disintegration theory of Rutherford and Soddy—most of these have a fascinating historical background of controversy and development.

The order of presentation of the subject calls, perhaps, for brief comment. The bulk of electrostatics is postponed till a late chapter but, in the interest of logical development, enough electrostatics (as far as induction) is presented at the outset to justify the use of the terms positive and negative and to give meaning to the conception of current as a flow of electrons. Magnetometry and the tangent galvanometer are also treated near the end of the book. Ohm's law, in view of its fundamental importance, is given at the beginning of current electricity, omitting at this stage the drop in potential difference of a cell on closed circuit.

Wireless telegraphy has been omitted, owing to the exigency of space. I have, however, devoted a chapter to signalling along wires, and another to the transmission of electricity and the Grid.

I am very much indebted to my former colleague, Mr R. E. Williams, who has read the manuscript and made numerous suggestions, and also to my pupil Mr J. W. G. Porter, who has worked out the answers to the examples.

My explanation of the action of a simple cell was suggested by Prof. W. L. Bragg's book, *Electricity*.

My thanks are due to Mr B. F. Brown, who has taken for me, in the Repton laboratories, photographs of filing fields of magnets and currents, the electric arc deflected by a magnet, cathode rays casting a shadow, the Wimshurst machine, etc.

Dr R. J. Van de Graaff, of the Massachusetts Institute of Technology, kindly sent me photographs of his electrostatic generator, and the Copper Development Association obtained for me a photograph to illustrate the refining of copper from the Ontario Refining Company. Messrs Geo. Newnes, Ltd. have given me permission to reproduce certain illustrations, Figs. 54, 56, 134, 180, 287, 288 and 289, from their publication, *The Principles of Electrical Engineering*.

The following also have assisted me by providing photographs and information: Professor Carl Størmer, the General Electric Co. Ltd., the British Thomson-Houston Co. Ltd., the Metropolitan-Vickers Electrical Co. Ltd., the English Electric Co. Ltd., Messrs Ferranti Ltd., the Central Electricity Board, the Western Union Telegraph Co., the Cambridge Instrument Co. Ltd., the American Telephone and Telegraph Co., the G.P.O., Messrs Kelvin, Bottomley and Baird Ltd., the Westinghouse Brake and Saxby Signal Co. Ltd., the Igranic Electric Co. Ltd., Messrs Crompton, Parkinson Ltd., Messrs Siemens Bros. and Co. Ltd., the Royal Institution, the Science Museum, Messrs W. Canning and Co. Ltd., the London Power Co. Ltd., the Royal Meteorological Society, the Southern Railway, the Director of the National Portrait Gallery, Prof. Blackett and the Royal Society, the Editor of *The Welder*, Messrs Watson and Sons (Electrical Medical Ltd.), Messrs Williams and Wilkins Co. of Baltimore, U.S.A., the Tella Co. Ltd., and Messrs Imperial Airways.

I have taken considerable care in devising and collecting problems and questions, as I regard these as a most important feature of a science text-book. I must express my thanks to the following Examining Bodies for permission to reproduce School Certificate questions: the Oxford and Cambridge Joint Board (O. and C.), the Northern Universities Joint Matriculation Board (N.), the University of London (L.), the Cambridge Local (C.) and the Oxford Local (O.) Examination Syndicates. The letters in brackets will be found printed after the questions to designate their source.

A. E. E. M.

Repton, November 1937

In this edition the term "resistivity" is used in preference to "specific resistance", and "oersted" instead of "gauss" as the unit of magnetic field strength.

July 1944

MAGNETISM & ELECTRICITY

Chapter I

MAGNETISM

It has been known from very early times that a certain stone, found in various parts of the earth, possesses the property of attracting iron. The stone is a black ore of iron, Fe_3O_4, and is known as *magnetite*, since it was found by the ancients in Magnesia, Asia Minor.

The phenomenon forms the basis of the strange (and incredible) story of the Black Mountain in the History of the Third Calender, The Arabian Nights Entertainments. This mountain was covered on its seaward side with the nails of ships, as a result of its terrible magnetic power. "About noon we were come so near that we found what the pilot had foretold to be true: for we saw all the nails and iron about the ships fly towards the mountain, where they fixed, by the violence of the attraction, with a horrible noise."

Magnetite possesses the further property that when suspended freely, it always sets itself in a particular direction. The Chinese, earlier than 2500 B.C., used it as a primitive compass: hence it is known as *lodestone* or leading stone.

A bar of iron, when stroked with lodestone, acquires the same properties. It will attract other pieces of iron, and will also set itself N. and S. when freely suspended. Such a bar is called an *artificial magnet*. Lodestone, since it occurs in nature, is called a *natural magnet*.

The attractive power of a magnet was the subject of much speculation by the Greeks. Thales of Miletus attributed it to a soul. During the Middle Ages gross and fantastic superstitions arose about the magnet's power. When held in the hand it was said to be a cure for gout. In the presence of garlic it was said to lose its power, which goats' blood would restore. Such were the current beliefs when this mysterious phenomenon was first subjected to scientific scrutiny by Dr Gilbert, "the father of magnetism".

Gilbert was court physician to Queen Elizabeth. His great work *On the Magnet and Magnetic Bodies and on the Great Magnet the Earth*, written in Latin, and published in 1600, was the first important scientific book to appear in England.

Gilbert's initial task was to expose, by experiment, the falsity of most of the statements made about magnets before his time. For example, Porta had asserted that iron rubbed by diamonds becomes magnetised and turns to the north. "We made the experiment ourselves", wrote Gilbert, "with 75 diamonds in presence of many witnesses employing a number of iron bars and pieces of wire, manipulating them with the greatest care while they floated on water supported by corks, but never was it granted me to see the effect mentioned by Porta."

A knowledge of the properties of magnets can best be obtained by following Gilbert's lead, and performing a few simple experiments. Indeed, the secret of Gilbert's greatness is revealed in the dedication of his book: "To you alone, true philosophers, ingenious minds, *who not only in books, but in things themselves look for knowledge.*"

Experiments with bar magnets.

1. A bar magnet will attract only iron and steel strongly. Most other substances—copper and silver coins, for example—it will not attract at all. (It attracts two other metals, nickel and cobalt, slightly.) A substance which is attracted by a magnet is said to be *magnetic.*

2. If a bar magnet is dipped into iron filings, the filings will be found to adhere mainly to the ends (see Fig. 1). The magnetism of the magnet, therefore, appears to be concentrated in regions at its ends, and these regions are termed the *poles* of the magnet. It is found that *poles always occur in pairs.*

Fig. 1

3. A bar magnet will settle down with its axis roughly N. and S. if it is suspended freely. A convenient method of suspending it is to place it in a stirrup of copper wire, hanging from a thread of unspun silk, which does not strongly resist being twisted. The pole of the magnet pointing N. is called the *N.-seeking pole,*

and the other the *S.-seeking pole*. These are generally referred to as the N. pole and the S. pole of the magnet respectively.

4. If the N. pole of another magnet is brought near to the N. pole of a magnet suspended as just described (see Fig. 2), the suspended N. pole will tend to move away, or, as Gilbert would say, "to be put to flight". Thus two N. poles repel. Similarly, it can be shown that two S. poles repel, whereas N. and S. poles attract. We can sum this up as follows: *like poles repel, unlike poles attract.*

If an unmagnetised iron bar is brought up to a suspended magnet, it will be found to attract both poles. Thus the only sure proof that a bar is magnetised is that it repels one end of a magnet or compass needle.

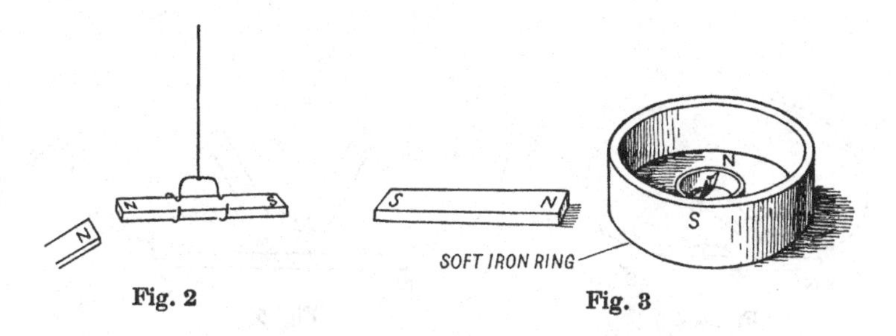

Fig. 2

Fig. 3

5. A magnet will attract iron through paper, wood, copper gauze, etc.—in fact, through any substance except iron. Thus iron can be used to shield a space from a magnet's influence. Expensive watches sometimes have a soft-iron case. Place a compass needle inside a soft-iron ring (see Fig. 3) and bring up a magnet as shown. The needle will be almost completely unaffected: to be completely shielded it would need to be completely surrounded by iron.

Methods of magnetising.

An iron bar can be magnetised by the *method of single touch.* Lay the iron bar to be magnetised on the table and stroke it with one end of a bar magnet. Move the magnet in the path shown in Fig. 4, say twenty times, making a wide return sweep after each stroke along the bar. That end of the bar where each stroke finishes will be found to have opposite polarity to the stroking pole.

If two bar magnets are available, the bar may be magnetised more rapidly by *the method of double or divided touch*. The bar is stroked from its middle outwards by N. and S. poles (see Fig. 5). Where the N. pole finishes its stroke, the bar acquires a S. pole, and vice versa.

Eventually a bar that is continuously stroked does not become any more strongly magnetised. It is then said to be *saturated* with magnetism.

The magnet which is used to do the stroking does not lose its magnetism during the process. Magnetism, however, can be destroyed by subjecting a magnet to rough treatment—dropping

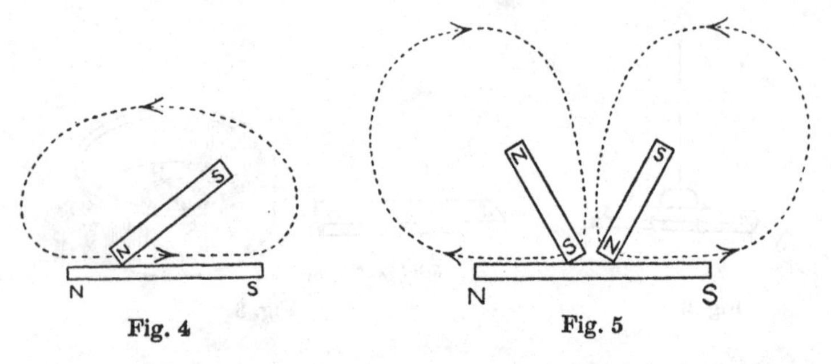

Fig. 4

Fig. 5

it on the floor or hammering. A magnet can be most completely demagnetised by heating it to red heat. Iron becomes non-magnetic at red heat. During the battle of Jutland the funnels of the *Lion* became red hot and therefore non-magnetic, thereby rendering useless the compass, which was corrected for their magnetic effect.

The powerful bar magnets used in the laboratory are not made by stroking with lodestone or artificial magnets. A much more efficient method of magnetising a bar of iron, by means of an electric current, was discovered in the early part of the nineteenth century. We shall describe this method in a later chapter.

Steel (a harder form of iron) is used for permanent magnets in preference to soft iron, since, though more difficult to magnetise, it retains its magnetism very much more tenaciously under rough treatment.

Strong permanent magnets are used widely in magnetos, telephones and electrical measuring instruments. Steel, alloyed with 10 per cent tungsten, has great retentive powers: it is used in magnetos. Steel containing 35 per cent cobalt is also retentive and can be magnetised very strongly; this is used for making bar magnets used in the laboratory. A nickel-iron alloy containing up to 50 per cent Ni is used for transformer cores (see p. 230), since it may be readily and strongly magnetised and, in contrast to the alloys just mentioned, is easily demagnetised.

Theories of magnetism.

When the simple phenomena of magnetism were clearly established the next step was to suggest some explanation of them; in other words, to devise a theory of magnetism. The theory has to explain the following:

(1) the difference between an unmagnetised and a magnetised bar of iron;

(2) how stroking can produce magnetism without any appreciable diminution of the magnetic power of the stroking magnet;

(3) why iron is magnetic whereas other metals such as copper and silver are not;

(4) why magnetism is concentrated at the ends of a magnet;

(5) why magnetic poles always occur in pairs (a N. pole never appears without a corresponding S. pole);

(6) why rough treatment and heat destroy magnetism;

(7) why a bar becomes saturated with magnetism and is incapable of being magnetised beyond a certain limit.

The first theories, put forward during the eighteenth century, were fluid theories. Fluid theories were fashionable among physicists at that time; both heat and electricity were regarded as invisible and weightless fluids. Coulomb imagined that there were two magnetic fluids, one of N.-seeking polarity, and the other of S.-seeking polarity. Aepinus, in a book published in 1759, developed a one-fluid theory, and suggested that at the two poles of a magnet, the concentrations of the fluids were greater and less than normal, respectively.

But grave difficulties confront a fluid theory when applied to magnetism, as will readily be appreciated by the reader if he tries to apply one to explain the facts we have enumerated above.

As a result of these difficulties fluid theories of magnetism gave place to the molecular theory of magnetism, which was propounded by Weber and developed by several others, notably Ewing. The fundamental assumption of this theory is suggested by the following experiment.

Magnetise a steel knitting needle by stroking it with a bar magnet. Break it in the middle with a pair of wire-cutting pliers. Where the needle is broken two new poles appear (see Fig. 6). Test these poles by repulsion with a compass needle. Now take one-half of the knitting needle and break it again. Once more two new poles appear. Every time the needle is broken, two new

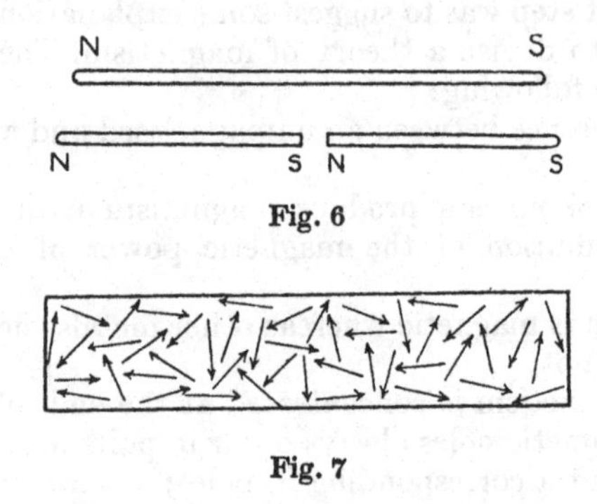

Fig. 6

Fig. 7

poles appear. (The needle must be thin enough to snap without undue force or it becomes demagnetised.)

Suppose it were possible to continue breaking the needle until the ultimate particle of which it is composed—the molecule of iron—were obtained. Would the molecule of iron have a pole at each end? The molecular theory of magnetism assumes that it would.

The fundamental assumption of the theory is, therefore, that *each molecule of iron is a tiny magnet.* The molecules of non-magnetic substances, such as copper and silver, are assumed not to be magnets. In an unmagnetised bar of iron the millions of molecules are pointing at random in all directions (see Fig. 7). They are not arranged, however, in a completely haphazard fashion. The N. pole of a molecule attracts the S. pole of its neighbour, with a result that small groups of molecules, known as closed chains, tend to be formed inside the bar (see Fig. 8). Thus each N. pole is very close to a S. pole, and they neutralise each other's external effect. Hence the bar appears to be unmagnetised.

When the bar is stroked the attraction and repulsion of the stroking pole breaks up these closed chains and the molecules are gradually set in alignment, as shown in Fig. 9. Saturation is reached when all the molecules are in alignment.

The theory explains why the magnetism is concentrated at the ends of the magnet. In the middle of the magnet N. and S.

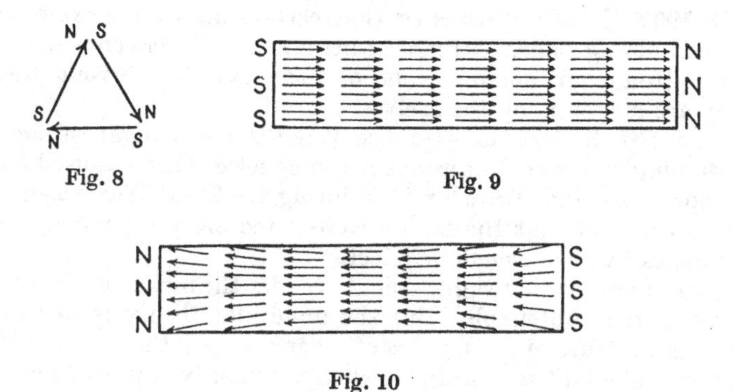

Fig. 8

Fig. 9

Fig. 10

poles are still in juxtaposition and neutralise each other's effects but at the ends there are rows of un-neutralised poles. The fact that a bar magnet's poles occupy a considerable area and are not confined to the extreme ends of the bar is explained by the mutual repulsion of the poles at the ends, which causes the molecules to spread out into the fan-shape shown in Fig. 10.

When a bar magnet is hammered the molecules are shaken out of alignment and hence its magnetism is destroyed. Again, the molecules of a substance are always vibrating slightly (except at the temperature of absolute zero). When a bar is heated the molecules vibrate more violently, until eventually they shake themselves out of alignment. On cooling, they set themselves in closed chains and the bar is demagnetised. The closed chains are stable owing to the attractions of the N. and S. poles, whereas open chains are not.

There is another significant phenomenon which lends some support to this theory. When a bar is magnetised its length increases slightly, by about $\frac{1}{400000}$th of its original length. Its width also decreases slightly. If this were all that occurred, it would be easy to explain. However, on being strongly magnetised, the bar suffers a contraction in length. All that we can say therefore is that the phenomenon points strongly to a rearrangement of the molecules inside the bar.

Modification of the molecular theory of magnetism.

In 1903 Heusler discovered that certain alloys, for example one containing 25 per cent Mn, 14 per cent Al, 61 per cent Cu, are quite strongly magnetic. None of the metals comprising this alloy are magnetic by themselves.

Again, Sir Robert Hadfield discovered that a manganese steel containing 78 per cent of iron is non-magnetic. This was used for the outer casing of British mines during the Great War, when it was discovered that the enemy were detecting the presence of the mines by their magnetic effect.

These facts, at first sight, seem to render the molecular theory of magnetism untenable. But the molecular theory is so convincing a picture of the process of magnetisation that we should be most reluctant to discard it: we must modify it in some way to accord with the new facts.

We have done this by assuming that the elementary magnets in our theory are not the molecules but aggregations of molecules called domains, about $\frac{1}{1000}$th inch in diameter. The directions of the magnetic axes of the domains are normally distributed at random and are brought into alignment during magnetisation. The full theory is too advanced to be given in detail here.

Magnetic induction.

Unmagnetised iron when placed in the vicinity of a magnet becomes weakly and temporarily magnetised. Hold an unmagnetised bar of iron near to a strong magnet (see Fig. 11): the iron will pick up iron filings while the magnet is near but loses this power when the magnet is taken away.

The phenomenon is called *magnetic induction*, and the iron (while the permanent magnet is near) is called an *induced magnet*. The induced poles are shown in Fig. 11: the N. pole of the magnet induces a S. pole in that end of the bar nearest to it.

It is not easy to demonstrate the polarity of the induced poles, owing to the proximity of the much more powerful poles of the permanent magnet. Here is one method, however:

Lay the permanent magnet on a table so that it is lying with its axis N. and S. but its N. pole pointing S. Place a small compass needle to the S. of it. If placed too near, the S. pole of the

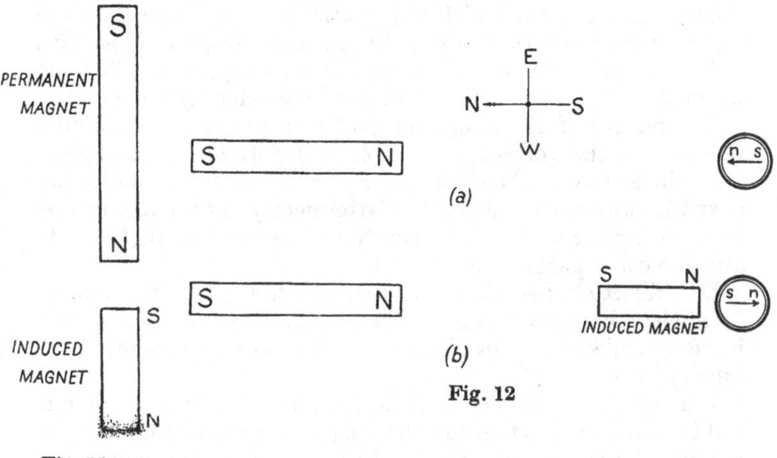

Fig. 11

Fig. 12

compass needle will be attracted by the N. pole of the magnet. Move the compass needle just far enough away so that it is uninfluenced by the magnet and its N. pole points N. (see Fig. 12 *a*). Now put the unmagnetised iron bar between the magnet and the compass (see Fig. 12 *b*). The S. pole of the compass needle will swing round owing to the attraction of the induced pole near to it, showing that the latter is a N. pole.

The attraction of unmagnetised iron by a magnet may be accounted for in terms of induction. As may be seen clearly in Fig. 11, the induced pole nearest to the magnet is opposite in polarity to the nearest pole of the magnet: and unlike poles attract. The attraction between the iron and the magnet is mutual. When both are floated on corks, either will move under this attraction if the other is held fixed.

On the molecular theory of magnetism we can explain magnetic induction as being a temporary alignment of some of the elementary magnets in the induced magnet.

The magnetic field.

Perhaps the most mysterious thing about a magnet is the fact that it can exert a force on a piece of iron or another magnet through empty space. It exerts, all around it, a magnetic influence. Any space in which a magnetic influence can be detected (by a compass needle, for example) is called a *magnetic field*. This is the modern term for what, in Gilbert's day, was called the magnetic "atmosphere" or "effluvia" surrounding a magnet.

The nature of the magnetic field of a magnet is strikingly revealed by the following simple experiment. Lay a large piece of cardboard over a bar magnet. Sprinkle iron filings uniformly over the cardboard and tap the latter gently. The filings will be found to set themselves in the pattern shown in Fig. 13(*b*). Note also the other patterns in Fig. 13.

Dr Gilbert studied these magnetic patterns, but it remained for Michael Faraday (1791–1867) to reveal their true significance. Faraday called the lines along which the filings set themselves, *lines of force*.

If a small compass needle is moved along the lines of force it will be found always to set itself along, or tangential to, the lines. Filings set themselves along the lines of force in the same way. They become temporary induced magnets and behave like tiny compass needles.

The lines of force of a magnet may also be plotted with a small compass* by making dots to represent the direction of the compass needle as the needle is moved away from the magnet and joining up the dots, see Fig. 14.

Now a filing or compass needle is acted upon by the magnetic field due to the magnet, and may be regarded as setting itself in the direction of the field, just as a weather vane sets in the

* Filings respond only to strong fields, while a compass responds to weaker fields. Thus filings show the magnet's field only, while a compass shows the resultant field due to the magnet and also to the earth.

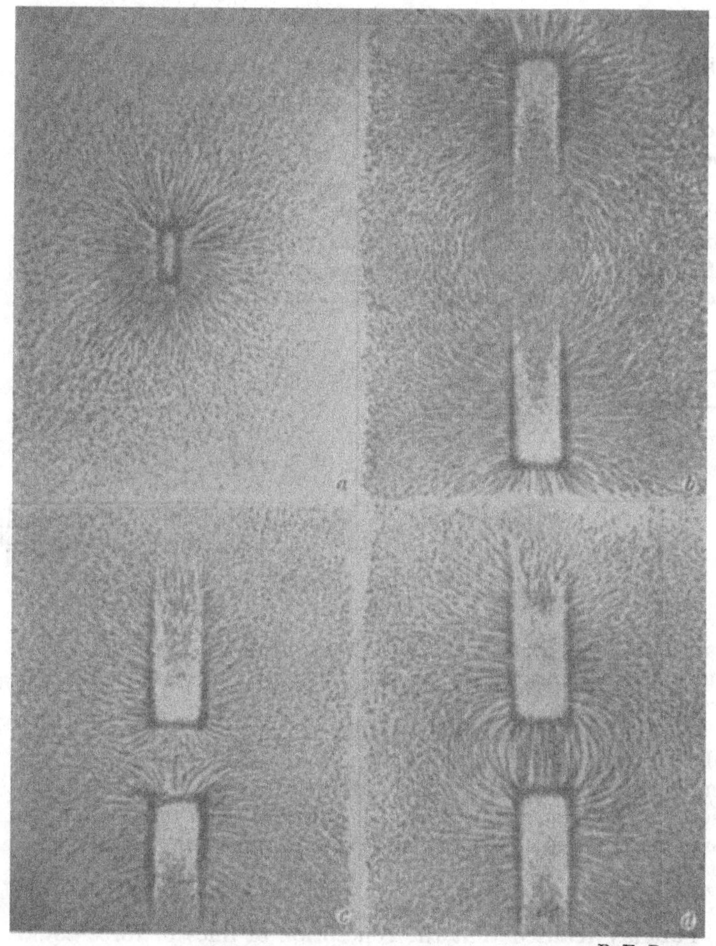

B. F. Brown

Fig. 13. The magnetic fields obtained by sprinkling iron filings on cardboard resting on (*a*) the upper pole of vertical magnet, (*b*) a horizontal bar magnet, (*c*) two like poles, and (*d*) two unlike poles.

direction of the wind. Hence the aptness of Faraday's term, "lines of force": the direction of such a line gives the direction of the resultant magnetic force at any point on it.

Why the lines of force of a bar magnet have their curved shape.

The whole of the magnetism of a magnet may be regarded as concentrated at the two poles, and the fact that there are two poles is responsible for the complex nature of the magnetic field.

Consider a point *A* (see Fig. 15) on a curved line of force. Under the influence of the N. pole alone a compass would set itself along *NA*: under the sole influence of the S. pole, along *AS*. Under the combined influence of both poles the compass sets along the intermediate direction *AR*. Note that *AR* is

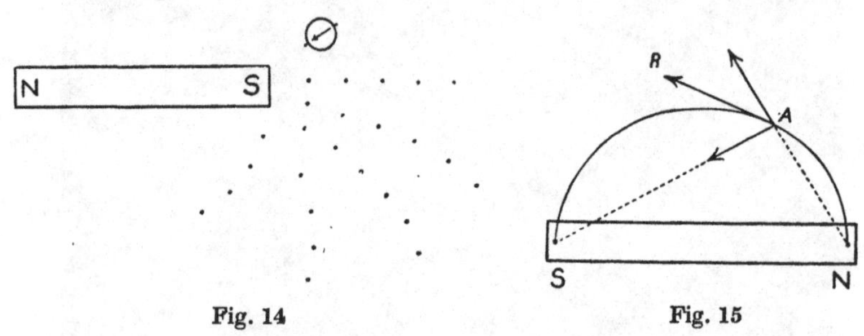

Fig. 14 Fig. 15

tangential to the line of force. Thus, by considering different suitable positions of *A*, the shape of the line of force can be explained.

For the sake of precision, Faraday suggested that the direction along a line of force in which the N. pole of a compass tends to move should be called the positive direction: it is usually designated by an arrowhead. A complete line of magnetic force may be regarded as starting on a N. pole and ending on a S. pole.

Properties of lines of force.

Faraday realised that the conception of lines of force could give a mental picture of what is happening in the space between two magnets. He proceeded to suggest properties for these hypothetical lines of force, which he imagined as sticking out of all magnets into the surrounding space. He saw that by assuming that the lines are in tension like pieces of stretched elastic, and that they tend to contract, he could account for the attraction of unlike poles, since the lines of force stretch straight across from one pole to the other (see Fig. 16(*a*)). But, unlike elastic, the tension of the lines of force must increase as they shorten, since the force between two unlike poles increases as they approach. The fact that the lines of force in Fig. 16(*a*) tend to curve outwards seems to suggest that they repel or tend to elbow each other, as it were, sideways. By the addition of this hypothesis Faraday accounted for the repulsion of like poles. The lines of force of the two poles (see Fig. 16(*b*)) may be thought of as trying to push each other away.

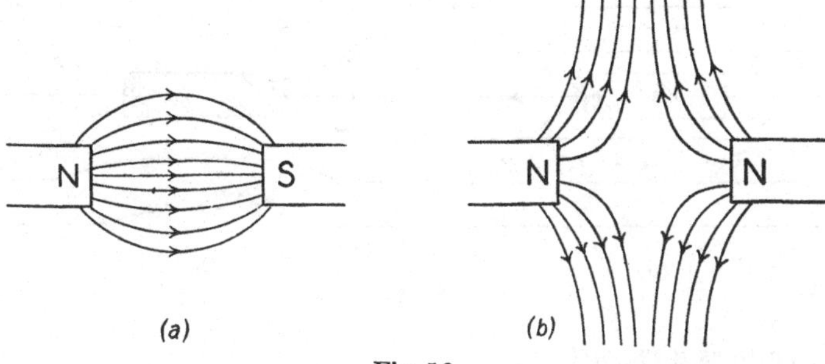

Fig. 16

The success of Faraday's picture or model, when applied to more difficult problems in magnetism, was astonishing. It provided something concrete to think about and suggested new possibilities and experiments. It has been universally adopted by physicists since Faraday's time, and throughout this book we shall make use of lines of force when dealing with magnetic forces acting through space. We shall, indeed, regard the properties of tension and lateral repulsion as axiomatic. For the theory has never been known to fail and has, moreover, been given exact mathematical expression by Clerk Maxwell.

The effect of unmagnetised iron in a magnetic field.

If a bar of unmagnetised iron is placed in a magnetic field it becomes an induced magnet whose lines of force are represented by the dotted lines in Fig. 17(*a*). The resultant field due to the induced magnet and the external field is shown in Fig. 17 (*b*). It will be seen that at the ends of the bar the external field is reinforced, because the lines of force due to the bar are in the same direction as those of the field. On the other hand, the field at the side is weakened, because the two sets of lines of force in Fig. 17 (*a*) are in opposite directions there.

There is another way of regarding this phenomenon. The lines of force of the external field seem to prefer to go through iron rather than through air. Iron is said to be more permeable than air, or to have the greater *permeability*.

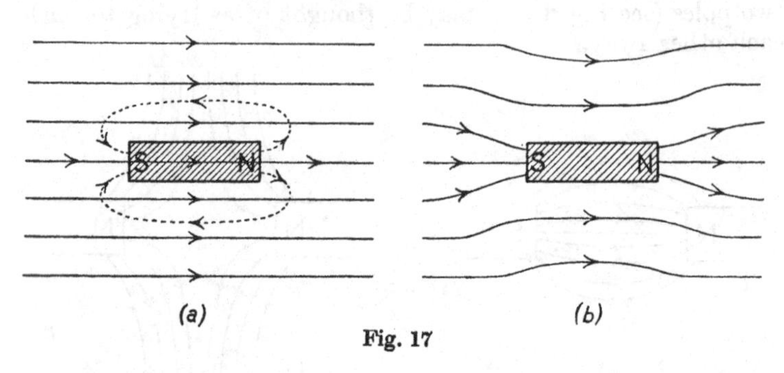

Fig. 17

The earth is a magnet.

Several explanations of the fact that a magnet tends to set itself N. and S. were put forward before Gilbert's time. The curious suggestion that the "coldness" of iron was responsible, Gilbert dismissed as "no better than old wives' gossip". Of the more serious theories, that there were points of attraction in the heavens, at the Celestial Poles, or, as Cardan unaccountably suggested, at the star in the tail of the constellation Ursa Major, Gilbert wrote: "The common herd of philosophers, in search of the causes of magnetic movements, called in causes remote and far away." He himself put forward the theory that the cause lay in the earth itself: the earth was a magnet with a pole of S.-seeking polarity (called the N. magnetic pole) near the geographical N. pole of the earth, and one of N.-seeking polarity (called the S. magnetic pole) near the geographical S. pole. (The earth's N. magnetic pole must be of S.-seeking polarity, since it attracts the N. poles of magnets.)

Gilbert's theory has been proved to be true, and the positions of the earth's magnetic poles have been located (see Chapter XVI). All round the earth there is a magnetic field due to the earth's magnetism. Magnets set themselves along the lines of force of the earth's field.

The direction (or more strictly, the vertical plane) in which a magnet sets itself at a particular place is called the *magnetic meridian* there. Similarly the *geographical meridian* is the vertical plane at a place which would pass through the geographic N. and S. poles.

Summary

A substance which is attracted by a magnet is said to be *magnetic*.

Iron is magnetic and so, to a slight extent, are nickel and cobalt.

A magnet has *two chief properties*: (1) it attracts iron; (2) it sets itself N. and S.

Its magnetism is concentrated at the ends, called the *poles*.

Poles never occur singly but in pairs.

Like poles repel, unlike poles attract.

An iron bar can be magnetised by stroking with a magnet, using the *method of single touch* or the *method of double touch*. Eventually the bar cannot be magnetised further and is said to be magnetically *saturated*. Its magnetism may be destroyed by heat or rough treatment.

Soft iron is more readily magnetised than steel, but steel retains its magnetism better.

The assumption of the *molecular theory of magnetism* is that iron is composed of elementary magnets, once thought to be molecules and now believed to be groups of molecules called domains. In an unmagnetised iron bar the elementary magnets are oriented at random: in a magnet they are in straight lines.

Iron placed near to a magnet becomes a temporary induced magnet.

Any space in which a magnetic influence can be detected is called *a magnetic field*.

A line of magnetic force is a line whose direction at any point represents the direction of the resultant magnetic field there: it is the direction in which a free N. pole tends to move.

A magnet sets N. and S. because *the earth is a magnet*: the magnet sets in the direction of the earth's lines of magnetic force.

QUESTIONS

1. How would you test whether a piece of metal is magnetic or non-magnetic? If you find that it is magnetic, how would you test whether it is magnetised or unmagnetised?

2. Given two bar magnets, the poles of which are not marked, how would you (1) locate the N.-seeking pole of each, (2) determine which magnet is the more strongly magnetised?

3. Suggest an experiment to show that the poles of a bar magnet are of equal strength.

4. (*a*) Devise an experiment to show that a piece of iron attracts a magnet just as truly as a magnet attracts iron.

(*b*) Explain, on the basis of induced magnetisation, the process by which a magnet attracts a piece of soft iron.

5. How would you magnetise a needle so that it should have a N. pole at each end? Would there be any other pole?

6. How would you proceed to magnetise a flat steel ring so that it may exhibit (*a*) polarity at opposite ends of a diameter, (*b*) no polarity?

7. The N. pole of a weak magnet is found to repel the N. pole of a compass needle, but the N. pole of a strong magnet is found to attract it. Explain this.

8. Do the facts of induction suggest to you any reason why a horse-shoe magnet retains its magnetism better when a bar of soft iron (called a keeper or armature) is placed across its poles than when it is not so treated? Explain fully.

9. How would you screen a region from the effect of a magnet's field? Give reasons for your method.

10. Write an account of the molecular theory of magnetism and the experimental evidence on which it is based.

11. Write an account of Faraday's symbolism—lines of force. What is its value?

Why are most of the lines of force due to a magnet curved?

12. Sketch the fields of force of

(*a*) Two equal bar magnets arranged in the form of a T with a gap between them.

(*b*) A horse-shoe magnet attracting a cube of iron.

13. Explain why iron filings, which are sprinkled on a sheet of cardboard over a bar magnet, take up definite positions when the cardboard is tapped lightly. Give a large diagram of the arrangement of the filings. (L.)

14. Three steel knitting needles, exactly similar in appearance, are supplied to you. One is known to be magnetised with opposite poles at its ends, another has consequent poles in the middle, and the third is unmagnetised. Describe how you would proceed to identify them if the only other apparatus available is a means for suspending them horizontally. (L.)

15. Describe and explain what happens when a test-tube loosely filled with iron filings is laid horizontally on a table and stroked with the N.-seeking pole of a magnet several times in the same direction. Give a short account of the theory regarding the magnetisation of iron which is supported by this experiment, and describe another experiment which supports the theory. (N.)

16. Describe the experiments you would make in order to show that steel is more difficult to magnetise than soft iron, but keeps its magnetism better. Mention two methods of demagnetising a piece of iron. (N.)

17. Indicate the difference in effects produced when the end of a soft-iron rod is held (*a*) near one end of an ordinary bar magnet, (*b*) near the middle of the magnet.

Hence suggest a method to enable you to distinguish between two similar-looking pieces of iron, one of which you are told is a magnet and the other a piece of soft iron, no apparatus of any sort being available. (N.)

18. Describe experiments which illustrate the law that under magnetic induction south polarity appears in a piece of iron where the lines of force enter it and north polarity where they leave it.

Also describe an experiment which shows that at a sufficiently high temperature iron loses its magnetic properties. (N.)

19. A number of steel needles are placed on a table parallel to and touching each other. One end of a strong magnet is brought near to

but not touching the points of the needles and after a while is removed. Describe and explain the behaviour of the needles.

What difference, if any, would have been observed if the needles had been of soft iron? (O. and C.)

20. What do you understand by a magnetic line of force? Draw diagrams to illustrate the field of force round

(*a*) a bar magnet with its N.-seeking pole pointing N.,

(*b*) a horse-shoe magnet (neglect the earth's field),

(*c*) a bar magnet lying within a soft-iron ring.

In each case explain the features of the diagram. (O.)

21. A strong bar magnet is laid on a block of wood on a table; a small compass needle is supported on the same level at a distance of a few inches from one end of the magnet and in line with its axis. What will be the effect of (*a*) interposing a slab of soft iron between the compass needle and the bar magnet, (*b*) placing the slab to one side of and parallel to the bar magnet? Explain as well as you can the reason for these effects. (N.)

22. What is meant by a line of magnetic force? Why do lines of magnetic force never cross? The N. pole of a magnet is opposite the S. pole of a second magnet and a soft-iron ring lies in the space between them, but does not touch the magnets. Draw a diagram of the lines of force between the poles. Draw also a diagram of the arrangement of the lines when the soft-iron ring is replaced by a brass ring. Give explanations of the diagrams. (Wales.)

23. What do you mean by a line of magnetic force?

Explain the method of plotting a magnetic field of force by means of a small compass needle.

A bar magnet lies horizontally in the magnetic meridian with its S. pole pointing northwards. At a certain point, due north of it, the magnetic force vanishes. Sketch the lines of force in the neighbourhood of this neutral point. (C.)

Chapter II

ELECTRICITY

Discovery of electricity.

As long ago as 600 B.C. the Greeks knew that amber, when rubbed with cloth or fur, would attract small pieces of chaff and straw. But they made no attempt to investigate the phenomenon, and it remained a strange, isolated fact for over 2000 years until the time of Gilbert.

Gilbert began to rub other substances, such as glass and resin, and found that they too underwent this curious change when rubbed, manifested by the power of attracting light objects. In order to describe their change of condition Gilbert coined a new word, *electricity*, from the Greek word for amber, *electron*. A rubbed rod which attracts light objects is said to be *charged with electricity*.

An ebonite fountain pen can readily be charged by rubbing it on the coat sleeve; it will then attract tiny pieces of paper. The phenomenon is utilised in ebonite clothes "brushes" which, when rubbed on the coat, attract bits of fluff and dust.

Two kinds of electricity.

At the beginning of the eighteenth century it was discovered that there are two kinds of electricity.

Rub an ebonite rod with fur and hang it in a stirrup supported by means of a silk thread (see Fig. 18). Rub another ebonite rod with fur and bring it near to the rod in the stirrup. The latter moves away, showing that the two rods *repel* one another.

Now rub a glass rod with silk and bring it near to the rubbed ebonite rod in the stirrup. The rods attract each other. Similarly, it can be shown that two rubbed glass rods repel each other.

We can only conclude from this difference in behaviour that there are two kinds of electricity. Monsieur du Fay (1699–1739), chief gardener to King Louis XV and a member of the French Academy of Sciences, was the first to observe and grasp the significance of this difference: he called the charge on glass and similar substances vitreous electricity, and that on ebonite or resin, resinous electricity. We now call them positive and negative electricity, respectively. There are very good reasons for this change of names, which we shall explain later (p. 25).

Glass rubbed with silk is said to be positively charged.

Ebonite rubbed with fur is said to be negatively charged.

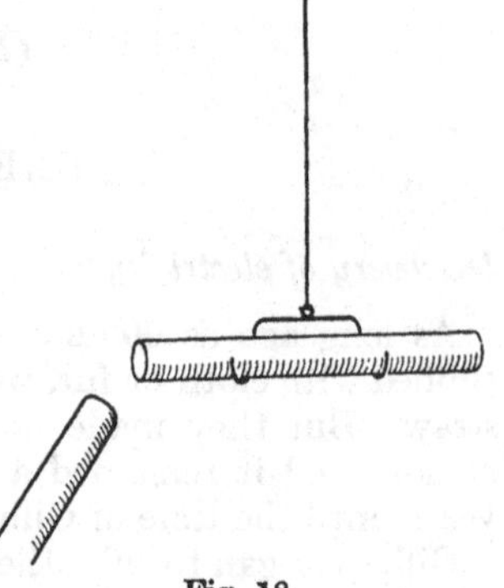

Fig. 18

We may state the law of attraction and repulsion between positive and negative charges as follows:

Like charges repel, unlike charges attract.

It should be noted that if ever we found a substance which could be charged so that it attracted both vitreous (or positive) and resinous (or negative) charges but repelled itself, we should have established the existence of three sorts of charges; in fact, every charge ever found attracts one and repels the other; so that we have never had to consider more than two sorts of charge.

The flow of electricity.

The next advance in the study of electricity was made by Stephen Gray (1696–1736), a pensioner at the Charterhouse, and a Fellow of the Royal Society. He made the important discovery that some of the attractive power of a rubbed glass rod can be transferred from it to another body.

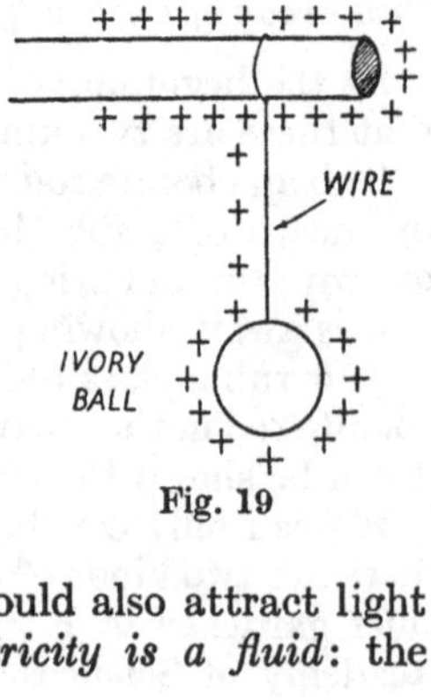

Fig. 19

He suspended an ivory ball by means of a wire from a rubbed glass rod (see Fig. 19), and found that the ivory ball would also attract light bodies. Hence arose the idea that *electricity is a fluid*: the electricity was regarded as flowing from the glass tube along the wire to the ivory ball.

Conductors and insulators.

Some substances allow electricity to flow through them and are called *conductors.* Carbon and the metals are good conductors. The best conductor of all is silver and the next best copper. Connecting wires in an electric circuit are made of copper rather than silver, as silver is too expensive.

Other substances such as silk and rubber do not allow electricity to pass through them and are called *insulators.* They are used to cover or insulate connecting wires and this insulation must be scraped away where contact is to be made in an electric circuit.

There is an intermediate class of substances which may be regarded either as poor conductors or imperfect insulators. The human body and water belong to this category. The human body, if inserted in the circuit carrying the current to light a domestic lamp, will not carry sufficient current to light the lamp, but merely enough to give an unpleasant electric shock.

The following table gives a list of conductors and insulators:

Good conductors	Poor conductors	Good insulators
The metals Carbon	Water Stone The human body Wood Cotton	Air Sulphur Paraffin wax Ebonite Glass Porcelain Silk India rubber Oils Dry paper

The first electroscopes.

When an ebonite rod is rubbed with fur, or sometimes even when the hair is combed with an ebonite comb, a slight crackling of sparks can be heard. Electricity passes from the ebonite, when it is "overcharged" so to speak, to the fur or hair in the form of a spark.

During the middle of the eighteenth century experiments in electrostatics (the study of electricity at rest) provided an attractive entertainment for genteel society. Friction machines, taking the form of a sulphur globe which could be rotated and rubbed by the hand, or a glass cylinder with a silk rubber, were invented in order to generate sufficient charge to produce quite large sparks.

The human body, if insulated from the earth by a stool with glass legs, can be readily charged either by flicking it with fur or connecting it to an electric machine, and sparks drawn from it. According to Priestley, this experiment made "a principal part of the diversion of gentlemen and ladies who come to see experiments in electricity".

It is a significant fact, which can be paralleled in all branches of science, that the next advance, made by the more serious students of electrostatics, was the invention of instruments for detecting and measuring electric charges called *electroscopes*.

The simplest type of electroscope consists of two thin wires hanging from an insulator. On being given a charge, the two wires repel each other and open like compasses. The angle through which they diverge is a measure of the charge. Volta used two dry straws instead of wires, and Cavendish a pair of cork balls on the ends of two straws or linen threads. Pith balls (made of the white substance found beneath the bark of an elder tree) were also used and are preferable to cork, since they are lighter (see Fig. 20).

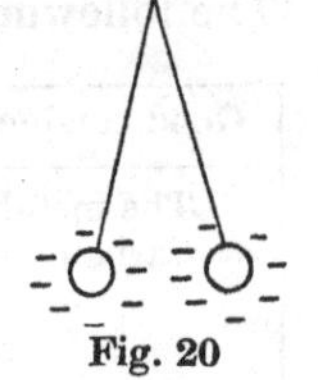
Fig. 20

The most satisfactory instrument, and one which is still in use, was invented by Bennet, a Derbyshire clergyman, and is known as the gold-leaf electroscope.

The gold-leaf electroscope.

It consists of a metal rod terminating at its upper end in a disc and having attached to its lower end a thin leaf of gold foil or Dutch metal. Only the upper horizontal edge of the foil is attached to the rod, so that the foil can move about this edge as about a hinge. The rod passes through an insulating plug, preferably of sulphur, in a metal case which has a glass front so that the leaf can be observed. The reader should note that the sulphur plug in Fig. 21 is shaded. The convention in electrostatic diagrams is to shade insulators and leave conductors unshaded.

Fig. 21

The electroscope may be charged by flicking its disc with fur, when it acquires a negative charge. The leaf shares some of this negative charge and since it is light, and the like charges on it and the lower end of the rod repel, it will be caused to diverge from the rod, as shown in Fig. 21.

If a positively charged glass rod is brought near to the negatively charged electroscope the leaf will tend to collapse. We can explain this by saying

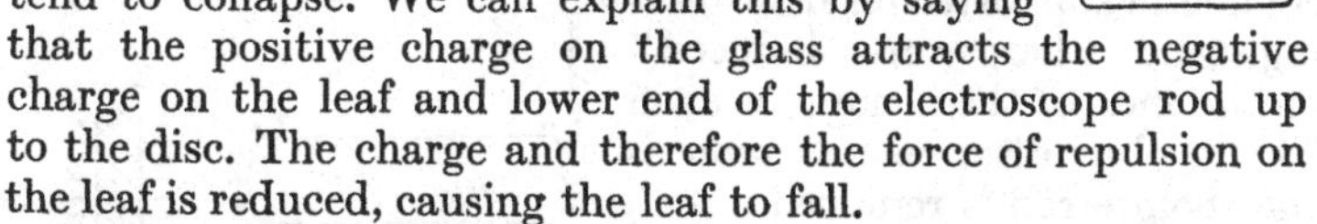

that the positive charge on the glass attracts the negative charge on the leaf and lower end of the electroscope rod up to the disc. The charge and therefore the force of repulsion on the leaf is reduced, causing the leaf to fall.

On the other hand if a negatively charged rod is brought near to a negatively charged instrument the leaf tends to diverge farther. The negative charge on the disc is now being repelled down to the leaf and the lower end of the rod, causing a larger charge there and stronger repulsion.

Thus a gold-leaf electroscope can be used to detect the sign of an unknown charge and also, from the extent of the divergence or fall of the leaf, to give an estimate of its size.

Testing insulators

The insulating properties of substances may be tested by means of a gold-leaf electroscope. Charge an electroscope. Touch the disc with a finger. The leaf will collapse at once, showing that the charge has been conducted away to earth through the body. Charge the electroscope again, and touch the disc with a piece of sulphur held in the hand. Nothing happens to the leaf, showing that the sulphur is a very good insulator. Silk, paper, wood and other substances can be tested in the same way.

The charge on the rubber.

When ebonite is rubbed with fur, the fur acquires a positive charge which is equal to the negative charge on the ebonite. This fact can be proved experimentally as follows. Rub an ebonite rod with a piece of fur mounted on an insulating handle. Bring the two together near to a negatively charged electroscope. They will have no effect on the leaf since their equal and opposite charges neutralise each other. When brought up separately, however, the fur can be shown to have a positive charge and the ebonite a negative, since the former makes the leaf fall and the latter makes it rise.

Electrostatic induction.

Bring a negatively charged ebonite rod near to a cylindrical conductor supported on an insulating stand. Charges are "induced" on the ends of the conductor as in Fig. 22, but disappear as soon as the ebonite rod is removed.

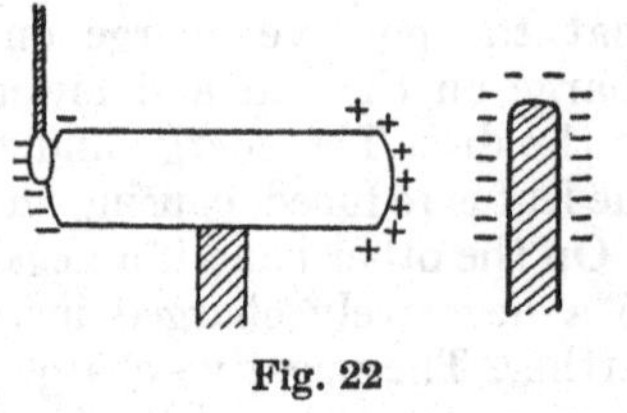

Fig. 22

The existence and nature of the induced charges can be demonstrated by laying a proof plane, consisting of a small metal disc on an insulating handle, on one end of the conductor, and taking away a sample of the charge to be tested with a charged electroscope.

If a positively charged glass rod is used instead of an ebonite rod, the induced charge on the nearer end of the insulated conductor is negative and on the farther end, positive.

The phenomenon is known as *electrostatic induction*, and was discovered by John Canton (1718–1772). The effect takes place only in conductors and not in insulators.

Fluid theories.

How are we to account for the appearance and subsequent disappearance of the two induced charges in electrostatic induction?

We have already mentioned that electricity was early regarded as a fluid because (unlike magnetism) it can flow. And as soon as it was clearly realised that there were two kinds of electricity it was assumed that there were *two sorts of fluids*.

The phenomenon of electrostatic induction suggests that all

uncharged conductors contain both fluids. When the two fluids are present in equal amounts they neutralise each other and the body is apparently uncharged. Hence the aptness of the terms positive and negative. When a conductor is charged positively or negatively it has an excess of positive or negative fluid, respectively. Its apparent charge is the algebraic sum of the charges it actually contains.

In the electrostatic induction represented in Fig. 22 the fluids are separated because the negative inducing charge attracts the positive fluid to the nearer end of the conductor and repels the negative to the farther end. Electrostatic induction does not occur in insulators, because the fluids cannot move freely therein.

On the other hand, insulators must contain the fluids, because on rubbing two insulators, such as ebonite and fur, both fluids appear. Moreover, the fur acquires a positive charge *equal* to the negative charge on the ebonite.

In the process of rubbing, then, any of the following may be happening:

(1) negative fluid may be rubbed out of the fur into the ebonite, leaving the fur with a deficiency of negative fluid and the ebonite with an excess of it;

(2) positive fluid may be rubbed out of the ebonite into the fur;

(3) both (1) and (2) may occur simultaneously.

Benjamin Franklin introduced a simplification here by replacing the two-fluid theory by a *one-fluid theory*. He suggested that only positive electricity could flow. Thus when ebonite is rubbed with fur, according to Franklin's theory positive electricity is rubbed out of the ebonite into the fur.

We shall see that this single-positive-fluid theory was used to depict an electric current flowing along a wire—a convention which we retain to this day although we now believe that Franklin chose the wrong fluid and that a current consists of a flow of negative electricity in the opposite direction.

In the further examples of induction which we shall discuss in the next few pages we shall regard both positive and negative electricity as being equally mobile: we shall, in fact, use the two-fluid theory. The reader may, if he wishes, translate our account in terms of a one-fluid theory.

Charging by induction.

A conductor can be charged by induction. Hold a charged ebonite rod near to a conductor and momentarily earth the conductor by touching it with the finger (see Fig. 23). The negative charge on the far end of the conductor is repelled away through the body to the earth. Remove the finger first and then the charged ebonite rod. The conductor has now a positive charge. We must state, without at this stage attempting an explanation, that wherever the conductor is touched, it is the charge on the far end which goes to earth.

To charge a conductor negatively by induction a positively charged glass rod may be used. On touching the conductor with the finger the positive electricity goes to earth (see Fig. 24).

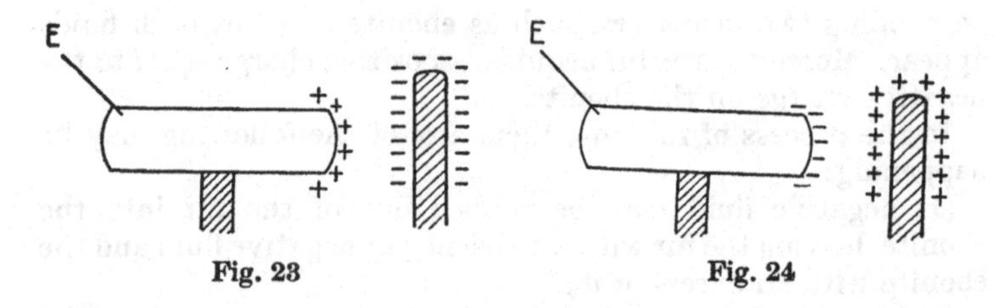

Fig. 23 Fig. 24

Explanation of the attraction of small objects.

When a charged ebonite rod is brought near to a small piece of paper, some of the negative charge in the paper is repelled to earth. The paper is therefore positively charged by induction, and since unlike charges attract, it is attracted by the negatively charged ebonite.

Charging an electroscope by induction.

A gold-leaf electroscope can be charged by induction.

Hold an ebonite rod near to the disc of an electroscope. The leaf will diverge although the disc has not been touched by the rod.

This is due to induction: a negative charge is induced on the leaf and lower end of the electroscope rod, and a positive charge on the disc (see Fig. 25i). Now touch the electroscope momentarily to earth, keeping the ebonite rod near while this is being done. The leaf collapses, since the induced negative charge is now repelled to earth (Fig. 25 ii). When the ebonite rod is taken away the leaf rises again, owing to the fact that the whole electroscope is positively charged (Fig. 25 iii).

Thus a negatively charged rod charges an electroscope positively by induction. Similarly, the electroscope can be charged negatively by induction by means of a positively charged rod. Draw diagrams similar to Fig. 25, showing the stages in this process.

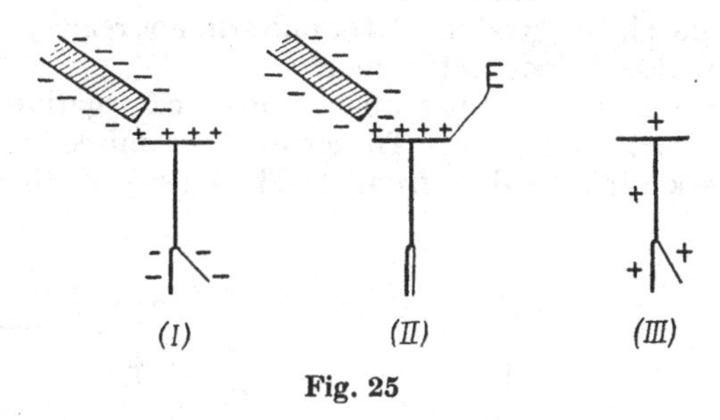

Fig. 25

Effect of an earthed conductor on a charged electroscope

If an earthed conductor such as the hand is held near to a charged electroscope, without touching it, the leaf collapses slightly. This is due to the fact that the hand is charged slightly by induction by the charge on the electroscope. The induced charge on the hand is of opposite sign to that on the electroscope and hence the leaf tends to collapse.

Thus in using a charged electroscope the only infallible test that a body is charged is increased divergence of the leaf, since an uncharged conductor will cause a slight collapse.

Alternatively, a sure way to test a body which is suspected of bearing a charge is as follows:

(1) Bring it near to an uncharged electroscope: this settles whether it is charged.

(2) If it is charged, bring it near to a charged electroscope: this settles the sign of the charge.

The electrophorus.

The *electrophorus* is an instrument for producing numerous electric charges from a single charge and is a good example of induction.

A disc of ebonite is given a negative charge by rubbing with fur (Fig. 26 (*a*)). A metal disc with an insulating handle, called the cover, is placed on the ebonite. A positive charge is induced on the lower side of the cover and a negative charge on the upper (Fig. 26 (*b*)). Practically no charge is lost by the ebonite by contact, since it is a good insulator and will not readily part with its charge: also the contact is poor.

The cover is earthed, whereupon its induced negative charge is repelled to earth (Fig. 26 (*c*)). The cover is now raised (Fig. 26 (*d*)), and a spark can be taken from it. The energy of the spark is

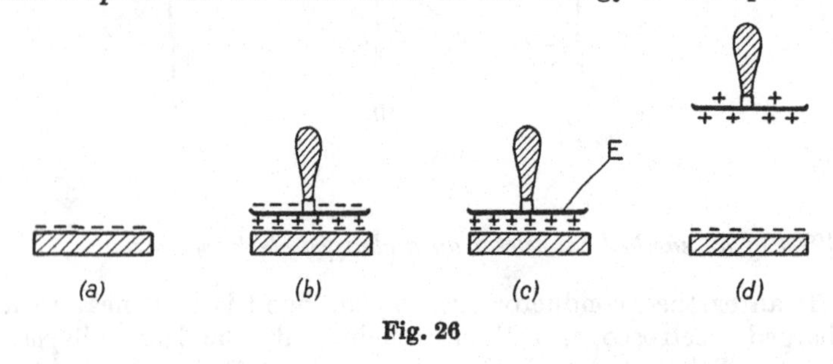

Fig. 26

derived from the work done in lifting the cover against the attraction of its positive charge and the negative charge on the ebonite.

The operation can be repeated many times without any further rubbing of the ebonite.

The instrument works better if the ebonite is placed on an earthed metal plate, called a *sole*. The negative charge on the ebonite induces a positive charge on the upper side of the sole and a negative charge on the lower side, which is repelled to earth.

The complete action of the sole is not easy to understand, but we may regard one of its chief functions as the preventing of the negative charge on the ebonite from leaking away, as a result of the attraction of its positive charge.

The modern era.

Towards the end of the nineteenth century physicists had accumulated a great store of systematised knowledge about the behaviour of electricity, particularly current electricity, i.e. electricity flowing along wires, but they were still very ignorant of its nature. It was as though they were experimenting with water in a pipe, measuring its rate of flow, its pressure at different points, and the gurgling noise it made as it passed through, without ever being able to open the pipe, and examine the actual nature of the water.

But in the last decade of the century they were able to examine electricity outside its conducting wires, as it passed through space. By experimenting on the discharge of electricity through rarefied gases and other phenomena, they were able to build up the modern theory of the nature of electricity, and with it, a theory of the constitution of matter. We cannot describe their researches but must content ourselves with a brief summary of their results.

The constitution of matter.

All matter is made up of some ninety elements, and the elements consist of minute constituent particles, called *atoms*, which cannot be subdivided without destroying their nature. The atoms of the elements combine with each other to form *molecules* of new substances, rather as the twenty-six letters of the alphabet may be combined to form words.

During the nineteenth century, atoms were thought of as being rather like minute marbles—indivisible and unsplittable. No one had any notion of their interior structure, nor indeed that their structure was soon to be revealed.

In 1897 Sir J. J. Thomson (and others) discovered and measured particles of negative electricity which were knocked out of atoms of gas during an electric discharge. These particles, called at first negative corpuscles by Sir J. J. Thomson, are now known as *electrons*. Electrons can now be obtained very readily by "boiling" them out of a metal wire by heating the wire with an electric current, as in a wireless valve. They are ejected from atoms under the action of X-rays, and in the spontaneous explosion of heavy atoms such as those of radium and uranium.

In whatever manner electrons are obtained they have always the same mass and the same charge. They are the units or ultimate particles of negative electricity.

The discovery of the electron was the beginning of a new epoch. Soon the unit of positive electricity, the *proton*, was discovered. A great simplification was introduced in our ideas of the structure of matter. The atoms of the ninety elements, instead of being made of ninety different kinds of stuff, were believed to be built up of two kinds of stuff only, protons and electrons. In recent years a new particle, the *neutron*, with a mass equal to that of the proton but no charge, has been discovered and it is thought that this, also, is a component of most atoms.

The hydrogen atom.

The lightest atom known, that of hydrogen, consists of a single proton, called the nucleus, round which revolves a single electron—rather like a planetary system, the proton representing the sun and the electron a planet (see Fig. 27). Since the electron is made of negative electricity, it is attracted by the positive proton (unlike charges attract), and this attraction keeps the electron revolving in its orbit.

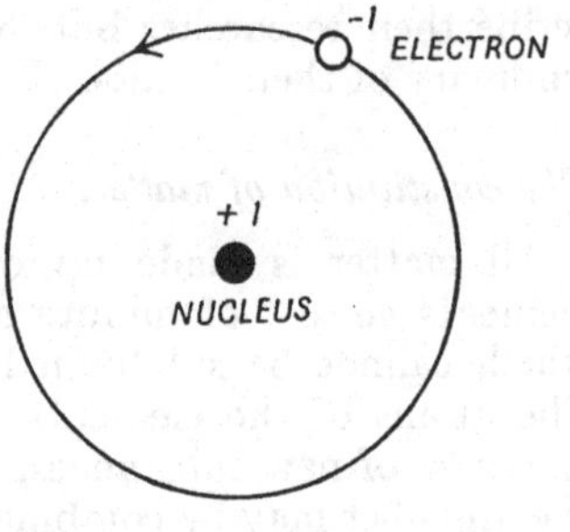

Fig. 27

The proton is about 1840 times as heavy as the electron. In it, therefore, most of the mass of the atom resides. Its charge is equal and opposite to that of the electron: hence the charges in the atom neutralise each other and the net charge on the atom is zero.

Atoms of the rest of the elements.

The hydrogen atom, as we have said, has a nucleus which consists of a single proton. The heavier atoms may have nuclei which consist both of protons and neutrons.

The second lightest atom known is that of helium, which is about four times as heavy as the hydrogen atom. The nucleus consists of 2 protons and 2 neutrons and this has a charge of +2 and a mass of 4 units. It has 2 revolving or "planetary" electrons, with a combined charge of −2 (see Fig. 28).

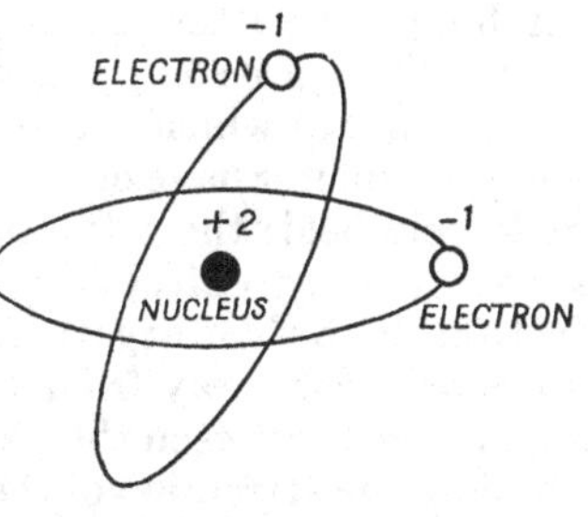

Fig. 28

The next heaviest atom is lithium, with 3 planetary electrons. Its nucleus has a net charge of +3 units. The beryllium atom has 4 planetary electrons, the boron atom has 5, the carbon atom has 6, and so on up to the heaviest atom, that of uranium, with 92. In each case the net positive charge of the nucleus is equal to the total negative charge of the planetary electrons. A heavy atom, with its host of revolving electrons, may be compared to a swarm of bees or gnats.

Electrification by rubbing.

The outermost planetary electrons of an atom, being farther away from the positive nucleus, are probably less strongly attracted than the innermost electrons. When two substances, such as glass and silk, are rubbed together, it seems likely that some of the outer electrons of the atoms composing the molecules of the glass are detached and pass into the silk. The glass is therefore left with a deficiency of electrons and hence a positive charge, while the silk has an excess of electrons and hence a negative charge.

If the glass, with its deficiency of electrons, is left on a table, electrons gradually enter it from the earth (which has a vast store of free electrons), until its deficiency is made up. Similarly, the excess of electrons on the silk eventually find their way to the earth, when the silk becomes uncharged once again.

This explanation of electrification by rubbing is essentially a one-fluid theory—but it is the negative electricity, consisting of millions of electrons, which moves. The massive positive nuclei of the silk atoms could not move without the silk disintegrating.

Conductors and insulators.

A lump of matter, although it looks so solid, is really very porous. There are comparatively large spaces between the atoms or molecules of which it is composed. Now the atoms of a *metal* or any conductor have one or two outer planetary electrons which are loosely held: the attraction of the positive nucleus for them is considerably enfeebled by the swarm of inner planetary electrons with their negative electric charges. Hence these outer electrons break away from their atoms and wander through the empty spaces between the atoms. Every now and again they are seized by the attraction of the nucleus of another atom which has lost one or two of its outer electrons, and there is a constant interchange of electrons between the atoms. But at any particular moment there is a large number of free electrons wandering unattached between the atoms. These free electrons are responsible for the conductivity of the metal.

If electrostatic induction is produced in the metal lump by, say, a positive charge held near, the free electrons are attracted to one end of the metal lump, and there is a general migration of the free electrons to that end. This flow of electrons in spaces between the atoms of the metal constitutes an electric current. We shall see shortly how the flow can be maintained, giving rise to a continuous current round and round a closed circuit. Any force which causes the electrons to move is called an Electro-Motive-Force (E.M.F.).

An insulator, on the other hand, contains no free electrons: all the planetary electrons of its atoms are firmly held by the nuclei. It is true that some of the outer planetary electrons may be removed by rubbing, but this is merely a surface phenomenon. No free electrons can pass through an insulator.

Galvani and Volta.

After our digression on the modern theory of the constitution of matter and the nature of electricity, we must pick up, once again, the thread of the historical development of the subject.

In the year 1780 an accidental discovery of Luigi Galvani, professor of anatomy at Bologna University, revealed a new type of electrical phenomenon and opened what may be termed the era of current electricity. While engaged in the dissection of a frog, Galvani noted that if the frog's leg was touched by two dissimilar metals, such as copper and zinc, the leg twitched. It is said that the discovery was made when a frog was hung up by a brass pin and touched, accidentally, an iron railing just below. Galvani realised that the twitching of the frog's leg was due to an electric shock, just as the muscles of the human body give an involuntary twitch under the action of an electric shock. He attributed the source of the electricity to the frog and called it animal electricity: his followers called it galvanism. It was, of course, known that certain fishes such as the torpedo could give powerful electric shocks.

Galvani's researches created great interest in the scientific world and his fellow-countryman, Alessandro Volta (1745–1827), also began to examine the phenomenon. Volta, who had already made an intensive study of the electricity produced by rubbing and invented the electrophorus, was dissatisfied with Galvani's theory that the source of electricity lay in the frog. He believed it probable that the electricity was identical with that obtained by rubbing, that it was derived from the two metals, and that the frog's leg served merely as a detector of the flow of electricity.

A keen controversy among scientific men arose on this question. In 1800, the year after Galvani's death, Volta proved his own view to be correct. He was able to dispense entirely with the frog, and invented what is called the simple voltaic cell.

The simple cell.

If a plate of copper and a plate of zinc are placed in dilute sulphuric acid and the two plates are connected by wires to a small pea lamp (see Fig. 29), the lamp lights. A current of electricity is flowing through the lamp. The copper, zinc and sulphuric acid constitute a *simple cell,* the source of the electric current.

Both the copper and the zinc plates contain free electrons, but when they are connected by a wire electrons tend to flow from the zinc to the copper through the wire. The circuit is completed through the sulphuric acid. We shall see in Chapter VIII that it is the chemical action inside the cell which keeps the electrons circulating.

Now in Fig. 29 arrows have been drawn depicting the current as flowing through the wire from the copper to the zinc. In the early days there was much confusion as to what constituted an electric current. Some people held the opinion that there was a simultaneous flow of both positive and negative electricity in opposite directions. To simplify the matter it was decided to regard the current as a flow of positive electricity only. Although we now believe the current to consist of a flow of negative electrons we have retained the old and universal sign convention (which assumes a flow of positive electricity) throughout this book. As far as outward effects are concerned, a flow of positive electricity in one direction amounts to the same thing as a flow of negative electricity in the opposite direction.

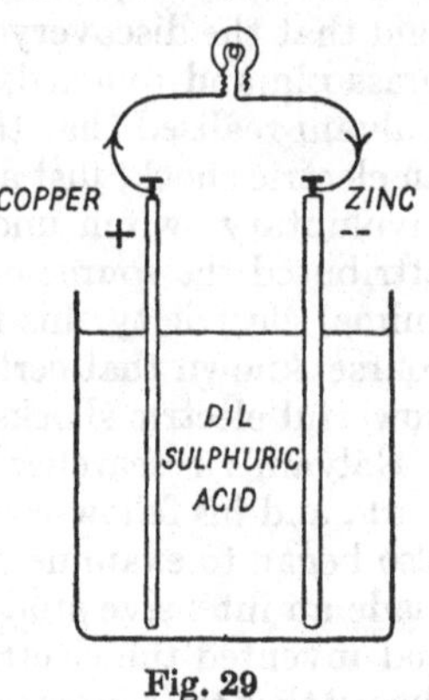

Fig. 29

Historical.

The first announcement of his cell was made by Volta in a letter to the President of the Royal Society, London, on 20 March 1800. It created the greatest interest and enthusiasm, because it provided a new and *continuous* source of electricity, one that could be set up by anyone for a few pence.

Volta's fame spread throughout Europe. He was invited to Paris to lecture to the Academy. When in 1804 he wished to resign his professorship at Pavia, Napoleon, who took a keen interest in electrical research, refused: "I cannot agree to Volta's resignation. If his activities as a professor are too great they must be limited. He may even give only one lecture a year, but the University of Pavia would be wounded to the heart if I were to allow so famous a name to be struck off the rolls of its members: furthermore, a good general must die upon the field of honour."

Volta, loaded with honours, was indeed more fortunate than his compatriot whose discovery he had investigated and developed. Galvani had refused to take the oath of allegiance to the Cisalpine Republic when Napoleon invaded Northern Italy in 1796: he died dispossessed of his professorship and in poverty.

Summary

There are two kinds of electricity, positive and negative.

Glass rubbed with silk is positively charged; ebonite rubbed with fur is negatively charged. Like charges repel, unlike charges attract.

Electrostatic induction. A charged body induces temporary charges in a neighbouring conductor: if the conductor is momentarily touched to earth it becomes permanently charged.

The *gold-leaf electroscope* is used to detect, distinguish and measure charges. It may be charged by flicking the disc with fur or by induction.

The *electrophorus* is an induction machine for generating electric charges.

The ultimate particle of negative electricity is called the *electron*; that of positive electricity is called the *proton*.

An electric current is a flow of electricity. Conventionally it is always regarded as a flow of positive electricity: in fact, it is believed to be a flow in the opposite direction of negative electricity, in the form of electrons.

Matter consists of atoms, or groups of atoms called molecules. Atoms consist of a positively charged nucleus containing protons (and usually some neutrons), round which revolve electrons.

When a substance is charged by rubbing, some of the outer electrons of the atoms of the rubbed substance (or the rubber) are removed, with the result that the substance has a deficiency (or excess) of electrons; hence its charge.

Conductors allow electricity to pass through them; *insulators* do not. The best conductors are the metals and carbon. *A conductor owes its conducting powers to the presence inside it of free electrons,* which have broken away from their atoms.

A simple cell, consisting of plates of copper and zinc dipping into dilute sulphuric acid, will maintain an electric current in a circuit, i.e. a closed conducting path. **The current (of positive electricity) is said to flow through the external circuit from the positive plate—the copper, to the negative plate —the zinc.**

QUESTIONS

1. (*a*) A small light pith ball is suspended by a silk thread. A charged ebonite rod is brought near. The ball is first attracted to the rod, but after touching it is repelled. Explain.

(*b*) Usually, unless the atmosphere is very dry, the force of repulsion on the pith ball gradually decreases (the silk thread comes back gradually to a vertical position), the pith ball is once more attracted, and then again repelled. Explain.

2. Two pith balls are suspended from the same point by means of long silk fibres. A glass rod which is charged positively is brought up slowly to the pith balls and finally touches them. Describe and explain what is observed. What happens when the glass rod is taken away? (L.)

3. Describe a gold-leaf electroscope and explain, with the aid of simple diagrams, how you would charge it positively by induction. How would you then use it for finding the sign of another charge? (N.)

4. Describe and explain what happens (*a*) when a positively charged insulated conductor is brought near the cap of a gold-leaf electroscope, (*b*) when the cap of the electroscope is momentarily touched by hand and the charged conductor is then removed, and (*c*) when the charged conductor is again brought up to the electroscope and made to touch the cap. (O.)

5. Describe the gold-leaf electroscope and show how you would use it to test the insulating properties of (*a*) a silken cord, (*b*) a piece of ebonite, (*c*) a wire gauze, (*d*) a piece of damp paper, (*e*) a rod of sulphur.

State the results you would expect to find in these cases. (L.)

6. Describe an experiment to show that in electrification by friction (*a*) two kinds of charges can be obtained, (*b*) equal quantities of positive and negative electricity are produced simultaneously.

Why can a charged ebonite rod attract small uncharged particles?

7. A charged ebonite rod is held near to the lower end of a burette from which water is slowly dropping into a can standing on the disc of an electroscope. The leaf of the electroscope slowly rises. Explain.

What would you expect to happen to the divergence of the leaf if the charged ebonite rod is now brought near to the electroscope?

8. Briefly describe an electrophorus and explain how it is used. How would you prove that the insulated metal plate is charged by induction and not by conduction?

If an electrophorus fails to give a spark what defect would you suspect? (L.)

9. Describe fully, in terms of the electron theory, what happens when a conductor is charged (*a*) positively, (*b*) negatively by induction.

10. Write an account of the modern theory of the constitution of matter. How does this theory account for the difference between conductors and insulators?

Chapter III

OHM'S LAW

The water analogy.

The flow of electricity through a wire consists of a moving stream of free electrons passing through the spaces between the atoms of the wire: it is often compared to the flow of water through a pipe.

Now what interests us about the flow of water through a pipe is the rate at which it is passing through, say in gallons per second, rather than the volume of water in the pipe at any instant: so in the case of electricity, we are mainly concerned with its rate of flow, which is called the **current**. An electric current is measured in **amperes.** A current of 1 ampere consists of a flow of 6,250,000,000,000,000,000 electrons per second. The small pea lamp in Fig. 29 takes a current of about $\frac{1}{8}$ ampere, and the headlamp of a motor car about 3 amperes. The current may be measured by an instrument called an *ammeter*, which is inserted in the circuit and through which the current flows.

Water will not flow along a pipe unless there is a difference of pressure between the ends. Similarly, we may say that electricity will only flow if caused to do so by a difference of electrical pressure, technically known as a **potential difference** (abbreviated to P.D.). In a cell this P.D. is maintained by the chemical action: the copper is said to be at a higher potential than the zinc, with the result that a current of positive electricity flows from the copper to the zinc when they are connected by a wire. It will be remembered that we have agreed to regard a current as a flow of positive electricity, although we believe it to be a flow of negative electrons in the opposite direction.

The water analogy is represented in Fig. 30(*b*). Note the diagrammatic representation of an electric *circuit* (a closed conducting path through which a current can pass) in Fig. 30(*a*) and, in particular, the symbols used to represent a cell, lamp and switch. In Fig. 30(*b*) a driven water wheel or pump corresponds to the cell, and pipes and a tap correspond to the connecting wires and switch respectively.

The difference in water pressure maintained by the pump may be measured by simple water gauges, called manometers, as represented in Fig. 30(*b*). The pressure of the water supports a vertical column of water in each manometer tube, and the difference in the heights of the columns is the difference of pressure maintained by the pump between the two sides of itself.

Similarly, the difference of electrical pressure or P.D. maintained by a cell may be measured by an electrical pressure gauge called a voltmeter, connected by wires to the two plates of the cell. Electric potential is measured in **volts**. The P.D. produced by a simple cell is about 1 volt.

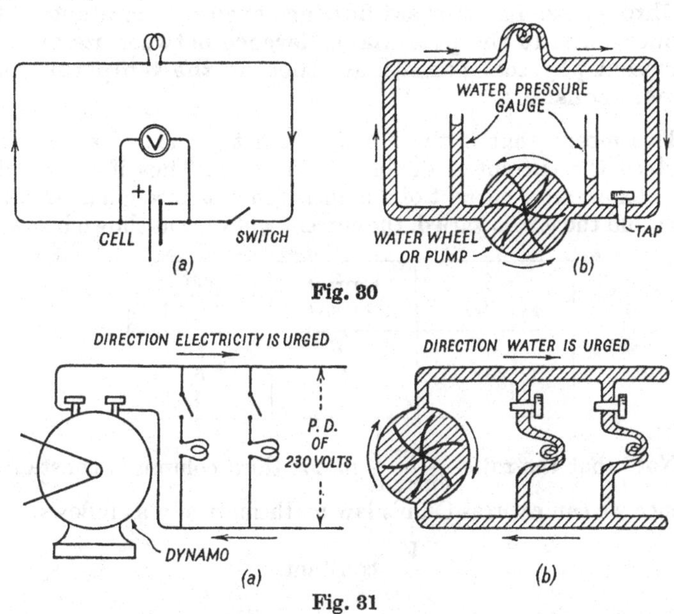

Fig. 30

Fig. 31

The electric supply mains in a house.

When we require water in our homes we turn on a tap. Water gushes out at a certain pressure which could be measured with a strong pressure gauge. Similarly, when we require electricity we close a switch and electricity is supplied at a certain pressure or voltage. The commonest voltage in this country is 230 volts. This voltage is maintained, not by cells, but by dynamos or generators in power stations. The water analogy in this case is represented in Fig. 31 (*b*). Both water and electricity are stationary in the figure: the arrows indicate the direction in which the electricity and the water are urged.

Ohm's law.

There is a simple and fundamental relation between the P.D. (in volts) required to drive the current (in amperes) through a wire or conductor. It is known as Ohm's law.

Ohm's law. The current flowing through a conductor is proportional to the potential difference between its ends, provided physical conditions, such as the temperature, remain constant.

This means that if the P.D. between the ends of a wire is doubled the current is doubled, and so on. Thus if a P.D. of 10 volts causes a current of 2 amperes to flow through a certain wire, and the P.D. is varied, the current will vary as shown below:

P.D. (volts)	Current (amperes)	$\frac{\text{P.D.}}{\text{Current}}$
10	2	5
20	4	5
5	1	5

Note that the ratio $\frac{\text{P.D.}}{\text{current}}$ in the third column is constant. Hence we can express Ohm's law mathematically as follows:

$$\frac{V}{i} = \text{constant},$$

where V = P.D. in volts, i = current in amperes.

Now electricity, on passing through a wire, experiences a certain amount of obstruction which is greater the longer and thinner the wire: similarly, the flow of water is impeded by a long narrow pipe. The free electrons flowing through the wire keep colliding with the atoms of the wire. This obstruction is called the *resistance* of the wire, and varies with the nature of the substance of which the wire is made. The resistance of an iron wire, for example, is about seven times as great as that of a copper wire of the same dimensions.

Thus if a P.D. of 10 volts causes a current of 2 amperes to flow through a copper wire, it will cause a current of only $\frac{2}{7}$ ampere to flow through an iron wire of the same dimensions.

In the case of the copper wire

$$\frac{V}{i} = \text{const.} = \frac{10}{2} = 5.$$

In the case of the iron wire

$$\frac{V}{i} = \text{const.} = \frac{10}{\frac{2}{7}} = 35.$$

The constant ratio V/i is seven times as great for the iron wire as the copper. It is clear, therefore, that we can use this ratio to express the resistance of the wire. The unit of resistance is the *ohm*, and it is *the resistance of a conductor through which a current of* 1 *ampere will flow when a* P.D. *of* 1 *volt is maintained between its ends.*

Ohm's law may be written as follows:

$$\frac{\mathbf{V}}{\mathbf{i}} = \mathbf{R},$$

where V = P.D. in volts, i = current in amperes, R = resistance in ohms.

Electrical engineers use an inelegant term, the mho, to represent the unit of conductance. Conductance is the reciprocal of resistance. Thus a resistance of 1 ohm has a conductance of 1 mho: a resistance of 2 ohms has a conductance of half a mho.

Ohm's great achievement was the clear distinction he drew between the quantities, V, i and R. He was led to his discovery by comparing the flow of electricity with the flow of heat. Heat flows along a wire from a point at a higher temperature to one at a lower, and the rate of flow is proportional to the difference of temperature. Similarly, electricity flows from a point at a higher potential to one at a lower, and its rate of flow is proportional to the difference of potential.

Although Ohm's law is of fundamental importance in electricity, yet owing to the fact that Ohm stressed the more speculative, theoretical side of his discovery rather than the experimental verification, its publication in Germany was received with hostile and sarcastic criticism. Not until the significance of his work had been appreciated in England and he had been awarded the Copley medal of the Royal Society was Ohm's true merit acknowledged by his fellow-countrymen. Fortunately recognition came just in time and he enjoyed, during his last five years, what had been his life-long ambition, a university professorship.

Small and large units.

The following submultiples of the ampere, volt and ohm are in common use:

1 milliampere $=\frac{1}{1000}$th ampere,
1 millivolt $=\frac{1}{1000}$th volt,
1 microampere $=\frac{1}{1000000}$th ampere,
1 microvolt $=\frac{1}{1000000}$th volt,
1 microhm $=\frac{1}{1000000}$th ohm.

A large unit of resistance,

1 megohm = 1,000,000 ohms.

Measurement of resistance.

Current and P.D. may be measured by direct-reading instruments called ammeters and voltmeters. Resistance, however, cannot be measured direct.

The simplest method of determining the resistance, R, of a wire is to pass a current, i, through it—measured by an ammeter, and to find the P.D., V, across its ends by means of a voltmeter. Then, by Ohm's law, $R = V/i$.

The circuit required is shown in Fig. 32. The current flows from the + terminal of the cell through the circuit and back to the − terminal of the cell. Some forms of ammeter and voltmeter have their terminals marked + and −, and the current should be led in at the + terminal and out at the − terminal. The current may be varied by means of a variable resistance or rheostat (see also Fig. 33), and hence several pairs of readings of V and i may be taken.

It is important to grasp clearly that the same current flows through the cell, unknown resistance, variable resistance and ammeter, just as the same volume of water must flow per second through any cross-section of a series of pipes connected together: otherwise there would be an accumulation of electricity (or water) at some point in the circuit. Thus the order in which these components of the circuit are connected together is quite immaterial. Anyone who doubts this should try connecting the ammeter at different places in the circuit, otherwise unaltered; it will always read the same: the current does not get "used up" as it goes round. On the other hand, it is essential to connect the

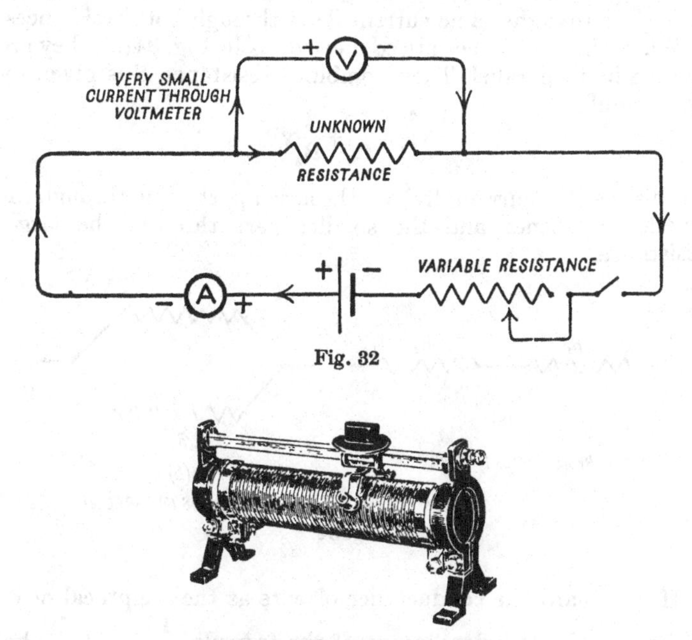

Fig. 32

Fig. 33. Variable resistance or rheostat.

voltmeter *across* the unknown resistance and not in the rest of the circuit. The voltmeter takes a very small current and acts as a kind of bypass through which a tiny fraction of the total current flows.

If a current of 1 ampere or more is passed for some minutes through the circuit, the unknown resistance may well become warm. Since the resistance of most metals increases considerably when the temperature rises, consistent results will not be obtained unless the duration of the current is reduced to a minimum.

Resistances in series and in parallel.

Two resistances R_1 and R_2, connected as in Fig. 34 (*a*), are said to be in series. Their combined resistance R is equal to their sum:

$$\mathbf{R} = \mathbf{R_1} + \mathbf{R_2}.$$

It is clear that the same current flows through both resistances.

When the resistances are connected, as in Fig. 34 (*b*), they are said to be in parallel. Their combined resistance R is given by the formula

$$\frac{1}{\mathbf{R}} = \frac{1}{\mathbf{R_1}} + \frac{1}{\mathbf{R_2}}.$$

In this case the current divides, the larger part going through the smaller resistance and the smaller part through the larger resistance.

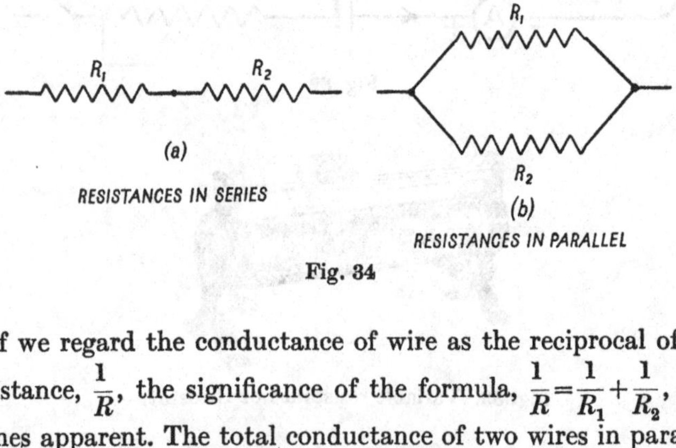

Fig. 34

If we regard the conductance of wire as the reciprocal of its resistance, $\frac{1}{R}$, the significance of the formula, $\frac{1}{R} = \frac{1}{R_1} + \frac{1}{R_2}$, becomes apparent. The total conductance of two wires in parallel is equal to the sum of their conductances.

Connecting the wires in parallel is comparable to increasing the thickness of one of the wires. Hence the resultant resistance is decreased. On the other hand, connecting two wires in series is comparable to increasing the length of one of the wires. Hence he resultant resistance is increased

Example. Two resistances of 2 and 3 ohms are connected in parallel. Find their combined resistance. If a total current of

1 ampere passes through the two resistances, find the current through each.

$$\frac{1}{R}=\frac{1}{R_1}+\frac{1}{R_2}$$

$$=\tfrac{1}{2}+\tfrac{1}{3}=\tfrac{5}{6}.$$

$$\therefore R=1\tfrac{1}{5} \text{ ohms.}$$

Applying Ohm's law to the combined resistances:

$$V=iR.$$

$\therefore$ P.D. between ends of each resistance $=1\times 1\frac{1}{5}=1\frac{1}{5}$ volts.

Applying Ohm's law to the 2-ohm resistance:

Current through 2-ohm resistance $=\dfrac{V}{R}$

$$=\frac{1\frac{1}{5}}{2}=\frac{3}{5} \text{ ampere.}$$

$\therefore$ Current through 3-ohm resistance $=1-\frac{3}{5}=\frac{2}{5}$ ampere.

It will be seen that the currents through the 2-ohm and 3-ohm resistances are in the inverse ratio of the resistances, i.e. 3 : 2.

Proof of the formulae.

1. $R=R_1+R_2$.

In Fig. 34(*a*), suppose a current i passes through the resistances.

P.D. between ends of $R_1=iR_1$,

" $R_2=iR_2$.

$\therefore$ Total P.D. across both resistances $=iR_1+iR_2$.

Now total combined resistance $=\dfrac{\text{Total P.D}}{\text{Current}}$

$$\text{i.e. } R=\frac{iR_1+iR_2}{i}$$

$$=R_1+R_2.$$

2. $\dfrac{1}{R}=\dfrac{1}{R_1}+\dfrac{1}{R_2}$.

In Fig. 34(*b*), suppose currents i_1 and i_2 pass through R_1 and R_2, respectively.

Let V = P.D. between the ends of each resistance.

Applying Ohm's law:

$$i_1=\frac{V}{R_1},$$

$$i_2=\frac{V}{R_2},$$

$$\text{Total current}=i_1+i_2=\frac{V}{R}.$$

$$\therefore \frac{V}{R}=\frac{V}{R_1}+\frac{V}{R_2},$$

$$\text{i.e. } \frac{1}{R}=\frac{1}{R_1}+\frac{1}{R_2}.$$

Ohm's law applied to a complete circuit.

A cell or dynamo is said to produce an **Electro-Motive Force (E.M.F.)**. An E.M.F. is a force which makes electricity move.

The E.M.F. of a cell is measured by the total P.D. that it can produce: **the P.D. that it maintains between its terminals when it is on open circuit**, i.e. not delivering a current. It may be measured (approximately) by connecting a voltmeter across the terminals of the cell. The E.M.F. of a simple cell is about 1 volt, of a single dry cell, about 1·5 volts, and of an accumulator, about 2 volts.

The E.M.F. of a cell driving a current through a circuit must overcome the complete resistance of the circuit: this comprises not only the external resistance but also the internal resistance of the cell. Thus applying Ohm's law to a complete circuit, we have

$$\frac{\mathbf{E}}{\mathbf{i}}=\mathbf{R}+\mathbf{r},$$

where E volts = E.M.F. of the cell, i amperes = current, R ohms = external resistance, r ohms = internal resistance.

Example. A cell of E.M.F. 2 *volts and internal resistance* $\frac{1}{10}$ *ohm is connected to an external resistance of* 5 *ohms. Calculate the current flowing in the circuit.*

Applying Ohm's law to the circuit:

$$\frac{E}{i}=R+r,$$

$$\frac{2}{i}=5+\frac{1}{10},$$

$$i=\frac{2}{5\cdot 1}=0\cdot 39 \text{ ampere.}$$

Arrangement of cells.

A number of cells is called a battery. Cells may be connected in series or in parallel.

When cells are connected in series the total E.M.F. of the battery is equal to the sum of their E.M.F.S: the internal resistance of the battery is equal to the sum of the internal resistance of the cells.

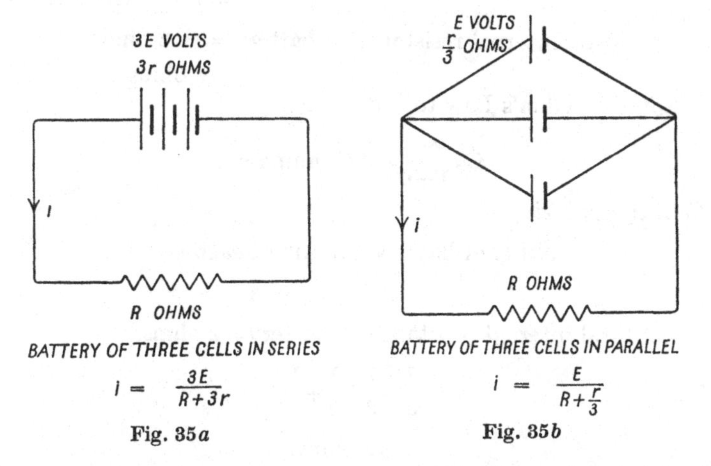

Fig. 35*a*

Fig. 35*b*

In Fig. 35*a* each cell has an E.M.F. of E volts and an internal resistance of r ohms: the current i, which the battery will send through an external resistance R ohms, is obtained by applying Ohm's law. In the general case, if there are n cells in series,

$$\frac{nE}{i}=R+nr.$$

When cells are connected in parallel as in Fig. 35*b*, the E.M.F. of the battery is equal to E, the E.M.F. of each cell: the internal resistance of the battery, in the case of three cells, is equal to that of three resistances, r in parallel, i.e. $r/3$. In the case of m cells in parallel,

$$\frac{E}{i}=R+\frac{r}{m}.$$

Example. What is the strongest current that three cells, each of E.M.F. 1·5 *volts and internal resistance* 2 *ohms, can send through an external resistance of* 1 *ohm?*

The cells may be connected either in series or in parallel (as in Figs. 35*a* and *b*).

Cells in series:

Total E.M.F. of battery $=3\times 1{\cdot}5$
$=4{\cdot}5$ volts.

Total internal resistance of battery $=3\times 2$ ohms
$=6$ ohms.

Applying Ohm's Law to the circuit:

$$i=\frac{4{\cdot}5}{1+6}=0{\cdot}64 \text{ ampere.}$$

Cells in parallel:

E.M.F. of battery = E.M.F. of each cell
$=1{\cdot}5$ volts.

Let total internal resistance of battery $=x$ ohms.

Then
$$\frac{1}{x}=\frac{1}{2}+\frac{1}{2}+\frac{1}{2},$$

$$x=\tfrac{2}{3} \text{ ohms.}$$

Applying Ohm's law to the circuit:

$$i=\frac{1{\cdot}5}{1+\frac{2}{3}}=0{\cdot}9 \text{ ampere.}$$

Hence the strongest current is 0·9 ampere and the cells must be connected in parallel.

This result may seem surprising. But, in this case, it is more advantageous to reduce the internal resistance of the battery by connecting the cells in parallel than to increase the E.M.F. by connecting them in series.

Generally speaking, it is advantageous to connect cells in parallel if their internal resistance is large compared with the external resistance, and to connect them in series if their internal resistance is small compared with the external resistance.

SUMMARY

A current, i, is measured in amperes. An electric pressure or potential difference, V, is measured in volts. A resistance, R, is measured in ohms.

Ohm's law. The current flowing through a conductor is proportional to the potential difference between its ends, provided physical conditions, such as the temperature, remain constant.

$$\frac{\mathbf{V}}{\mathbf{i}}=\mathbf{R}.$$

Resistances in series: $\mathbf{R}=\mathbf{R}_1+\mathbf{R}_2$.

Resistances in parallel: $\frac{\mathbf{1}}{\mathbf{R}}=\frac{\mathbf{1}}{\mathbf{R}_1}+\frac{\mathbf{1}}{\mathbf{R}_2}$.

The E.M.F. of a cell is the P.D. between its terminals when it is on open circuit.

Ohm's law applied to a complete circuit:

$$\frac{\mathbf{E}}{\mathbf{i}}=\mathbf{R}+\mathbf{r}.$$

Grouping of cells.

(i) n cells in series:

$$\frac{nE}{i}=R+nr.$$

(ii) m cells in parallel:

$$\frac{E}{i}=R+\frac{r}{m}.$$

QUESTIONS

Draw diagrams of the circuit in each example.

1. (*a*) What current flows when a P.D. of 2 volts is maintained across a resistance of 6 ohms?

(*b*) What P.D. is required to send a current of 3 amperes through a resistance of 4 ohms?

(*c*) What is the resistance of a lamp when taking a current of ½ ampere at a P.D. of 230 volts?

(*d*) What E.M.F. is required to send a current of 1·2 amperes through a circuit of resistance 6 ohms?

2. (*a*) An ammeter, placed in series with an electric radiator, reads 10 amperes, and a voltmeter, placed across it, reads 230 volts. What is the resistance of the radiator?

(*b*) The resistance of a circuit is doubled, and the current is decreased to one-third. What was the change in the E.M.F.?

3. A cell sends a current of 0·25 ampere through two resistances of 5 and 3 ohms arranged in series. What is the potential drop across the ends of each resistance?

4. What do you understand by the electromotive force of a battery? A dry cell of E.M.F. 1·4 volts sends a current of 0·12 ampere through an 11-ohm resistance. What is the internal resistance of the cell? (C.)

5. (*a*) A Leclanché cell, of E.M.F. 1·5 volts and internal resistance 1 ohm, is short-circuited, i.e. its plates are connected by a wire of negligible resistance. What current flows?

(*b*) An accumulator, of E.M.F. 2 volts and internal resistance $\frac{1}{10}$ ohm, is short-circuited. What current flows?

6. What current will flow when a 2-megohm grid-leak is connected between the poles of a 4½-volt flash-lamp battery? Why should it not be advisable to connect the poles by a small resistance, say 2 ohms? (L.)

7. (*a*) Find the combined resistance of a 2-ohm and a 4-ohm resistance connected (i) in series, (ii) in parallel.

(*b*) The 2-ohm and 4-ohm resistances in parallel are connected to a cell of E.M.F. 2 volts and negligible internal resistance. What

current is taken from the cell and what currents flow in the two resistances?

(*c*) Is it true to say that in a divided circuit the current "takes the line of least resistance"? Explain fully.

8. Explain how the terms "series" and "parallel" are applied to conductors carrying currents.

Two conductors, respectively of 10 and 20 ohms' resistance, are connected in parallel, and joined by thick copper wire to the terminals of a cell of E.M.F. 2 volts and internal resistance 1 ohm. What is (*a*) the total current flowing in the circuit, (*b*) the current in each conductor? (N.)

9. A cell of 1·5 volts' E.M.F. and 5 ohms' internal resistance is connected to two resistances of 100 ohms and 10 ohms, respectively, in parallel. How much current will pass through the cell and through each resistance? (O. and C.)

10. Three wires of resistances 1, 2 and 3 ohms, respectively, are connected (*a*) in series, (*b*) in parallel. Calculate their combined resistance in each case.

11. What resistance must be placed in parallel with an 11-ohm coil to reduce its resistance to 10 ohms? (L.)

12. Two coils have a combined resistance of 12 ohms when connected in series, and 1 ohm when connected in parallel. Find their respective resistances. (L.)

13. State Ohm's law.

It is required to test the value of a resistance coil marked 5 ohms (2 amperes). Describe briefly how you would conduct the necessary test in the laboratory. Can you suggest a reason why the value 2 amperes is specified? (L.)

14. A searchlight requires a current of 100 amperes at 60 volts. It has to be worked off 250-volt mains. Find the extra resistance required, and show on a diagram how this resistance is connected in the circuit. (O.)

15. Four wires, each of 10 ohms' resistance, are joined in the form of a square. Two opposite corners are connected through a resistance of 40 ohms to the electric mains, the supply voltage being 100 volts. Calculate the equivalent resistance of the square and the fall of potential across it. (N.)

16. A battery of E.M.F. 1·5 volts and internal resistance 2 ohms is joined up to three other resistances as shown in Fig. 36. Calculate the current flowing in each of the resistances. (O. and C.)

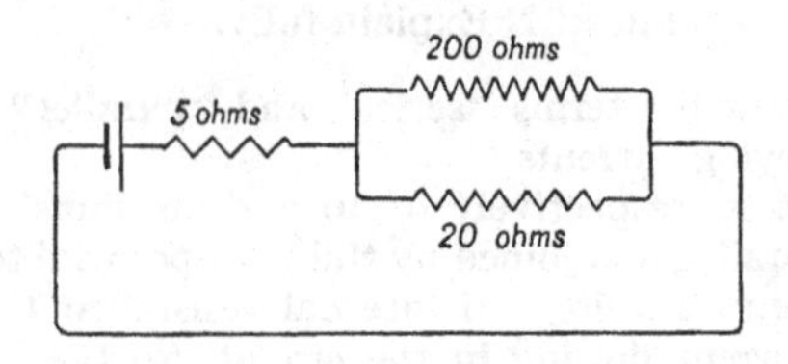

Fig. 36

17. A resistance coil is in series with an ammeter of negligible resistance and a lamp of 200 ohms' resistance, the voltage of the supply being 110. The potential difference, measured by a voltmeter between the terminals of the coil, is 2 volts. What should the ammeter reading be, and what is the resistance of the coil? Draw a diagram showing the connections of ammeter and voltmeter, marking + and −. (L.)

18. Three resistances of 6, 8 and 10 ohms, respectively, are connected in series in a circuit. A voltmeter is connected across the ends of the 8-ohm resistance and reads 4 volts. Find (*a*) the current flowing through the circuit, (*b*) the voltage across the 10-ohm resistance, (*c*) the total voltage across all three resistances.

19. Two resistances, one of 20 ohms, and the other unknown, are connected in series with an ammeter which reads 2·5 amperes. If the total voltage across the two resistances is 60 volts, find (*a*) the unknown resistance, (*b*) the voltage across the 20 ohms, (*c*) the voltage across the unknown resistance.

20. What current would you expect a battery of three cells in series, each of E.M.F. 1·5 volts and 0·5 ohm resistance, to pass through a 1-ohm coil? (N.)

21. It is required to send a current of 0·75 ampere through a resistance coil of 3 ohms. Each of the cells used in the battery has an E.M.F. of 1·5 volts and a resistance of 1 ohm. Show that three cells are required and that they must be fitted in series. (L.)

22. An electric bell has a resistance of 5 ohms and requires a current of 0·25 ampere to work it. Assuming the resistance of bell wire is 1 ohm per 15 feet, and that the bell push is 90 feet distant from the

bell, how many cells, each of E.M.F. 1·4 volts and internal resistance 2 ohms, will be required to work the circuit? (O. and C.)

23. The poles of a battery, consisting of three cells each of 1 ohm resistance and 1·6 volts E.M.F. joined in series, are connected by two wires in parallel. The wires have resistances of 5 ohms and 4 ohms, respectively. Calculate the current flowing in each wire. How would the intensities of these currents be affected if the cells of the battery were joined in parallel instead of being in series? (N.)

24. Two cells, *A* and *B*, are joined in series with a resistance and an ammeter. The ammeter reads 2·0 amperes. The terminals of *B* are reversed, and then the ammeter reads 0·40 ampere. If the E.M.F. of *B* is 1·0 volt, find that of *A*.

25. Two voltaic cells, *A* and *B*, are connected in series with an ammeter. The current indicated is 2·4 amperes. The cell *A* is then reversed so as to oppose *B* and the current observed is 0·6 ampere in the same direction. Calculate the ratio of the electromotive force of *A* to that of *B*. (L.)

26. The resistance of the telegraph wires joining two stations is 20 ohms, and the receiving instrument has a resistance of 80 ohms, and requires a current of $\frac{1}{20}$th of an ampere to work it. What is the smallest number of cells, each of E.M.F. 1·1 volts and resistance 2 ohms, which will suffice to send a signal from one station to the other? (O. and C.)

27. Six cells, each of E.M.F. 1·5 volts and internal resistance 1 ohm, are arranged in two rows, in parallel, of three cells in series. What current will this battery send through a resistance of 2 ohms?

28. State Ohm's law.

The following values of current and resistance were obtained with a circuit containing a battery of dry cells, an ammeter and a variable resistance *R*:

Resistance, *R*	5	10	15	20 ohms
Current, *i*	0·36	0·24	0·18	0·145 ampere

Plot the relation between *i* and *R* and explain why the graph does not pass through the origin. (L.)

Chapter IV

THE MAGNETIC EFFECT OF AN ELECTRIC CURRENT

The resemblances between the phenomena of magnetism and electrostatics (i.e. electricity at rest) led scientific men to look for some connection between them.

Several well-authenticated, isolated facts, unfortunately not reproducible, lent strong support to this possibility. When a ship at sea was struck by lightning the compass was often demagnetised or even magnetised in the opposite direction. It was reported in the *Transactions of the Philosophical Society* for 1735 that a number of knives, belonging to a certain tradesman of Wakefield, after being struck by lightning, picked up nails and behaved like magnets.

Oersted's experiment.

The honour of being the first to demonstrate experimentally this fundamental and most important connection between magnetism and electricity belongs to Hans Christian Oersted, professor of physics at the University of Copenhagen. At the end of one of his lectures in the winter of 1819–20 he placed a compass needle parallel to a wire in a circuit. When he switched on the current, he saw the magnetic needle turn at right angles to the wire; "he was quite struck with perplexity" for "he had not before any more idea than any other person that the force should be transversal" (i.e. at right angles).* He confessed that he had stumbled upon his discovery by accident, but as Lagrange has said, "Accidents only happen to those who deserve them."

The directions in which the N. pole turns when the compass needle is placed above and below the wire, respectively, are shown in Fig. 37. If the current is reversed the N. pole turns in the opposite direction in each case.

* Taken from a letter from one of Oersted's pupils to Faraday.

The magnetic field due to a current in a straight wire.

Since it affects a compass needle, an electric current must set up a magnetic field in its neighbourhood. We can examine the nature of this field by sprinkling iron filings on a card through a hole in which passes a wire carrying a current (see Fig. 38). *The filings set themselves in concentric circles round the wire.* To influence the filings the current must be large, about 5 amperes.

The positive directions of the lines of force, i.e. the direction in which the N. pole of a compass tends to move, can best be remembered by a rule invented by Maxwell.

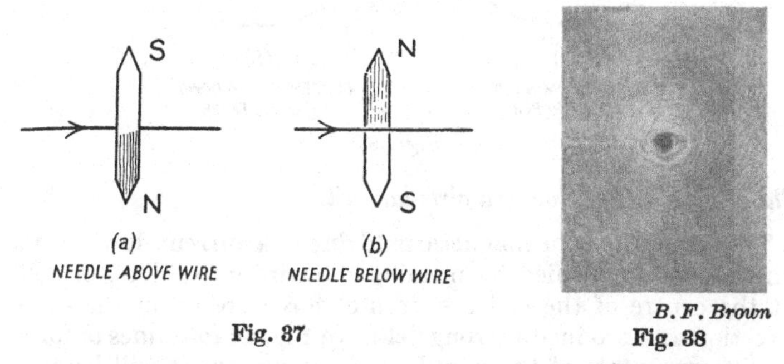

Fig. 37

Fig. 38

Fig. 38. A wire carrying a current passes through the hole at right angles to the paper and the iron filings set themselves in concentric circles.

Maxwell's corkscrew law. Imagine a corkscrew being screwed along the wire in the direction of the current. Then the direction in which the thumb rotates indicates the positive direction of the lines of force.

A compass needle sets itself tangentially to one of these lines, with the N. pole pointing in the positive direction. The reader should use Maxwell's rule to verify that Fig. 37 has been drawn correctly.

The lines of force due to a straight wire can be represented as in Fig. 39: $\odot$ represents the cross-section of a wire at right angles to the paper, carrying a current up out of the paper; the dot mav be regarded as the tip of an arrow. Similarly, $\otimes$ represents the cross-section of a wire carrying a current down into the paper;

the × may be regarded as the tail of the arrow. The lines of force are put closer together near the wire, since the field is stronger there.

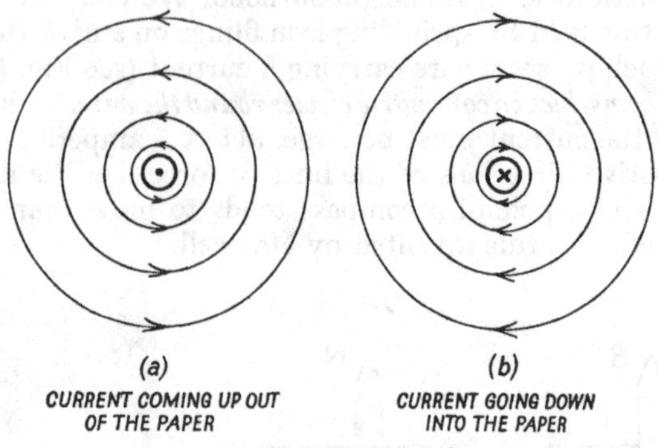

Fig. 39

The magnetic field due to a circular coil.

The strength of the magnetic field due to a current flowing in a wire can be intensified by bending the wire into a loop or coil. At the centre of the coil the lines of force are all in the same direction and produce a strong field. In Fig. 40 four lines of force at different parts of the wire have been drawn: it will be seen that in the centre of the coil they are all in the same direction at right angles to the plane of the coil.

We can examine the magnetic field due to a coil by the iron filing method (see Fig. 41) or by plotting the lines with a compass needle (see Fig. 42).

One of the most interesting things about *the magnetic field due to a current flowing in a coil* is that it is *identical with that of a very short or flat magnet*. Compare Fig. 13(*b*) and Fig. 41. The coil *behaves as though* one face were a N. pole and the other a S. pole. A simple way of remembering which face is N. and which S. is shown in Fig. 43. The reader should verify the correctness of this figure by the use of Maxwell's corkscrew law. Screw along one side of the coil. The direction in which the thumb moves is the direction in which a N. pole tends to travel; hence the face of the coil into which the thumb tends to move must be the S. pole; (unlike poles attract).

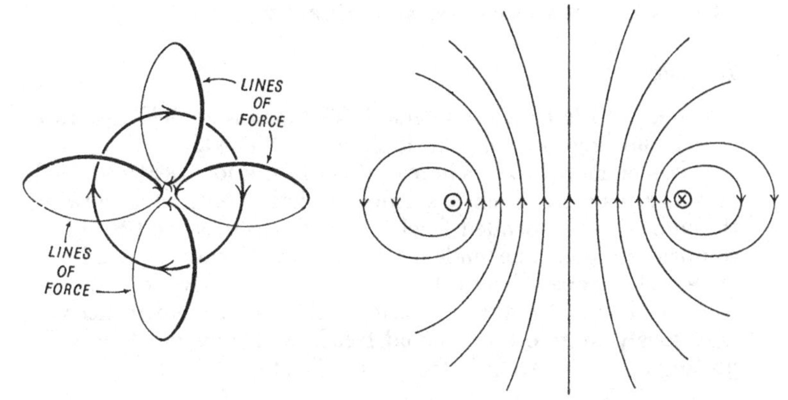

Fig. 40

Fig. 42. Lines of force due to current in circular coil. Compare with Fig. 41.

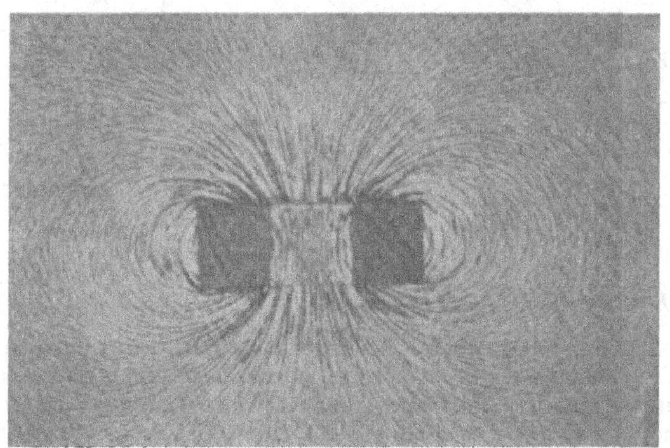

B. F. Brown

Fig. 41. The lines of force due to a circular coil carrying a current. Before the photograph was taken the coil was removed: the sides of the coil passed at right angles through the dark rectangular holes in the cardboard. The small square piece of cardboard in the middle is detachable and had, of course, to be carefully removed before the coil could be lifted out.

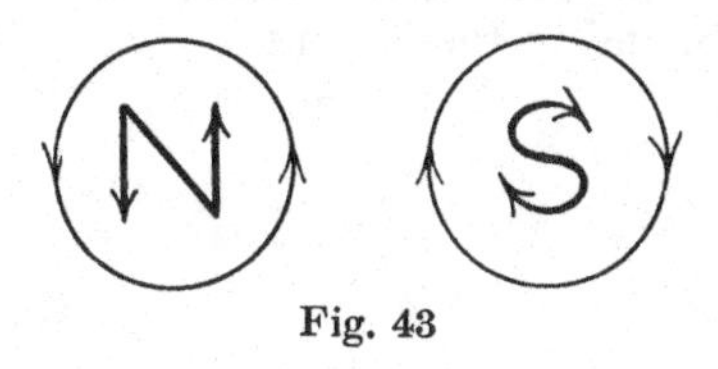

Fig. 43

The solenoid.

A long coil is called a solenoid. The magnetic field due to a current flowing in a solenoid is shown in Fig. 44.

It is even more obvious here that the field is identical with that of a bar magnet. Such a solenoid (when a current is flowing through it), if suspended freely, will set itself N. and S. like a compass needle. The ends of a current-bearing solenoid are attracted or repelled by a bar magnet. De la Rive, of Geneva, who was one of the first to perform this experiment, contrived ingeniously to give his solenoid freedom of movement by supporting it on a floating battery (see Fig. 45).

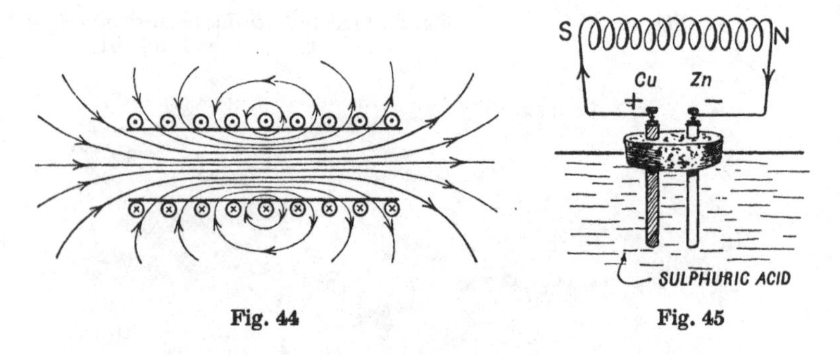

Fig. 44 Fig. 45

The magnetic field inside a solenoid.

An important difference between a current-bearing solenoid and a bar magnet is that the solenoid is hollow and has an intense magnetic field along its axis, while a magnet is solid. It should be noted that in Fig. 44 the lines of force inside the solenoid run from the end of S. polarity to the end of N. polarity. Thus a free N. pole, which when placed outside the solenoid or in the vicinity of a bar magnet is attracted by the S. pole and repelled by the N. pole of the solenoid or magnet, moves, inside the solenoid, away from the S. pole towards the N. pole. The positive direction of the magnetic field inside a solenoid is, therefore, from the S. end to the N. end.

Magnetising action of a solenoid.

If a bar of iron is placed inside a solenoid carrying a strong current it becomes magnetised. The end of the bar at the end of N. polarity of the solenoid acquires a N. pole and the other end a S. pole. This is a much quicker and more effective method of magnetising than stroking with a bar magnet. The direction of magnetisation clearly depends on the direction of the current and the way the solenoid is wound. *The reader should verify that the poles have been marked correctly in Fig.* 46, *using the method of Fig.* 43.

Now draw a diagram similar to Fig. 46 but with the direction of winding the other way, i.e. with the wire passing over the front of the bar at the left-hand end. With the current still flowing in at the left, show that the poles are reversed.

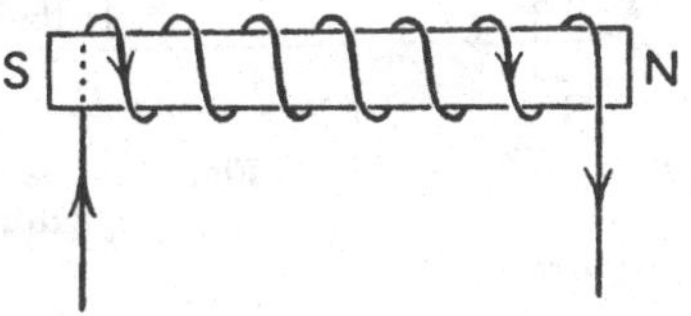

Fig. 46

To sum up the properties of a current-carrying solenoid:

(*a*) *outside* it resembles a bar magnet; hence its "setting" action;

(*b*) *inside* it has a strong magnetic field the "wrong way" (i.e. from S. end to N. end); hence it is valuable for magnetising.

The electromagnet.

We have seen that a solenoid carrying a current behaves like a magnet. If a bar of iron is inserted in the solenoid the strength of the magnet is much increased, since the lines of force due to the magnetised iron are added to those due to the current in the solenoid. Such an arrangement is called an electromagnet: it was invented by Sturgeon in 1825.

The strength of the magnet is proportional to the product of the strength of the current in amperes and the number of turns per centimetre. This product is often called the ampere turns. If many turns of wire are used a weak current will produce as strong a magnet as a stronger current flowing through fewer turns.

So long as soft iron is used and not steel the electromagnet loses practically all of its magnetism when the current is switched off. This is often a great convenience. A simple form of lifting electromagnet is shown in Fig. 47.

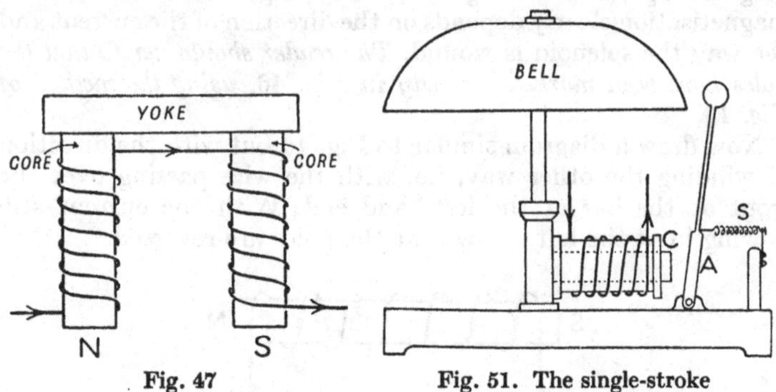

Fig. 47

Fig. 51. The single-stroke signalling bell.

Uses of the electromagnet.

Electromagnets are used for lifting heavy masses of iron and steel, such as girders, scrap iron, or a cargo of nails, for separating iron and copper scrap, for picking out tins and iron from a city's refuse, and, in hospitals, for drawing out pieces of iron lodged in the human eye (see Figs. 48–50).

Tramcars are fitted with a slipper electromagnetic brake. The slipper consists of an iron block carried just above the rail; the upper part of the block is wound as an electromagnet. When the current is switched on, the block is attracted to the rail and acts as a very efficient brake.

Lamps are made with an electromagnet in the base which clings to iron when the lamp is switched on. This is a convenience when looking at steel machinery, since the lamp will hang itself up in any convenient place.

The single-stroke signalling bell.

The single-stroke signalling bell, used extensively in railway cabins and stations, consists (see Fig. 51) of an electromagnet, with a soft iron core C, and a short bar of iron A, called the

By courtesy of the Igranic Electric Co. Ltd

Fig. 48. An electromagnetic crane lifting steel girders. The electromagnet is immediately below the chains. Inside the flat, circular, iron casing are a central core and coils through which passes an electric current.

By courtesy of the General Electric Co. Ltd.

Fig 49. An electromagnet dropping a cracker ball to break up scrap metal.

By courtesy of the Igranic Electric Co. Ltd.

Fig. 50. A magnetic separator for removing stray iron from bulk material. The material passes on a moving belt over the pulley, which is magnetised by passing an electric current through coils in its interior. The non-magnetic material is projected beyond the pulley but any iron or steel is attracted and carried under the pulley. The picture does not show the separator in action: iron filings have been placed on paper, placed where the belt normally passes, to reveal the magnetic field.

armature, which is pivoted at its lower end and attached to a light spring *S* at its upper end. When a current is passed through the coil of the electromagnet the armature is attracted and causes the hammer to make a single stroke upon the bell. When the current is switched off the spring *S* pulls the armature back to the position shown in the figure.

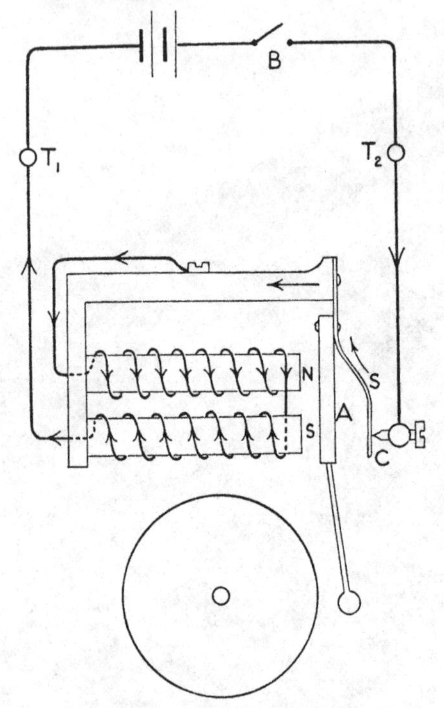

Fig. 52. The trembler bell.

The trembler bell.

The ordinary household trembler bell consists of an electromagnet, armature and spring similar to the one we have just described, but it has, in addition, a make-and-break mechanism which causes the armature to be attracted and pulled back several times per second.

The electromagnet comprises two coils wound on two arms of a soft-iron framework (see Fig. 52). When the bell-push *B* is pressed, the electromagnet attracts the armature *A*, to which is attached a small hammer, and the hammer hits the bell. The armature is also attached to a piece of springy steel, *S*, which makes contact with an adjustable screw at *C*. When the armature is attracted, the contact at *C* is broken; the circuit is no longer complete and the current ceases to flow. The springy steel *S* now pulls back the armature (which is no longer attracted by the electromagnet), and contact is made once more at *C*. The armature is again attracted by the electromagnet and thus *A* begins to "tremble", causing the bell to be struck repeatedly. The speed of trembling depends on the inertia of the armature and the springiness of *S*. The bell may be made to work most efficiently by adjusting the movable screw which makes contact at *C*.

There is usually a certain amount of sparking at *C* and the contact must be made of a metal which has a high melting point and which does not readily oxidise, in order to ensure a good electrical connection. Silver is the cheapest metal which fulfils these conditions and is often used. The ideal metal, platinum, is far too expensive.

Indicators.

In a large house the electric bell in the kitchen may be rung from many different rooms. To indicate the room, the wire from the room is connected to the bell through the coil of an electromagnet: when the current flows the electromagnet attracts an armature carrying a small disc or flag; when the current stops flowing the armature is released and the disc waggles to and fro. There is an indicator corresponding to each room.

The relay.

The relay is an instrument by means of which a small current may switch on a large current. Thus a tiny current may cause a ship to be launched or control distant switches in a Grid substation by operating a relay. A relay therefore may be regarded as an electric trigger.

Fig. 53 represents a simple magnetic relay. The feeble current passes through the coil of an electromagnet, having a large number of turns, and causes a light, delicately pivoted, armature

A to be attracted. The armature then makes contact at *C* and thereby completes a circuit into which a strong current is supplied by another source of electricity. The armature is pulled back by a weak spring, *S*, when the current in the coil of the electro-magnet ceases to flow.

We shall see (p. 256) that the relay is much used in telegraphy.

Relays are also used as cut-outs. If a current greater than a certain safe value flows in a circuit, the armature of the relay is attracted and it is arranged that the current is thereby switched off. The reader should design a cut-out on the lines of Fig. 53:

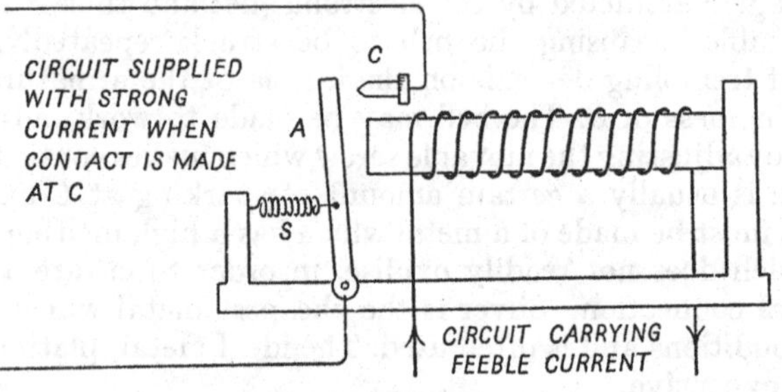

Fig. 53. The relay.

remember that there is one current only, in this case, and it must be switched off when it exceeds a certain maximum value.

All large electrical systems must be equipped with protective gear which operate, almost invariably, by means of relays. If, for example, one of the cables carried by the Grid pylons breaks, the protective relays come into operation and the wires are dead before they reach the ground.

Automatic railway signalling.

An interesting example of the use of a relay is the automatic operation of railway signals.

The line is divided into sections and the rails of each section are insulated from those of adjacent sections by placing fibre insulation pieces between the fish plates and the rails.

Fig. 54 represents a section which is insulated from the adjacent sections at the points *i*. A current from a battery *B* flows through the variable resistance *R*, the rails, and through a relay, when there is no train in the section. The relay is energised and operates a motor which lowers the signal arm. When a train enters the section the current can now flow from one rail to the other through the wheels and axles of the train. The relay is said to be short-circuited and practically no current flows through it. It is no longer energised and hence the signal arm (controlling entry to that section) moves to the stop position. It should be noted that, in this system, if the battery fails or there is a break in its circuit, the relay ceases to be energised and the signal automatically goes to the stop position.

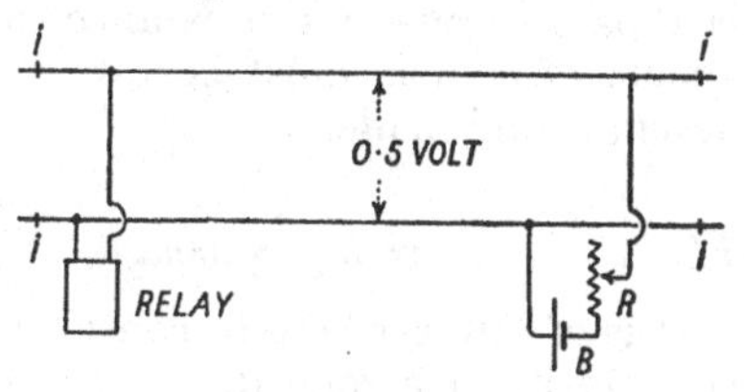

Fig. 54

By courtesy of the Westinghouse Brake and Saxby Signal Co. Ltd.

Fig. 55. Automatic signal apparatus, relays, etc., on the London to Brighton line of the Southern Railway.

The battery used has an E.M.F. of about 1 volt. The function of the resistance R is to prevent the battery being completely short-circuited and hence being made to supply a very large current, when the rails are connected through a train in the section.

Now the ground between the rails is an appreciable conductor. Its resistance decreases when the weather is wet, but we can take, as an average value, 10 ohms per 1000 feet of track. There is thus a permanent resistance, or "shunt", across the relay, which, of course, must be taken into consideration when it is designed and made.

The full block overlap system.

It is of interest here to consider how the signals, operated by the relays, are arranged.

Suppose, in Fig. 56, *A*, *B*, *C*, *D* represent four sections of the line. The length of each of these sections depends on the frequency and speed of the trains: a usual length is half a mile, but on suburban and underground lines, where trains follow each

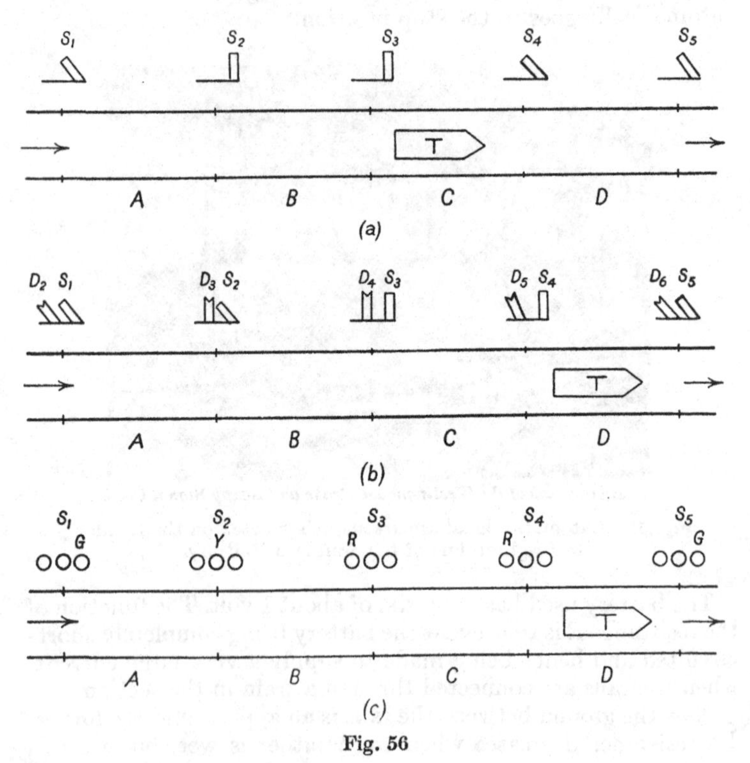

Fig. 56

other in quick succession, the length may be as small as 300–400 yards. Each section is controlled by a signal, S_1, S_2, S_3, etc. (see Fig. 56 (*a*)), and when a train is in any particular section the two signals behind it are automatically at the stop position. The system is known as the "full block overlap" system, since there must, obviously, be a complete section or block between two trains.

In Fig. 56 (*b*), beside the "stop" signals S_1, S_2, S_3, etc. are shown "distant" signals D_2, D_3, D_4, etc. The distant signals are

By courtesy of the Westinghouse Brake and Saxby Signal Co. Ltd.

Fig. 57. Automatic signals, Up Line between Three Bridges and Balcombe Tunnel. Note the "A" light, mounted below the main signal, which is illuminated when the red aspect is showing above it. This distinguishes an automatic signal from one which is cabin-controlled.

painted yellow, have a V-shaped notch, and are usually placed below the stop signals. The function of the distant signals is cautionary and informs the engine driver of the state of the next signal. Thus in Fig. 56 (*b*), D_2 is down, showing that S_2 is down, while D_3 is up, showing that S_3 is up.

The latest practice is to dispense with signal arms entirely and to use coloured lights, red—stop, yellow—caution, green—proceed, see Fig. 56 (*c*) and also Fig. 57.

Summary

The lines of magnetic force due to a current flowing in a straight wire are **concentric circles round the wire. The positive direction of these lines of force is the direction in which the thumb rotates when a corkscrew is screwed along the wire in the direction of the current. (Maxwell's corkscrew law.)**

The magnetic field due to a flat coil is similar to that of a short, flat, bar magnet; one face of the coil behaves as a N. pole and the other as a S. pole (see Fig. 42). A *solenoid* is a long coil; its poles may be determined also by Fig. 43. A current-bearing solenoid will magnetise an iron bar placed within it.

An *electromagnet* consists of a solenoid containing a soft-iron core. It forms the essential component of an electric bell and a magnetic relay.

QUESTIONS

1. (*a*) In what direction will the N. pole of a compass needle be deflected when held over a current flowing from north to south?

(*b*) A compass needle is held below a wire running north and south and the N. pole of the needle is deflected to the east. What is the direction of the current in the wire?

2. A long straight wire is set in the magnetic meridian and a pivoted magnetic needle is placed (1) vertically below the wire, (2) vertically above the wire. Explain in each case what happens to the needle when a current is passed through the wire from S. to N. and gradually increased. What would be the behaviour of the needle in cases (1) and (2) if the current were sent through the wire placed perpendicular to the magnetic meridian in the direction from E. to W. (N.)

3. Imagine a man swimming over a wire carrying a current, in the direction in which the current is flowing. Towards which hand (right or left) of the swimmer will the N. pole of a compass, also above the wire, tend to be deflected? Explain how you obtain your result. (A rule similar to this was devised by Ampère before the invention of the more convenient corkscrew rule of Clerk Maxwell.)

4. Describe the experiments you would carry out to investigate the distribution of the lines of magnetic force near (*a*) a long straight wire, (*b*) a circular coil, carrying an electric current. Sketch in each case the lines of force due to the current alone. (C.)

5. Explain fully:

(*a*) What is the cause of the increase in the magnetic field when a bar of soft iron is placed inside a solenoid?

(*b*) Why does a coil, connected to a battery and floating above it on water, tend to set itself in a particular direction? Draw a diagram showing how it does set, putting in the direction of the current and also the direction of the magnetic N. (i.e. the direction of the earth's magnetic field).

(*c*) The large spiral filament of an old-type lamp when carrying direct current may be attracted or repelled by a magnet, but if the current is alternating the filament vibrates violently when the magnet is brought near.

6. Fig. 58 represents a solenoid with widely spaced turns of wire, the × marks representing sections of turns where the current is flowing down into the paper and the ○ marks where the current is flowing up out of the paper. Copy the diagram (making it as large as possible) and indicate clearly the lines of magnetic force due to the solenoid and their direction.

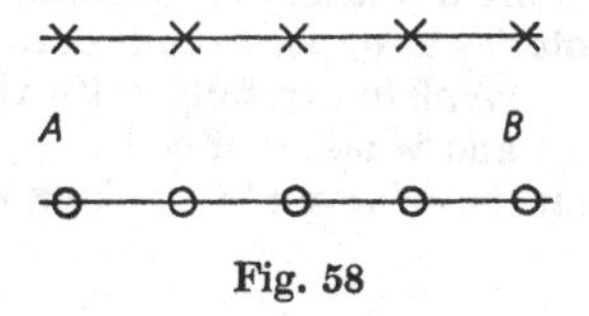

Fig. 58

A rod *CD* of soft iron is placed inside such a solenoid with the end *C* at *A* and the end *D* at *B*, and after some time is removed. What are now the magnetic properties of *CD*? What differences, if any, would have been observed if *CD* had been a steel rod? Give your reasons. (O. and C.)

7. Describe, with diagrams, how you would magnetise a long piece of hard steel so that it had a N. pole at each end, and a S. pole in the middle, (*a*) by means of a permanent magnet, (*b*) by means of an electric current.

8. Describe how you would make and use a small electromagnet of horse-shoe type, and mention any purpose for which it might be employed. Make a careful diagram, showing the direction of the current and the nature of the poles produced.

Explain why the core of an electromagnet is made of soft iron, while permanent magnets are made of hard steel. (N.)

9. Describe the construction of an electric bell, explaining how it acts. How would a two-room indicator be connected with it to show from which of the two rooms the bell had been rung? (N.)

10. Draw a circuit containing one battery whereby an electric bell may be rung from both the front door and the back.

11. Describe the action of the magnetic relay. Explain some practical use for which it is employed.

12. (*a*) On looking towards the poles of a U-shaped electromagnet does the current appear to circulate round the two poles in the same direction? Is the direction clockwise or anti-clockwise in the case of the S. pole?

(*b*) Two electromagnets have identical iron cores which are wound, the one with 1000 turns, and the other with 200 turns of wire. What current in the 200 turns will produce as strong a magnet as that obtained by passing 0·12 ampere through the 1000 turns?

13. How would you use a solenoid to magnetise a steel ring so that (1) it might exhibit polarity at opposite ends of a diameter, (2) so that it exhibits no polarity. Explain carefully, with the aid of a diagram, the direction of current and winding of coils.

Could the same results be achieved by stroking with magnets? If so, how?

14. Explain: If two coils of wire carrying currents are suspended so as to face each other, it is found that when the currents in the coils are in the same direction the coils attract each other, and when in opposite directions repel each other. (O. and C.)

15. Two circular loops of wire are placed facing and parallel to each other. Draw the lines of force when currents flow in the coils (*a*) in the same direction, (*b*) in opposite directions.

16. Explain why a solenoid tends to contract when a current passes through it. (Roget's spiral.)

17. *A* and *B* are the poles of an electromagnet standing on a piece of glass. Two short pieces of soft-iron wire of equal length are lying side by side on the line *AB*. On switching on the current of the magnet the two wires roll apart. Explain this. (C.)

18. A circular coil of wire, free to move, rests near one end of a bar magnet as shown in Fig. 59. What will happen if a current is made to flow in the coil in the direction shown? Give reasons for your answer. (L.)

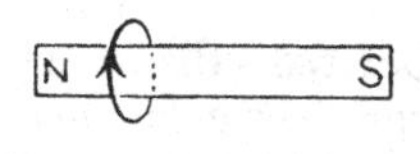

Fig. 59

19. A hollow steel tube is magnetised so as to have poles near its ends and placed with its axis east and west. A small compass needle is then moved slowly from a distant point on the prolongation of the axis so that it passes completely through the tube and eventually reaches a distant point beyond the tube. Describe and account for the behaviour of the compass. The steel tube is then replaced by a cardboard cylinder round which is a closely wound spiral of insulated wire carrying an electric current, and the compass needle is moved as before. Describe and account for the behaviour of the needle in this case. (O.)

20. A model railway track, consisting of a large oval loop in four insulated sections, is to be wired for automatic signalling using a battery and pea lamps only (no relays). Draw a wiring diagram. [Start with red lamps only, one controlling each section and arrange for them to light up at the correct times. You can then try including green and amber lights also.]

Chapter VI

GALVANOMETERS, AMMETERS AND VOLTMETERS

The first galvanometer.

Within two years of Oersted's discovery of the magnetic effect of an electric current, the first *galvanometer*, an instrument for detecting and measuring electric currents, had been invented. Schweigger, a German physicist, looped a wire into a rectangular coil with a large number of turns (see Fig. 79), thereby causing an electric current to produce an intensified magnetic field at the centre.

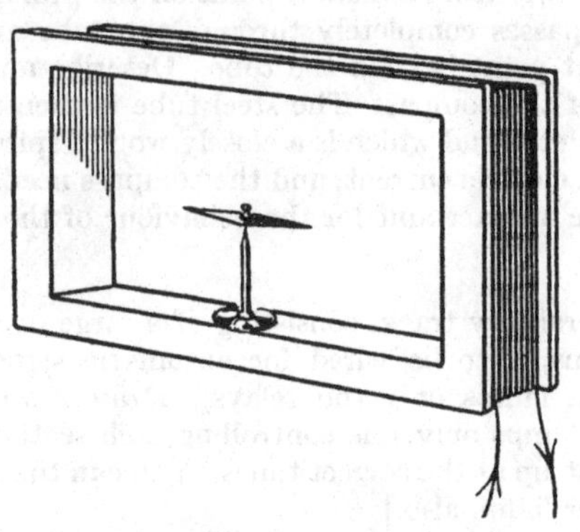

Fig. 79. Schweigger's galvanometer.

A pivoted compass needle at the centre of the coil tends, when a current is passed, to set itself along the lines of force, at right angles to the coil. The coil is therefore set with its plane N. and S., so that the compass needle lies in its plane while no current flows.

When the current flows the forces due to its magnetic field begin to rotate the needle, and as soon as this happens forces due to the earth's field try to pull the needle back. Finally, the *at right angles (see Fig. 61), in such a way that the* **f***irst finger is pointing in the direction of the* **f***ield, and the se***c***ond finger in the direction of the* **c***urrent: then the thu***m***b represents the direction of* **m***otion.* The letters in black type suggest an easy method of remembering this useful rule. The reader should apply Fleming's left-hand rule to Fig. 60 to verify that it gives correctly the direction of motion of the wire.

If the wire is inclined obliquely instead of at right angles to the field, the force is smaller but is still at right angles to both wire and field. The direction of the force can be obtained by Fleming's left-hand rule by considering the direction of the "component" of the field which is at right angles to the current.

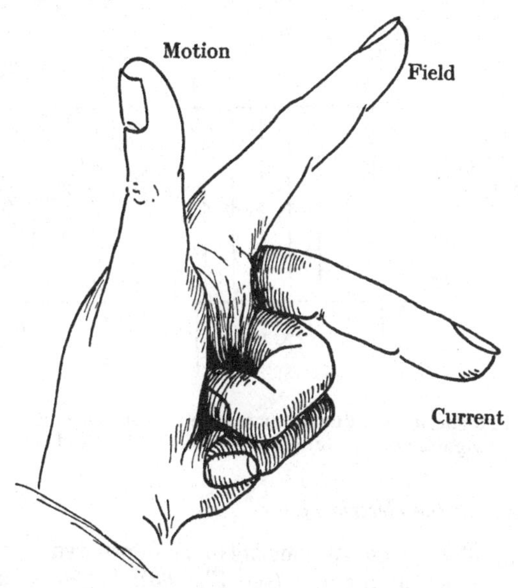

Fig. 61. Fleming's left-hand rule.

Explanation by means of lines of force.

By drawing the magnetic lines of force we can obtain a picture or model to show how the force on the wire in Fig. 60 is exerted.

Fig. 62 (*a*) shows the lines of force due to the current in the wire, and Fig. 62 (*b*) the lines of force of the external field due to the magnetic poles. Fig. 62 (*c*) shows the combined field of (*a*) and (*b*) when the wire is placed between the poles.

Note that, in Fig. 62 (*a*) and (*b*), the lines of force on the left of the wire are in the same direction as those of the external field, while those on the right of the wire are in the opposite direction. Consequently in the combined field of Fig. 62 (*c*) the field to the left of the wire is strong—there are a large number of lines, while the field to the right is weak.

If we assume, with Faraday, that the lines of force are in tension and trying to shorten (see p. 13), we should expect the wire to be urged to the right. This is precisely what we find by experiment.

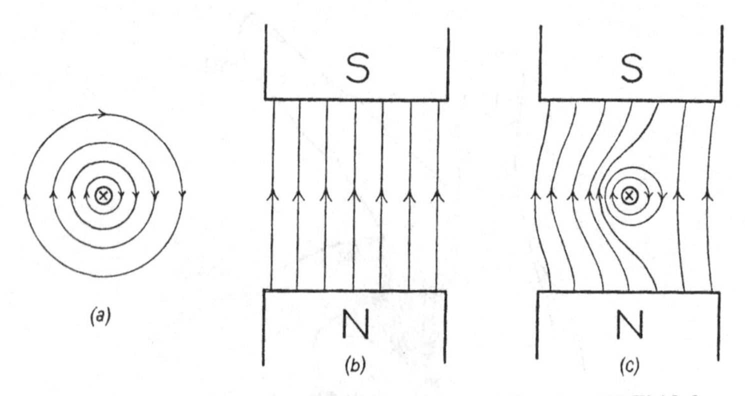

Fig. 62. (*a*) Magnetic field due to current in straight wire. (*b*) Field due to magnetic poles. (*c*) Combined field of (*a*) and (*b*).

The principle of the electric motor.

The simple electric motor consists of a coil pivoted between the poles of a permanent magnet (see Fig. 63). When a current is passed through the coil in the direction indicated in the figure we can show, by applying Fleming's left-hand rule, that the left-hand side of the coil will tend to move down and the right-hand side to move up. (Remember that the direction of the field due to the permanent magnet is from the N. to the S. pole.) Thus the coil will rotate in a counter-clockwise direction to a vertical position.

Suppose, when the coil is vertical, the current in it is reversed suppose also that the coil has sufficient momentum to overshoot the vertical position slightly. Once again the side of the coil which is now on the left carries a current into the paper and is urged downwards. Similarly, the right-hand side of the coil is urged upwards. The coil, therefore, continues to rotate until it is vertical again. Thus if the current is reversed every time the coil is vertical the latter will rotate continuously. This is the principle of the electric motor.

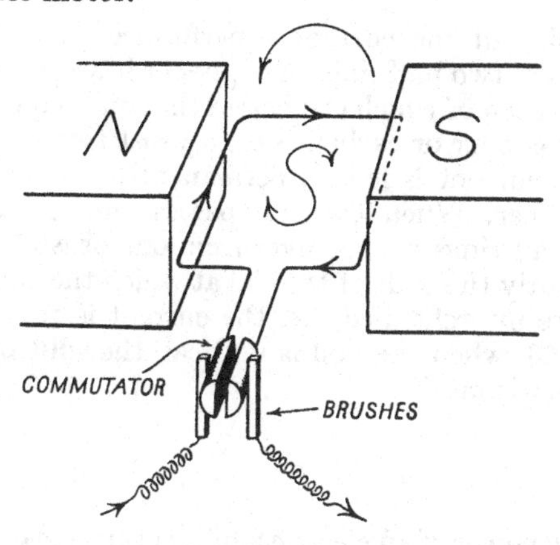

Fig. 63. Principle of the electric motor.

The lines of force of the resultant field due to the current in the coil and the permanent magnet are shown in Fig. 64. Compare this with Fig. 62*c*. By regarding the lines of force as being in tension it is easy to explain why the left-hand side of the coil moves down and the right-hand side moves up.

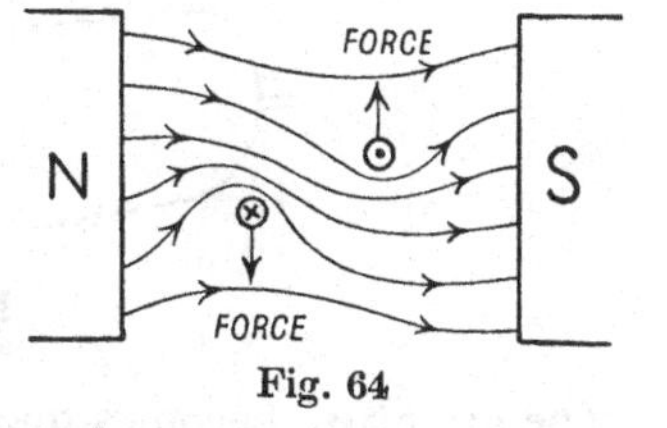

Fig. 64

There is another way of explaining the rotation of the coil. We have seen (p. 56) that a coil carrying an electric current behaves like a flat bar magnet, one face acting as a N. pole and the other as a S. pole. In Fig. 63 the upper face is a S. pole, and owing to its attraction by the N. pole of the permanent magnet the coil will tend to turn until it is vertical. The lower face of the coil is, of course, a N. pole, and will similarly be attracted by the S. pole of the permanent magnet. When the current is reversed the S. face of the coil now becomes a N. face, which is repelled by the N. pole of the magnet and consequently the coil continues to rotate.

The commutator.

The reversion of the current is performed by a *commutator*, which consists of two half rings of copper or brass, insulated from each other, to each of which one end of the coil is connected. Two "brushes" of copper or carbon press against this split ring as it revolves. The current is led in, permanently, at one brush and out at the other. When the coil passes through its vertical position the half rings change over from one brush to the other, and consequently the ends of the coil at which the current enters and leaves are interchanged, i.e. the current is reversed. Note that, in Fig. 63, when the coil is vertical, the split between the rings will be horizontal.

The armature.

The turning power of an electric motor is very greatly increased by winding the coil on a soft-iron core, called an *armature* (see Fig. 65), and also by increasing the number of turns of wire.

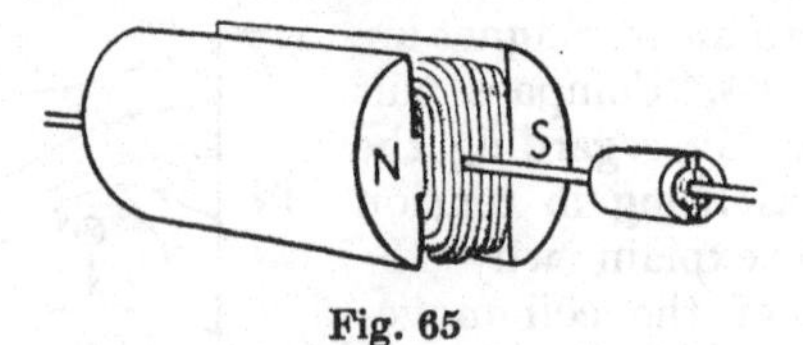

Fig. 65

The armature becomes magnetised by the current in the coil as indicated in the figure, with the result that the armature and coil comprises a much stronger magnet than the coil alone. Hence there is a greater force of attraction or repulsion between its poles and those of the permanent magnet than in the case of the coil alone.

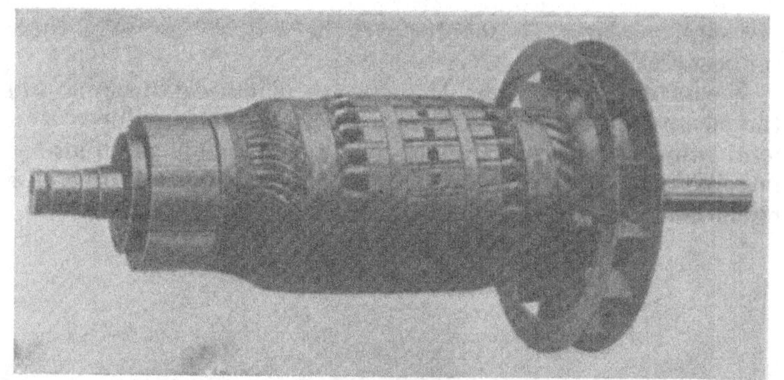

By courtesy of the British Thomson-Houston Co. Ltd.

Fig. 66. The armature of a D.C. motor. It consists of an iron core on which there are a number of coils, thereby making the torque much more uniform than is the case with a single coil, (see p. 211). This necessitates a commutator (on the left), with a large number of segments, instead of two. Note also the ventilating fan on the right. See also Fig. 68.

The field magnet.

The magnet between the poles of which the armature turns is usually an electromagnet, since such a magnet may be made more powerful than a permanent magnet. Its coils are known as the *field coils*, and may be connected in series (series-wound), see Fig. 67, or in parallel (shunt-wound) with the armature coil.

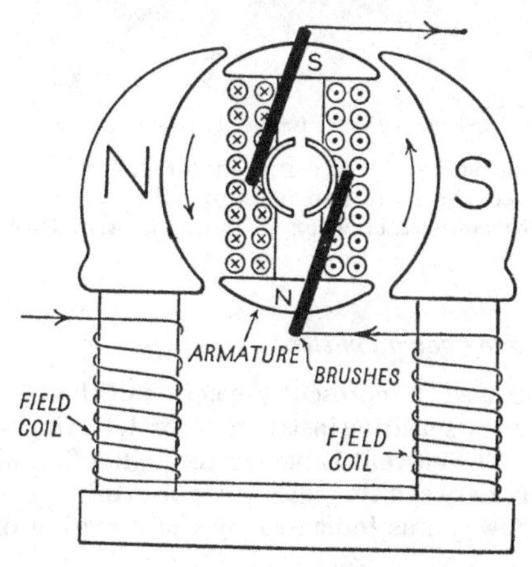

Fig. 67. Series-wound electric motor

We shall see later, in Chapter XII, the relative merits of these connections.

The inside faces of the pole pieces are cylindrical in shape, and the clearance between them and the armature is reduced to a minimum. In this way the magnetic field is made as intense as possible: the air gap is kept very small, since lines of force pass much more readily through iron than through air.

By courtesy of the British Thomson-Houston Co. Ltd.

Fig. 68. The frame and poles of a D.C. motor—armature and end-shields removed. Note that there are two pairs of poles. There are also smaller, subsidiary poles to counteract what is known as armature reaction. See also Fig. 66.

The Einthoven string galvanometer.

Fig. 60 may be used to represent the action of the Einthoven string galvanometer, a very sensitive instrument for detecting and measuring electric currents. The current is passed through a fine wire, kept taut by a weak spring, between the poles of a powerful electromagnet. The wire is urged sideways, as indicated by the direction of the force in Fig. 60.

In order to observe the motion of the wire a microscope is placed in a hole bored in one of the poles of the electromagnet (see Fig. 69). In the eye-piece of the microscope is a fine etched glass scale on which the position of the image of the wire is read.

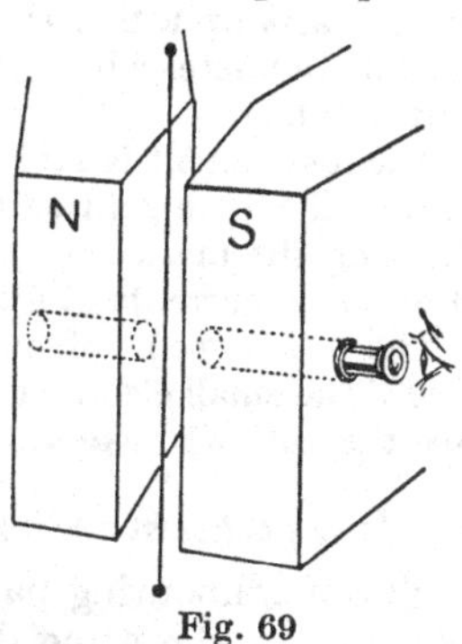

Fig. 69

Instead of the microscope a bright light and optical system, which projects an image of a tiny length of the wire on to a moving photographic film, may be used.

Fig. 70 is a record of the beating of the human heart taken by this method. A man places each hand in a bath of salt solution and the two baths are connected

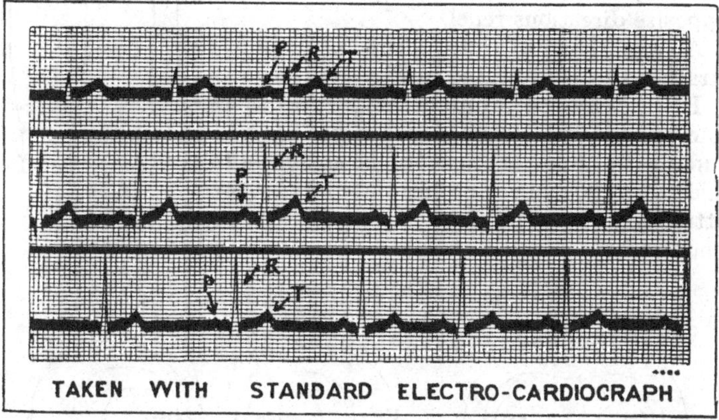

By courtesy of the Cambridge Instrument Co. Ltd.

Fig. 70. Electric currents set up by the beating of the heart. These three simultaneous records, obtained by means of a three-stringed Einthoven galvanometer, show the currents flowing between right arm and left arm, right arm and left leg, left arm and left leg. The patient sits with his two hands and left foot immersed in salt solutions, from which leads are taken to the galvanometer.

The heart pumps the blood by alternately contracting and dilating. It has two receiving chambers called auricles, and two pumping chambers called ventricles. The peaks marked P occur during the contraction of the auricles, and R and T during the contraction of the ventricles. Any irregularity in the beating of the heart map be detected by this method.

to the Einthoven string galvanometer. Each time the man's heart beats it sets up a tiny electric current, causing a slight movement of the galvanometer wire (or string) and a corresponding "kick" on the moving film.

The instrument is said to be "dead beat" since, when the current ceases, the wire returns at once to its zero position without overshooting the mark and vibrating. Thus it can be used for recording alternating currents of frequencies up to 200 per second.

The Einthoven galvanometer is also used in sound ranging to record the small electric current set up by the arrival of a sound wave (see the author's *Sound*).

The forces between two wires carrying currents.

Two neighbouring parallel wires carrying currents in the same direction attract one another, while wires carrying currents in opposite directions repel.

An experimental arrangement for demonstrating this is indicated in Fig. 71.

If we draw the resultant magnetic fields due to the two currents (see Fig. 72), assuming the lines of force to be in tension and to repel one another laterally, the attraction or repulsion may be readily understood.

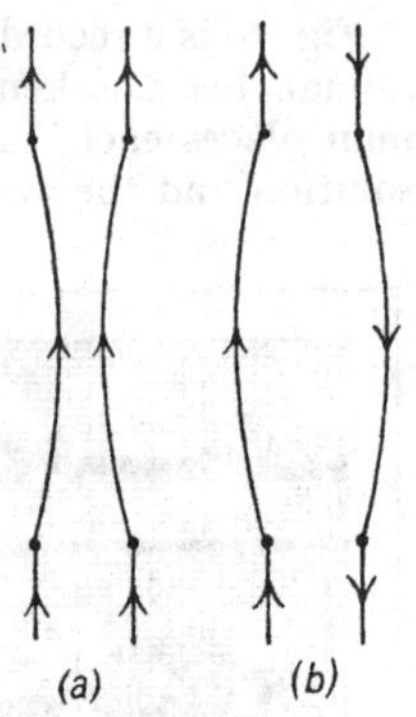

Fig. 71

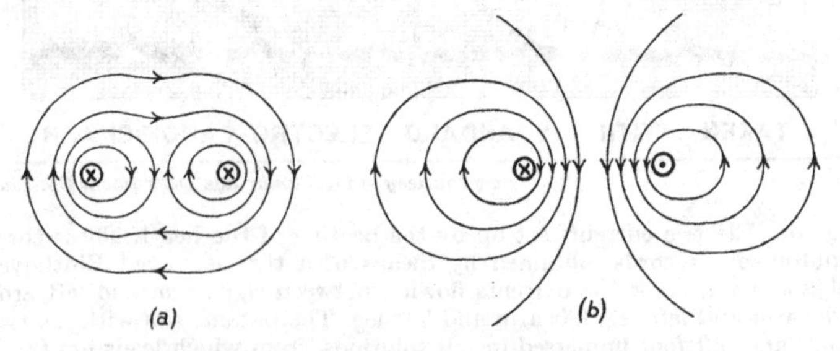

Fig. 72. (*a*) Lines of force due to two "like" currents.
(*b*) Lines of force due to two "unlike" currents.

The true nature of the phenomenon, discovered by André Marie Ampère (1775–1836), was at first completely misunderstood by some physicists. They saw in it only the attraction and repulsion of electric charges, instead of the reaction between the magnetic fields of two currents. But, as Ampère pointed out, like charges repel, whereas like currents (i.e. currents in the same direction) attract. One critic, seeking to disparage Ampère's achievement, remarked that since two wires carrying currents are each subject to forces under the action of a magnet, it was obvious that they must therefore exert forces on each other. To this Ampère's friend Arago replied by taking two keys from his pocket, and saying: "Each of these keys attracts a magnet: do you believe that they therefore also attract each other?"

Ampère's fundamental researches carried out in the latter part of 1820 and 1821 were of such importance and completeness that he has been termed the Newton of electrodynamics.

The news of Oersted's discovery had reached Paris in July 1820, and it stimulated intense activity among the brilliant members of the French Academy, men like Ampère, Biot, Arago, and their president, the celebrated Laplace. Eager for fame and priority, they seized upon the opportunity of investigating the new phenomenon, but it was Ampère who after the amazingly short space of a few months claimed the lion's share of success.

Ampère was a man of penetrating and profound intellect. His temperament was strange and intensely religious, as is evidenced by his self-chosen epitaph—*Tandem felix* (Happy at last). No doubt his terrible experience at the age of eighteen years during the French Revolution, when his father was guillotined, was partly responsible. It is said that for a year after that event he wandered over France aimless and distraught.

SUMMARY

A wire carrying a current in a magnetic field is acted upon by a force whose direction may be determined by **Fleming's left-hand rule** (see p. 75). The force on the wire may be accounted for by the tension of the lines of force of the combined magnetic fields due to the current and the original field.

The *simple electric motor* consists of a coil pivoted between the poles of a magnet: the current in the coil is reversed by means of a commutator each time the plane of the coil is opposite (i.e. parallel) to the poles of the magnet.

Two neighbouring wires carrying currents exert forces on each other; like currents attract, unlike currents repel.

QUESTIONS

1. A current flows downwards in a vertical wire. In what direction does the wire tend to move under the action of the earth's magnetic field?

2. What do you know of the action of a magnetic field on a conductor carrying a current?

The wire *BA* hangs vertically between the poles of a large electro-magnet (see Fig. 73). It is free to turn in any direction about *B*. State and explain what happens when a current is passed through the wire in the direction *AB*. (*A* is free to move.)

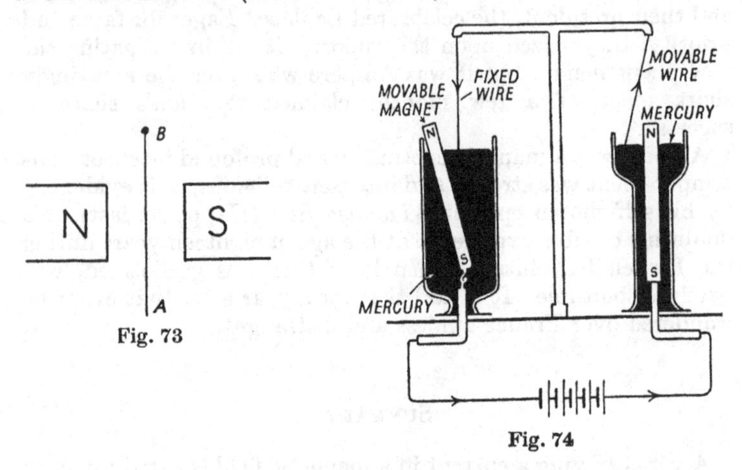

Fig. 73

Fig. 74

3. In Fig. 74 the glass vessels contain mercury, which is a good conductor and which also allows a magnet or the end of a wire to move in it. In the left-hand vessel the top of the magnet rotates round the fixed wire: in the right-hand vessel the wire rotates round the fixed magnet. If the directions of the currents in the wires are as shown, what is the direction of the rotations? Explain your reasoning fully.

(This experiment was first performed in 1821 by Faraday after several unsuccessful attempts by other physicists.)

4. In Fig. 75 *AB* is a metal rod which is free to roll upon, and makes contact with, insulated metal rails, between the poles of a magnet. In which direction will *AB* roll when the current passes in the direction shown? Explain fully.

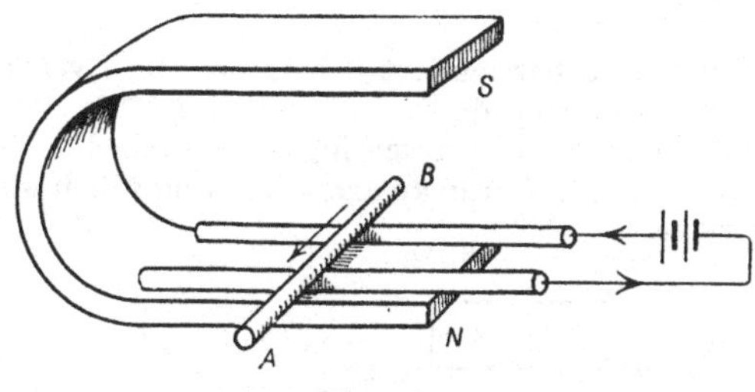

Fig. 75

5. In Fig. 76 a current passes from the axle of a wheel down a spoke which dips into mercury; the wheel is pivoted between the poles of a permanent magnet. In which direction will the wheel revolve? Explain fully. (This is known as Barlow's wheel.)

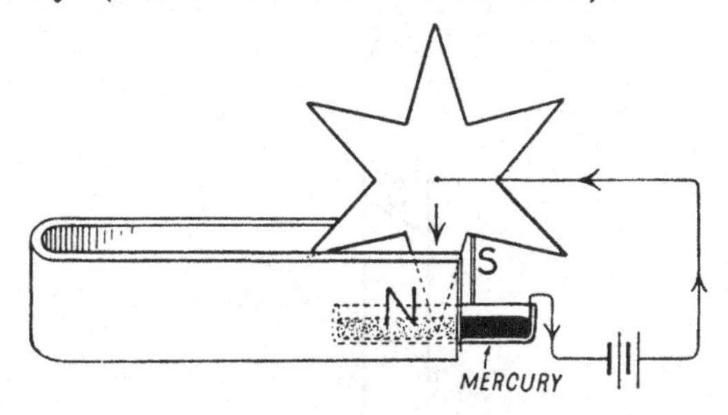

Fig. 76. Barlow's wheel.

6. Describe an experiment to show that when a conductor, through which an electric current is flowing, is situated in a magnetic field, a mechanical force is in general exerted on the conductor by the field.

How is this employed in the production of continuous rotation in the armature of a direct-current electromotor? (C.)

7. Explain the principle of the electric motor, by considering a single coil pivoted between the poles of a permanent magnet.

Describe with the aid of diagrams the nature and use of the following components of a practical motor:

(*a*) the commutator,

(*b*) the armature,

(*c*) the field magnet.

8. The N. pole of a bar magnet is held near to a vertical wire in which a current is flowing downwards (see Fig. 77). Draw a large diagram of the lines of force representing the combined field due to the magnet and the current. Put in an arrow showing which way the wire will tend to move.

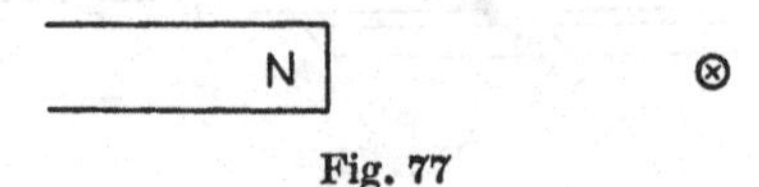

Fig. 77

9. A piece of very thin wire hangs straight down near to a bar magnet. When a current is passed through the wire, the latter twines itself round the magnet as in Fig. 78. Draw a diagram, putting in the direction of the current, and explain the phenomenon fully.

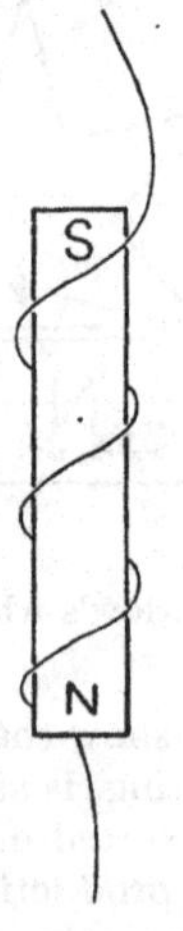

Fig. 78

Chapter V

ELECTRODYNAMICS

A wire carrying a current in a magnetic field tends to move. There is a force exerted on it. The study of such forces is called *electrodynamics.*

The force on a wire carrying a current in a magnetic field.

The existence of the force on a wire carrying a current in a magnetic field may be demonstrated experimentally by setting up apparatus similar to Fig. 60: the wire moves in the direction

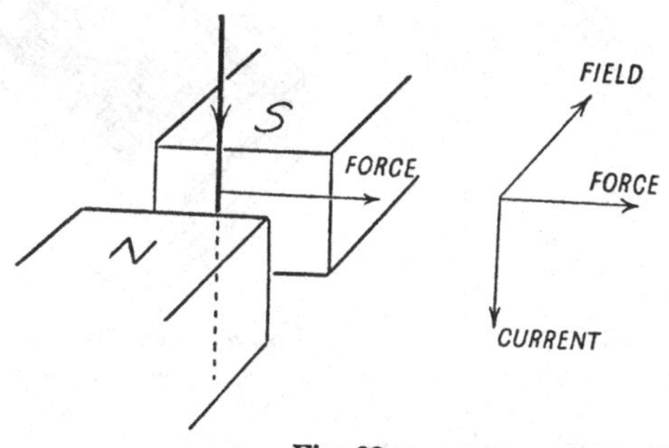

Fig. 60

indicated when the current is switched on. Note that the wire is at right angles to the magnetic field and that the force is at right angles to both wire and field.

If the current in the wire, or the direction of the magnetic field, is reversed, the direction of the force is reversed in each case.

Fleming's left-hand rule.

There is a rule, invented by Fleming, which enables one to predict the direction of the force. It is called *Fleming's left-hand rule. Hold the thumb and first two fingers of the left hand mutually* needle sets itself roughly N.E. or N.W. according to the direction of the current.

Thus two forces act on each pole of the compass needle (see Fig. 80): a controlling force, due to the earth's field, trying to restore the needle to a N. and S. direction, and a deflecting force, due to the magnetic field of the current in the coil, trying to make the needle set itself E. and W., at right angles to the coil.

It is of interest to note that the turns of wire in Schweigger's galvanometer were insulated with resin or sealing-wax: silk- and cotton-covered wires were not introduced until later.

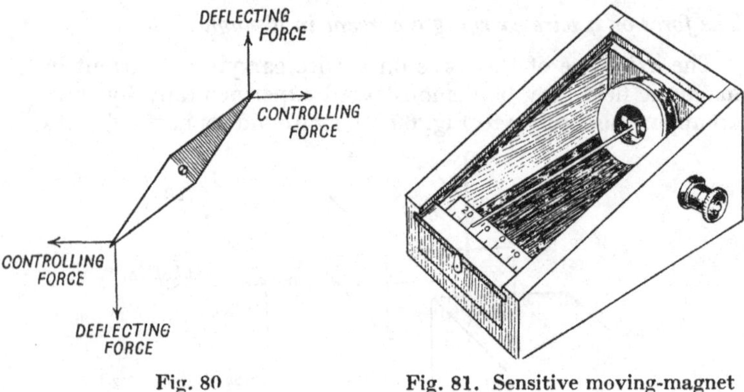

Fig. 80

Fig. 81. Sensitive moving-magnet galvanometer.

Other moving-magnet galvanometers.

The galvanometer we have just described is known as a moving-magnet galvanometer, since the magnet moves while the coil is stationary.

Another and very sensitive form of moving-magnet galvanometer, which will detect currents as small as $\frac{1}{100}$th milliampere, is shown in Fig. 81. The instrument is contained in a wooden box with a glass top to protect it from draughts. A slight movement of the small magnet at the centre of the coil is registered by a very light pointer, which is attached to the magnet and moves over the scale shown.

Such a galvanometer is suitable only for detecting (rather than measuring) tiny currents.

A form of moving-magnet galvanometer which can be used to *measure* currents is known as the tangent galvanometer, and is shown in Fig. 258, p. 322. The full theory of this instrument is given in Chapter XVII.

Moving-magnet galvanometers require to be set before use, i.e. they must be placed with the plane of their coils lying N. and S., so that the magnet lies in the plane of the coil. Since, ordinarily, the earth's field is used as the controlling force in these galvanometers, they are very susceptible to the effect of stray magnetic fields, caused, for example, by bringing magnets into their vicinity.

The moving-coil galvanometer.

The most sensitive type of current-measuring instrument is the moving-coil galvanometer, which uses a moving coil and fixed magnet.

A small light coil consisting of a number of turns of fine copper wire is pivoted or suspended between the poles of a powerful permanent magnet.

By courtesy of the Cambridge Instrument Co. Ltd.

Fig. 82. A moving-coil galvanometer, uni-pivot type. The coil rotates between the poles of a permanent magnet. The soft-iron ball at the centre of the coil is fixed and does not rotate with the coil: its function is to give an even scale.

When a current passes through the coil the galvanometer begins to act like an electric motor (see p. 76). The coil begins to rotate between the poles of the permanent magnet. If there were no resistance to this rotation then the smallest current would make the coil turn until its faces were opposite and parallel to the poles of the permanent magnet: but this is prevented by a controlling force exerted by either a hair spring or the torsion (resistance to being twisted) of the suspending fibre. The bigger the current flowing through the coil the stronger the turning effect or tendency to rotate and hence the greater the deflection.

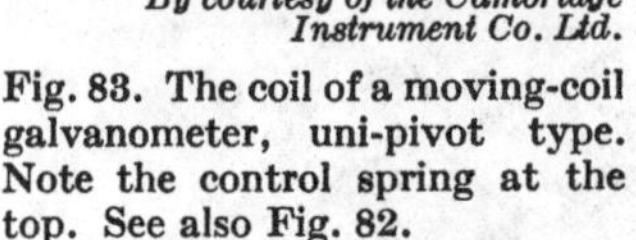

By courtesy of the Cambridge Instrument Co. Ltd.

Fig. 83. The coil of a moving-coil galvanometer, uni-pivot type. Note the control spring at the top. See also Fig. 82.

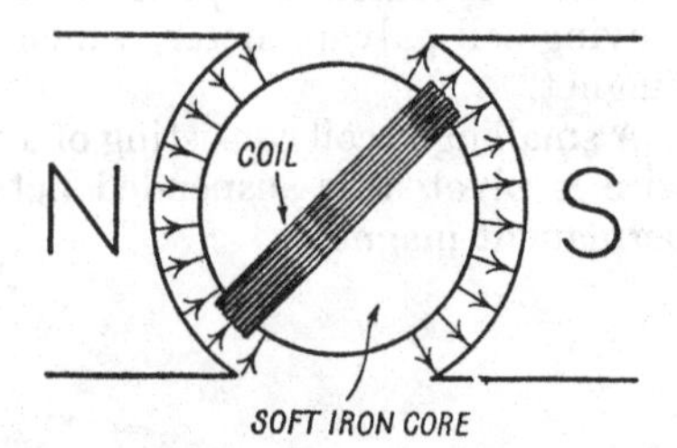

Fig. 84

There are many different forms or designs of this instrument: one form, known as the unipivot type (the coil is supported on a single pivot), is shown in Figs. 82 and 83. The permanent magnet is in the form of a ring with a gap between the poles. Note the fixed soft-iron ball or core at the centre of the coil, which does not turn with the coil, and which has, near its centre, a jewelled bearing to support the vertical spindle of the coil. This core serves to make the magnetic lines of force which enter it from the poles of the permanent magnet radial (see Fig. 84): the lines enter the core everywhere at right angles to its surface and hence, whatever the positions of the coil, its vertical sides move at right angles to the lines of force. This ensures an even scale, the distances between the markings on the scale over which the pointer moves all being equal.

Note also the fine spring attached to the top of the coil, in Fig. 83. When the coil turns this fine spring is twisted or coiled slightly and tries to uncoil, thereby providing the controlling force.

The moving-iron galvanometer.

If two bars of soft iron are placed inside a solenoid and a current is passed through the solenoid, the two bars spring apart. The bars are both magnetised in the same direction, and since their N. poles and also their S. poles are adjacent, they repel each other. This is the principle of the moving-iron galvanometer. It should be noted that even if an alternating current (one that keeps reversing in direction, usually 50 times per second) is passed through the solenoid the bars still spring apart. The direction of magnetisation of each bar, although it keeps changing, is always the same as that of the other bar. Hence a moving-iron galvanometer, unlike a moving-magnet or moving-coil galvanometer, is suitable for measuring alternating currents.

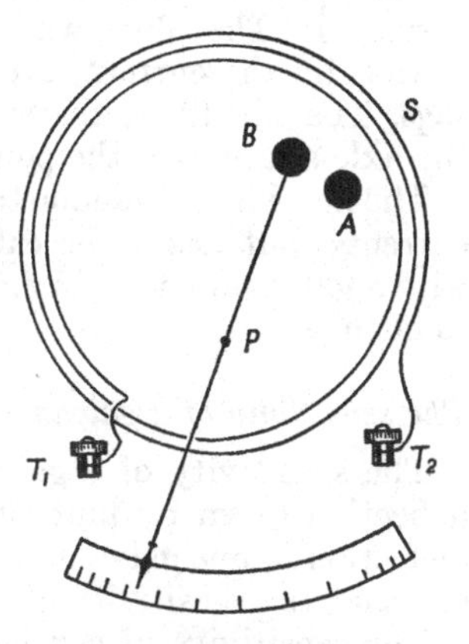

Fig. 85. Moving-iron galvanometer.

Fig. 85 represents the instrument. S is the solenoid which the current enters and leaves by the terminals T_1 and T_2. A is the end view of a fixed bar of soft iron. B is a movable bar of soft iron attached to a lever pivoted at P, which terminates in a pointer. Note that the scale, over which the pointer moves, is uneven, i.e. the distance between the marks on it are not equal.

The controlling force in this instrument is gravity, which tends to restore the lever to its original position. In some instruments the controlling force is provided by a light spring attached to the lever.

The hot-wire galvanometer.

The galvanometers we have described hitherto have made use of the magnetic field set up by an electric current. The hot-wire galvanometer utilises the heating effect of an electric current.

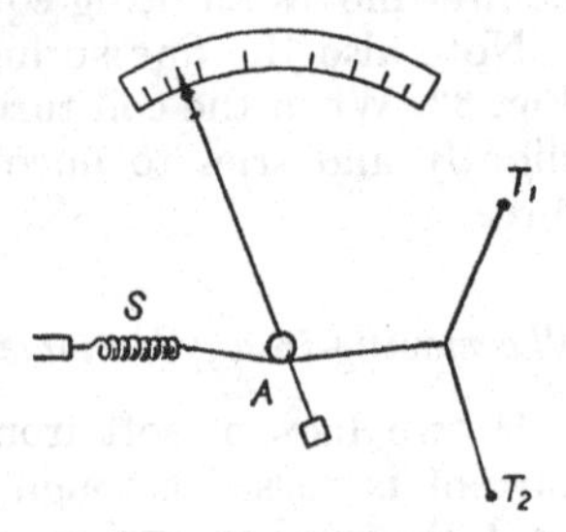

Fig. 86. Hot-wire galvanometer.

The current passes through a fine wire stretched between T_1 and T_2 (see Fig. 86), which is kept taut by a silk fibre passing round the axle A of the pointer and attached to a spring S. The fine wire becomes heated by the current, expands and sags. The silk fibre, therefore, turns the axle and moves the pointer.

The hot-wire galvanometer may be used to measure alternating currents, but has to be calibrated frequently by comparing it with a galvanometer of a more accurate type, since its zero is apt to change.

The sensitivity of a galvanometer.

The sensitivity of a galvanometer may be measured by the deflection given by unit current. Thus if a current produces a deflection in one galvanometer double that in another we should say that the sensitivity of the first is twice that of the second.

The sensitivity of a galvanometer depends on the deflecting force and the controlling force (see Fig. 80). For a high sensitivity the former must be large and the latter small. Thus a very sensitive moving-coil galvanometer will have a powerful permanent magnet and a light coil with a number of turns giving a large deflecting force, but its hair spring will be weak, in order that the controlling force shall be small.

Decreasing sensitivity by means of a shunt.

The sensitivity of a galvanometer may be decreased by putting a suitable resistance, called a *shunt*, across its terminals. Since the shunt and the galvanometer are in parallel, only part of the current passes through the latter, and hence the deflection is smaller than if the whole current had passed through it.

Example. The sensitivity of a galvanometer is to be reduced to $\frac{1}{4}$ *of its normal value. If the resistance of the galvanometer is* 50 *ohms, find the resistance of the requisite shunt.*

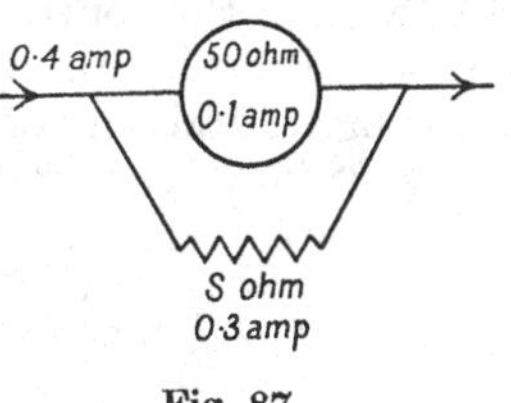

Fig. 87

When a current of 0·4 ampere is to be measured only 0·1 ampere must pass through the galvanometer. Hence 0·3 ampere passes through the shunt (see Fig. 87).

Applying Ohm's law to the galvanometer:

$$V = iR$$

$$= 0{\cdot}1 \times 50 = 5 \text{ volts.}$$

This is the P.D. across the galvanometer terminals and hence it is the P.D. across the shunt.

Applying Ohm's law to the shunt:

$$R = \frac{V}{i}$$

$$\therefore = S\frac{5}{0{\cdot}3} = 16\tfrac{2}{3} \text{ ohms.}$$

Hence the resistance of the shunt must be $16\frac{2}{3}$ ohms.

Ammeters.

An ammeter is a galvanometer specially modified to read directly in amperes. The galvanometer is, for this purpose, made very much less sensitive by placing a thick copper shunt across its terminals.

Let us suppose that a moving-coil galvanometer, which requires a current of 15 milliamperes, i.e. 0·015 ampere, to produce a full-scale deflection, is to be converted into an ammeter reading to 5 amperes. Only 0·015 ampere must pass through the moving coil and hence *a thick wire shunt is placed in parallel with the coil*, of such resistance that it will take the rest of the current,

$$5{\cdot}000 - 0{\cdot}015 = 4{\cdot}985 \text{ amperes.}$$

In Fig. 88 (*a*) a moving-coil galvanometer is shown with a thick copper wire shunted across its terminals. (The current is not, in practice, led in and out of the moving coil by loose wires as shown, but through the controlling hair springs at the front and back. The way in which the axle of the coil is supported is not shown in the figure and only one hair spring, that at the back, is drawn.)

Fig. 88 (b) is a diagram of the circuit.

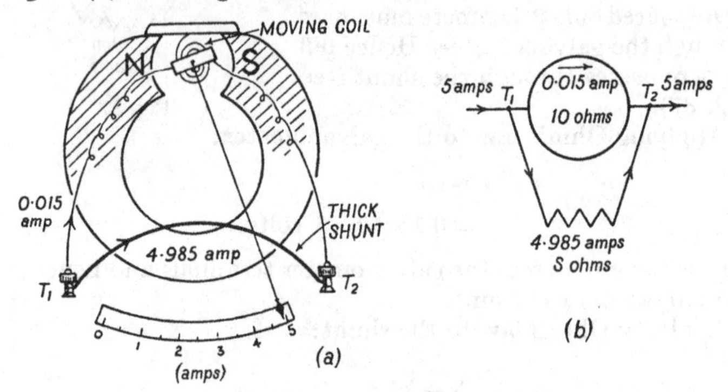

Fig. 88. A moving-coil ammeter.

We will assume that the resistance of the moving coil is 10 ohms and then calculate the necessary resistance of the shunt.

Applying Ohm's law to the moving coil:

$$V = iR.$$

∴ P.D. between ends of coil

$$= 0{\cdot}015 \times 10$$

$$= 0{\cdot}15 \text{ volt.}$$

But the ends of the coil are connected to the ends of the shunt, and therefore the P.D. between the ends of the shunt is also 0·15 volt.

Applying Ohm's law to the shunt:

$$R = \frac{V}{i} = \frac{0{\cdot}15}{4{\cdot}985}$$

$$= 0{\cdot}0301 \text{ ohm.}$$

Thus the requisite resistance of the shunt is 0·0301 ohm.

The total resistance of the ammeter is bound to be slightly less than that of the shunt, since the moving coil is in parallel with the shunt. It is indeed essential that *the resistance of an ammeter should be low*, since otherwise, if it were placed in a circuit to measure the current, it might well reduce the current considerably.

Any type of galvanometer may be converted into an ammeter by the use of a suitable shunt. Moreover, the range of the ammeter—up to 5 amperes, 50 amperes or 0·5 ampere—can be altered by using a different shunt. Fig. 89 shows a moving-coil galvanometer which gives a full-scale deflection with a current of 15 milliamps, shunted so as to read up to 1·5 amps.

By courtesy of Messrs Crompton Parkinson, Ltd.

Fig. 89. A sensitive moving-coil galvanometer which can be converted to an ammeter reading to different ranges by means of the shunts as shown.

Voltmeters.

A voltmeter is an instrument for measuring a P.D. direct in volts.

Since the current passing through a galvanometer is, by Ohm's law, proportional to the P.D. across its terminals, we can measure voltage by means of a specially modified galvanometer.

In order to convert our moving-coil galvanometer (of resistance 10 ohms, and requiring a current of 0·015 ampere to produce a full-scale deflection) into a voltmeter reading, say, to 5 volts, we must place, this time, *a high resistance in series with the moving coil.* The arrangement is shown in Fig. 90(a) and (b).

We will calculate the necessary value, S ohms, of the high resistance.

Suppose a P.D. of 5 volts is placed across the terminals T_1 and T_2 of the instrument. In order to produce a full-scale deflection a current of 0·015 ampere must flow through the moving coil and also through S, since it is in series with the moving coil.

Applying Ohm's law:

$$\frac{V}{i}=R,$$

$$\text{i.e. } \frac{5}{0{\cdot}015}=S+10,$$

$$5=0{\cdot}015S+0{\cdot}15.$$

$$\therefore S=\frac{4{\cdot}85}{0{\cdot}015}$$

$$=323\tfrac{1}{3} \text{ ohms.}$$

Hence the requisite resistance is $323\frac{1}{3}$ ohms.

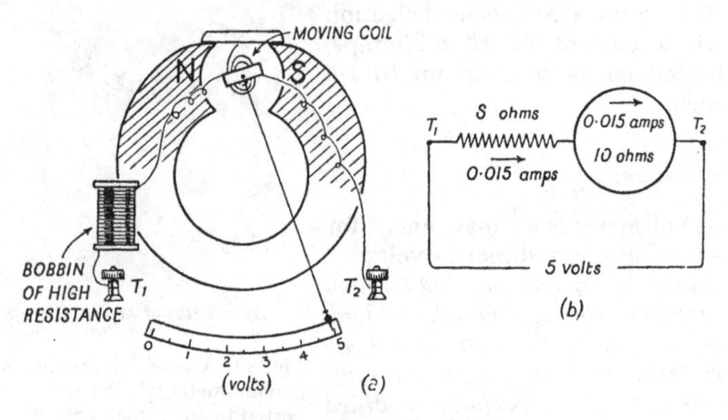

Fig. 90. A moving-coil voltmeter.

The total resistance of the instrument is $333\frac{1}{3}$ ohms. It is essential that *the resistance of a voltmeter should be high*, since otherwise, when placed across part of a circuit to measure the P.D., it will take too large a current, and lower appreciably the P.D. it is required to measure. If the resistance in the above example had turned out to be, say, 20 ohms, this would have shown that our moving-coil galvanometer was not suitable for conversion to a voltmeter.

Any type of galvanometer (which does not take too large a current) can be converted into a voltmeter by means of a suitable high resistance, and its range can be varied by altering the resistance. Fig. 91 shows the identical moving-coil galvanometer which was converted into an ammeter by means of a low-resistance shunt in parallel (see Fig. 89), now converted into a voltmeter by means of a high resistance in series. It is possible to buy such an instrument with a set of shunts, which will read either volts or amperes over considerable ranges (0 to 15 milliamps up to 0–30 amperes and 0 to 75 millivolts up to 0–150 volts).

By courtesy of Messrs Crompton Parkinson, Ltd.

Fig. 91. The same instrument as in Fig. 89, converted into a voltmeter by placing a high resistance in series with it.

SUMMARY

A *galvanometer* is an instrument for detecting and measuring an electric current.

Types of galvanometers

1. *Moving-magnet*, consisting of a small pivoted magnet at the centre of a fixed coil: the controlling force is supplied by the earth's field.

2. *Moving-coil*, consisting of a small pivoted coil between the poles of a permanent magnet; the controlling force is supplied by a hair spring or the torsion of a fibre.

3. *Moving-iron*, consisting of a fixed and a movable iron bar inside a solenoid; the controlling force is supplied by gravity or a spring.

4. *Hot-wire*, consisting of a fine wire which expands when heated by a current.

The *sensitivity* of a galvanometer is the deflection produced by unit current. It may be decreased by a shunt.

An ammeter is a galvanometer with a thick shunt in parallel. It must have a low resistance so that it will not reduce the current it is required to measure.

A voltmeter is a galvanometer with a high resistance in series. It must have a high resistance so that it takes a very small current. It can measure volts because, by Ohm's law, the current through it is proportional to the P.D. in volts across its terminals.

QUESTIONS

1. A small compass needle is placed at the centre of a circular vertical coil of insulated wire. How would you arrange the coil so that when a current is passed through it the direction of the compass needle (*a*) is unchanged, (*b*) has a maximum deflection? Give diagrams showing the direction of the current and the magnetic fields. What modification would you make in the apparatus in order to use it to detect small currents? (N.)

2. Copy Fig. 92 and put in an arrow showing which way the coil (free to turn about a vertical axis) tends to move. Explain fully your reasoning.

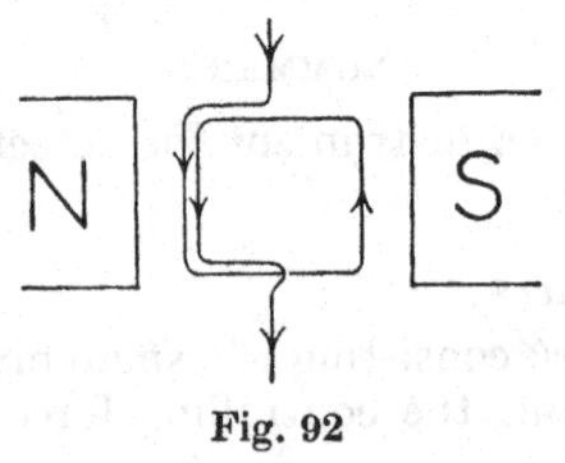

Fig. 92

3. Describe how you would determine, without breaking the circuit, the direction in which a current is flowing in a wire.

A rectangular coil hangs vertically in a magnetic field with its plane parallel to the lines of force. Describe and explain what happens when a current is passed round the coil and illustrate your answer by clear diagrams. (N.)

4. Describe the construction and action of a moving-coil galvanometer. At least one diagram is essential.

5. If two iron bars are placed side by side in a solenoid and a current is passed through the solenoid the bars spring apart. Explain fully why this occurs. Describe a galvanometer whose operation depends upon this phenomenon.

6. In Fig. 93 the two magnets are of very nearly equal strength: they are rigidly connected and are suspended by a fine wire. Copy the diagram and show by an arrow which way they will tend to turn when a current flows in the coil in the direction shown. Do both magnets tend to turn in the same direction? Explain fully. What is the direction of the restoring action of the controlling earth's field on each magnet? This arrangement is called an astatic galvanometer. Can you suggest why it is more sensitive than an ordinary moving-magnet galvanometer?

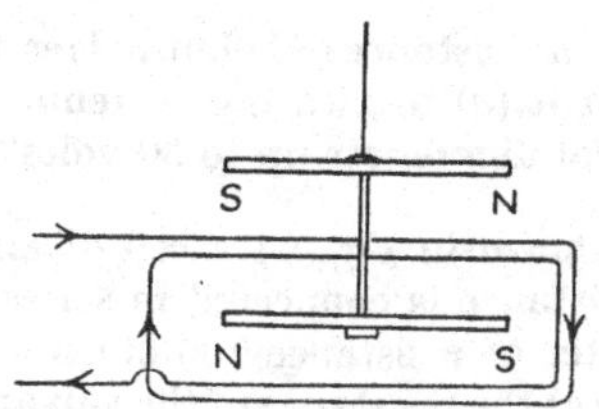

Fig. 93. Astatic galvanometer.

7. Describe a moving-magnet galvanometer.

Such a galvanometer is set with its coil in the magnetic meridian. How is its sensitiveness likely to be affected by placing a bar magnet on the table behind and parallel to the coil, (*a*) with its N. pole pointing north, (*b*) with its S. pole pointing north?

8. Explain the use of a galvanometer shunt.

A galvanometer of 250 ohms resistance, a cell of E.M.F. 1·4 volts and internal resistance 2 ohms, and a coil of resistance 70 ohms, are connected in series. Calculate the current through the galvanometer. If the galvanometer is now shunted with a coil of resistance 10 ohms, what current passes through the galvanometer? (L.)

9. Explain the use of a shunt. A galvanometer of 100 ohms resistance is to be provided with a shunt such that one-fifth of the whole current shall pass through the galvanometer. Find the resistance of the shunt. (O. and C.)

10. Explain how, by means of external resistances, an instrument of 5 ohms internal resistance, which gives a full-scale deflection for a current of 0·15 ampere, can be used (*a*) as an ammeter, reading up to 1·5 amperes, (*b*) a voltmeter, reading up to 3 volts. (C.)

11. An ammeter, whose resistance is 0·6 ohm, shows its maximum deflection when a current of 1 ampere is flowing. What resistance must be used as a shunt to the instrument so that the maximum deflection will be obtained when the current flowing in the circuit is 5 amperes? (L.)

12. How would you convert a 100-ohm galvanometer, reading up to 120 microamperes, into a voltmeter reading up to 2·4 volts?

13. A voltmeter connected across a battery of low resistance reads 4 volts. A resistance of 200 ohms is placed in series with it, when it reads 3 volts. Find the resistance of the voltmeter.

14. An ammeter has a resistance of 1 ohm and reads up to 0·1 ampere. How would you adapt it (*a*) to measure currents up to 10 amperes, (*b*) to measure potential differences up to 50 volts? (O. and C.)

15. In the circuit shown in Fig. 94 a cell of E.M.F. 1·5 volts and negligible internal resistance is connected in series with a 1000-ohm coil and a galvanometer of resistance 100 ohms, shunted through a 5-ohm coil. Calculate (*a*) the resistance of the galvanometer and shunt in parallel, (*b*) the total current from the battery, (*c*) the current in the galvanometer. (O. and C.)

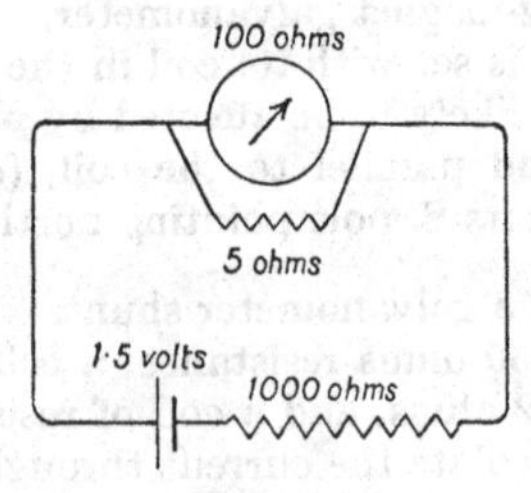

Fig. 94

16. Explain, with a diagram, the construction and mode of action of a moving-coil ammeter. Explain one factor on which the sensitiveness of the instrument depends.

A certain ammeter has a resistance of $\frac{1}{10}$ ohm and a maximum reading of 0·15 ampere. Explain, with a diagram, how it can be converted into a voltmeter of maximum reading 15 volts. (O. and C.)

17. A moving-coil voltmeter and an ammeter are sometimes used to show that the potential difference between the ends of a wire is proportional to the current through it. Criticise this method of verifying Ohm's law. (C.)

18. What do you understand by the sensitivity of a galvanometer? Describe the construction of a sensitive galvanometer that you have seen and explain how the sensitiveness is obtained or controlled. (C.)

19. A certain milliammeter has a resistance of 15 ohms. Explain, with an illustrative diagram, how you could convert the instrument into a voltmeter so that the indicated milliampere reading would correspond to the same number of volts. (O. and C.)

20. In Fig. 95 *ABCDE* represents a "universal shunt", consisting of resistances of 10, 90, 900 and 9000 ohms, and of a movable arm *OC* which can make contact with any of the studs *A, B, C, D, E*. Explain why the sensitiveness of the galvanometer increases as the arm moves from *A* to *E*; and show that it increases ten times each time the arm moves from one stud to the next.

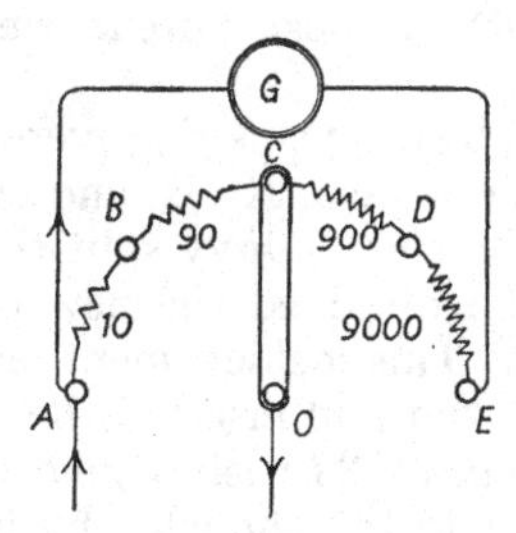

Fig. 95. Universal shunt.

Chapter VII

ELECTROLYSIS

A month after Volta's letter to the Royal Society in 1800, announcing the discovery of his cell (see p. 84), a large voltaic battery was set up in England. Experiments with an electric current began to be performed in this country: currents were passed through all kinds of substances and an important discovery resulted immediately.

Nicholson and Carlisle found that when a current was passed through acidulated water, where the two wires from the rest of the circuit dipped into the liquid, bubbles of gas appeared. The gases were found to be hydrogen and oxygen. Now some years before, Cavendish had made water by exploding hydrogen and oxygen: hence it was realised that the electric current had split up water into its constituent elements.

The discovery created something of a stir among chemists, for it provided them with a most potent weapon for chemical research.

Sir Humphry Davy, by its means, discovered two new elements. Until that time caustic potash, KOH, and caustic soda, NaOH, were believed to be elements. Davy split them up by means of an electric current and obtained the new elements potassium, K, and sodium, Na. This achievement earned him a prize offered by Napoleon, who had inserted the following characteristic notice in the *Moniteur*: "I wish to give as an encouragement a sum of 60,000 francs to the one who, by his experiments and discoveries in galvanism and electricity, will make a step forward comparable to that made in these sciences by Franklin and Volta, my particular aim being to direct and encourage the attention of physicists to that branch of the science, which is, in my opinion, the path to great discoveries." The prize was awarded to Davy while France and England were at war, an instance of the international spirit of science.

We have already noted two other effects of an electric current, the setting up of a magnetic field round a wire, and a heating effect in the wire. The third effect, the chemical splitting up of a liquid, will be studied in this chapter.

Terminology.

A liquid which allows a current to pass through it and is thereby decomposed is called an *electrolyte*. The process is called *electrolysis*. The two wires or plates dipping into the liquid are called *electrodes*: that by which the current enters the liquid is called the *anode*, and that by which it leaves is called the *cathode*. These words were coined by Faraday after consultation with a classical scholar. Anode means the way up (Greek ὁδός, a way), and cathode the way down.

Electrolytes consist of acids, bases and salts. It was found as a result of experiments that **hydrogen in acids, and the metals in salts and bases, are always liberated at the cathode, while oxygen and the non-metals are liberated at the anode.**

The electrolysis of a solution of hydrochloric acid.

When an electric current is passed through a solution of hydrochloric acid, HCl, in water, the following occurs:

$$2HCl \rightarrow \underset{\text{cathode}}{H_2} + \underset{\text{anode}}{Cl_2}.*$$

The hydrochloric acid is thus split up into its chemical constituents. In accordance with the rule we have just stated, the hydrogen is liberated at the cathode and the (non-metal) chlorine is liberated at the anode. These gases bubble out of the solution, which gradually becomes weaker or more dilute.

Chlorine is very active chemically, and hence it is desirable not to use a metal anode but one of carbon, which is not attacked by chlorine. The metal platinum is often employed for electrodes, since it is not attacked by most substances although it is acted upon, slowly, by chlorine.

* Each molecule of hydrogen and chlorine consists of two atoms and hence we write H_2 and Cl_2.

The electrolysis of copper sulphate solution.

(*a*) *With copper electrodes.*

When an electric current is passed through a solution of copper sulphate, $CuSO_4$, in water, using copper electrodes, it is found that copper is deposited on the cathode and the anode is eaten away.

We can account for this by assuming that the copper sulphate splits as follows:

$$CuSO_4 \rightarrow \underset{\text{cathode}}{Cu} + \underset{\text{anode}}{SO_4}.$$

(Note that again this accords with our rule that the metal is liberated at the cathode and the non-metal at the anode.)

When the SO_4 reaches the anode it attacks it,

$$Cu + SO_4 = CuSO_4.$$

This is known as a *secondary reaction.*

But a substance SO_4 is unknown to the chemists: it has never been isolated. We must regard the above, therefore, as pure hypothesis; but we shall find that it fits in very well with our theory of electrolysis.

The concentration of the copper sulphate solution does not diminish, since as fast as copper is being deposited on the cathode it is being dissolved from the anode.

The method is used for refining copper (see Fig. 96). Plates of impure copper are used as the anode and pure copper is deposited on the cathode, the impurities sinking to the bottom of the tank as mud.

The conductivity of copper is considerably reduced by small traces of impurities such as arsenic and phosphorus, and of the vast quantity of copper used annually in the manufacture of wire and electrical apparatus, a large part is refined electrolytically.

(*b*) *With platinum electrodes.*

When a solution of copper sulphate is electrolysed, using platinum electrodes, we find that, as before, copper is deposited on the cathode, but the platinum anode is not attacked: instead, oxygen bubbles off there. We assume, therefore, that when the SO_4 reaches the anode, since it cannot act upon the platinum, it attacks a water molecule instead.

By courtesy of the Copper Development Association and the Ontario Refining Co.

Fig. 96. The refining of copper by electrolysis. The copper cathodes of refined copper are shown being lifted from between the impure copper anodes in the solution of copper sulphate.

Electrolysis:

$$CuSO_4 \rightarrow \underset{\text{cathode}}{Cu} + \underset{\text{anode}}{SO_4}.$$

Secondary reaction:

$$SO_4 + H_2O = H_2SO_4 + O.$$

The oxygen liberated consists of pairs of atoms, O_2, called molecules, and hence we usually write the above equation as follows:

$$2SO_4 + 2H_2O = 2H_2SO_4 + O_2.$$

As we should expect, the solution gradually loses its blue colour, characteristic of copper sulphate solution, and becomes acid, H_2SO_4.

The electrolysis of acidulated water.

Apparatus for demonstrating the electrolysis of water is shown in Fig. 97. It is designed to enable the two gases liberated, hydrogen and oxygen, to be collected separately. The volume of hydrogen liberated is found to be double the volume of oxygen:

$$2H_2O = \underset{\text{2 vols.}}{2H_2} + \underset{\text{1 vol.}}{O_2}.$$

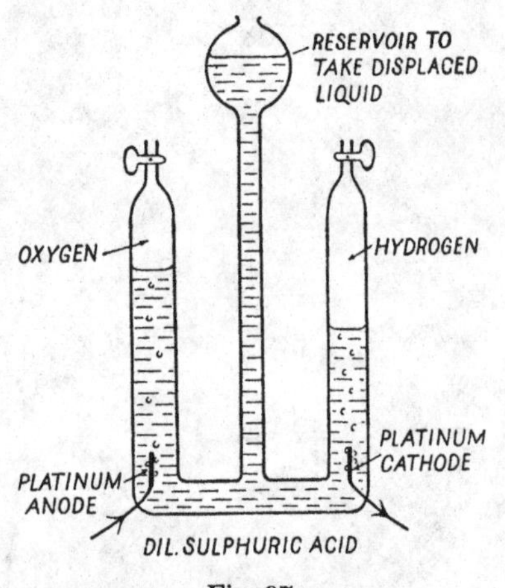

Fig. 97

Water, however, is not split up directly during electrolysis. Pure water is not an electrolyte: it must be rendered conducting by the addition of a little acid (base or salt), such as dilute sulphuric acid. It is the sulphuric acid which is really electrolysed. The water is decomposed in a secondary reaction.

Electrolysis:

$$H_2SO_4 \rightarrow \underset{\text{cathode}}{H_2} + \underset{\text{anode}}{SO_4}.$$

Secondary reaction:

$$2SO_4 + 2H_2O = 2H_2SO_4 + \underset{\text{anode}}{O_2}.$$

The electrodes must be made of platinum. Otherwise the SO_4, which, as we have said, is so active chemically that it cannot exist in a free state, will attack the anode instead of the water.

The electrolysis of sodium sulphate solution.

If sodium sulphate solution containing litmus is contained in a vessel similar to that in Fig. 98, and an electric current is passed through it, the litmus above the anode turns red, and that above the cathode blue.

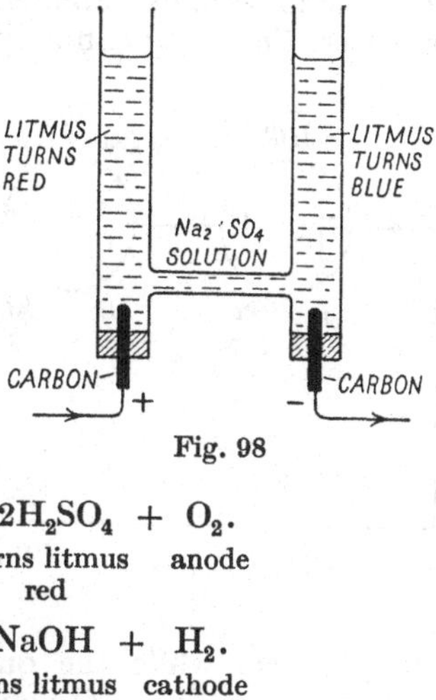

LITMUS TURNS RED

LITMUS TURNS BLUE

Na_2SO_4 SOLUTION

CARBON +

– CARBON

Fig. 98

Electrolysis:

$$Na_2SO_4 \rightarrow \underset{\text{cathode}}{2Na} + \underset{\text{anode}}{SO_4}.$$

Secondary reactions:

At anode: $2SO_4 + 2H_2O = \underset{\text{turns litmus red}}{2H_2SO_4} + \underset{\text{anode}}{O_2}.$

At cathode: $2Na + 2H_2O = \underset{\text{turns litmus blue}}{2NaOH} + \underset{\text{cathode}}{H_2}.$

Electrolysis may be used to distinguish between the positive and negative terminal of a supply of direct current. Pole-finding paper consists of blotting-paper impregnated with sodium sulphate and litmus. The paper is moistened with water and the bare ends of two wires, connected to the terminals, are placed on it. The paper goes red at the positive pole and blue at the negative.

Faraday's laws of electrolysis.

The quantitative investigation of electrolysis was performed by Faraday. He set out to discover the effect of (1) the size of the electrodes, (2) the strength of the electrolyte, (3) the size of the current, and (4) the time for which the current was passed. He discovered that the weight of a substance liberated depended only on the product of the current and the time, i.e. the quantity of electricity passed: the size of the electrodes and the concentration of the electrolyte were immaterial.

The unit in which quantity of electricity is measured is the coulomb. *One coulomb is the quantity of electricity which passes when* 1 *ampere flows for* 1 *second.* The number of coulombs, therefore, is given by the product of the current in amperes and the time in seconds. Comparing an electric current with a flow of water, the current in amperes corresponds to rate of flow in

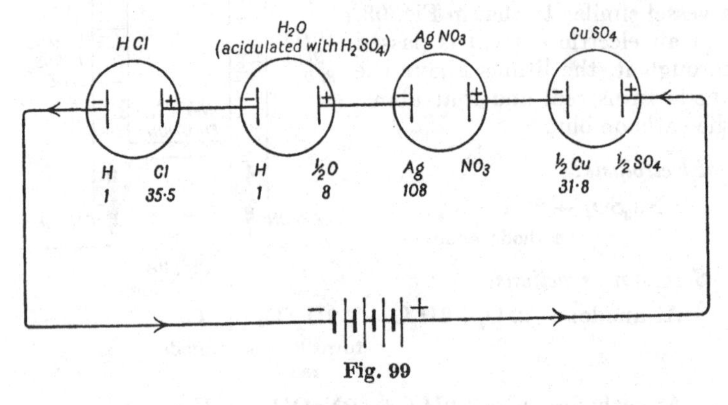

Fig. 99

c.c. per sec., while the quantity of electricity in coulombs corresponds to the total volume which has passed in c.c.

Faraday's discovery is summed up in what is known as **Faraday's first law of electrolysis.**

The weight of a substance liberated in electrolysis is proportional to the quantity of electricity passed.

Faraday then passed the same quantity of electricity through different electrolytes (see Fig. 99) and compared the weights of different substances liberated. Again he made a simple and this time an arresting discovery, summed up in **Faraday's second law of electrolysis.**

When the same quantity of electricity is passed through different electrolytes the weights of substances liberated are in the ratio of their equivalent weights.

The equivalent weight of an element is the weight that combines with or displaces 1 gm. of hydrogen. It is equal to the atomic weight divided by the valency:

Element	Hydrogen	Oxygen	Silver	Copper
Atomic weight	1	16	108	63·6
Equivalent weight	1	8	108	63·6 monovalent 31·8 divalent

The quantity of electricity which liberates 1 gm. of hydrogen, therefore, will also liberate 8 gm. of oxygen, 108 gm. of silver, and 31·8 gm. of divalent copper (see Fig. 99). This quantity of electricity, liberating the equivalent weight in grams of all elements, is sometimes called 1 Faraday, and is equal to 96,500 coulombs.

Example. A current which passes through solutions of copper sulphate and silver nitrate, connected in series, deposits 1·00 gm. of copper. Calculate the weight of silver deposited.

By Faraday's second law,

$$\frac{\text{Weight of silver deposited}}{\text{Weight of copper deposited}} = \frac{\text{Equivalent weight of silver}}{\text{Equivalent weight of copper}},$$

i.e.
$$\frac{\text{Weight of silver deposited}}{1} = \frac{108}{31{\cdot}8}.$$

$$\therefore \quad \text{Weight of silver deposited} = 3{\cdot}40 \text{ gm.}$$

The electrochemical equivalent.

The electrochemical equivalent of a substance is the weight liberated in electrolysis by 1 coulomb.

This is a quantity which can be determined by experiment and which is useful for purposes of calculation. It can also be calculated by assuming Faraday's second law and using the experimental fact that the equivalent weight in grams of all substances is liberated by 96,500 coulombs:

$$\text{Electrochemical equivalent} = \frac{\text{Equivalent weight}}{96{,}500} \text{ gm. per coulomb.}$$

Experimental determination of the electrochemical equivalent of copper.

In order to determine the electrochemical equivalent of copper a *copper voltameter** is used. This consists of a vessel containing a solution of copper sulphate into which dip two electrodes of copper (see Fig. 100). The middle plate is used as the cathode and the two outer plates are connected together to form the anode, thereby enabling a deposit of copper to be formed on both faces of the cathode.

The voltameter is connected in series with a battery, an ammeter, rheostat and switch.

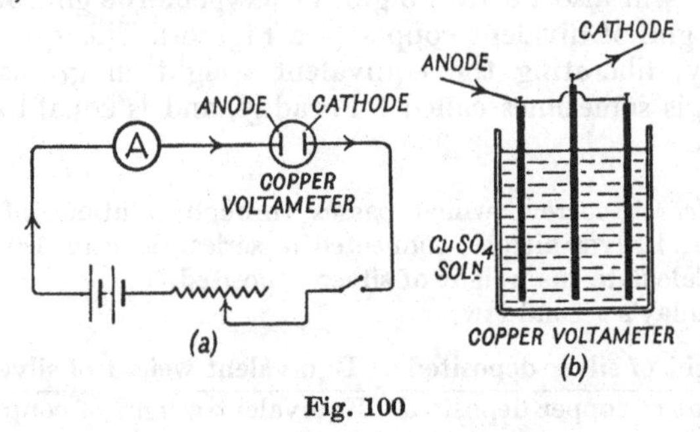

Fig. 100

The cathode is first dried, cleaned with emery paper and weighed as accurately as possible. By using the wire rider on the arm of a sensitive balance it can be weighed to the third decimal place in grams.

The current is switched on and adjusted immediately to a suitable value. The current density should not be greater than 1 ampere per 50 sq. cm., since if it is too great the deposit of copper will not stick to the cathode and may wash off.

The time at which the current is switched on is noted and the reading of the ammeter taken every minute.

It is desirable that the current should be passed for as long a period as possible in order to obtain a large deposit of copper. If 1·5 amperes are passed for a quarter of an hour a deposit of about 0·4 gm. is obtained.

When the current is switched off the time is noted. The

* This has nothing to do with a *voltmeter*.

cathode is taken out, washed, dried and reweighed. The washing should take place under a gently running tap, and then in distilled water. For rapid drying the wet cathode may be dipped in methylated spirits to remove as much of the water as possible. Methylated spirits is a more volatile liquid than water and hence evaporates more rapidly. It is inadvisable to dry the cathode too near a flame, since the freshly deposited copper may oxidise.

The following is a typical set of readings:

Mass of copper deposited $= 0{\cdot}439$ gm.

Average reading of the ammeter $= 1{\cdot}48$ amperes.

Time for which current is passed $= 15$ minutes

$= 900$ seconds.

$\therefore$ Quantity of electricity passed $= 1{\cdot}48 \times 900$

$= 1332$ coulombs.

$\therefore$ Electrochemical equivalent of copper

= Mass of copper deposited by 1 coulomb

$$= \frac{0{\cdot}439}{1332}$$

$= 0{\cdot}000330$ gm. per coulomb.

Measuring current by means of a voltameter.

Once the electrochemical equivalent of a substance like copper has been determined accurately the voltameter may be used as a very accurate instrument for measuring current. Indeed, in Faraday's day, it was the only accurate method of measuring current. The actual measuring instruments required are, of course, a balance and a watch. The weight of copper deposited by the current in a measured time is found and hence the current may be calculated.

If the electrochemical equivalent of copper had been known, the experiment described in the previous paragraph could have been used to calibrate the ammeter. Our calculation would then have been as follows:

Electrochemical equivalent of copper = 0·000330 gm. per coulomb.

∴ Quantity of electricity required to deposit 0·439 gm. of copper

$$= \frac{0{\cdot}439}{0{\cdot}00033}$$

$$= 1332 \text{ coulombs.}$$

Since current was passing for 900 seconds,

$$\text{Current} = \frac{1332}{900} = 1{\cdot}48 \text{ amperes.}$$

Since the average reading of our ammeter was 1·48 amperes, we should conclude that the ammeter was reading correctly.

Faraday and Joule calibrated their galvanometers in a similar way, using a water voltameter.

In 1884 Lord Rayleigh and Mrs Sidgwick made a very accurate determination of the electrochemical equivalent of silver. Their value was 0·001118 gm. per coulomb, and an identical result was obtained by Kohlrausch in Germany two years later. A silver voltameter is more accurate than a copper one, since a given quantity of electricity deposits more than three times the weight of silver than copper.

Although the unit of current, the ampere, was theoretically defined in terms of its magnetic effect (see p. 322), a practical need was felt for a definition also in terms of its electrolytic effect. At an international conference held in London in 1908 **the international ampere** was defined as **the unvarying current which, passing through a solution of silver nitrate in water in accordance with certain specifications, deposits silver at the rate of 0·001118 gm. per sec.**

Theories of electrolysis.

What happens inside an electrolyte when a current is flowing? A theory of electrolysis must explain the following:

(1) how the molecules are split up;

(2) why hydrogen and the metals go to the cathode, while oxygen and the non-metals go to the anode;

(3) how the current passes through the liquid;

(4) Faraday's laws.

The anode may be regarded as positively charged and the cathode as negatively charged. Hence it was early assumed that the molecule of an electrolyte consists of two parts, one of which is positively charged and the other negatively. The positively charged part is attracted to the cathode and the negatively charged part to the anode.

Thus in the first theory, that of Grotthus, the molecules were assumed to be charged positively at one end and negatively at the other. Those near to the electrodes were torn asunder: the positively charged anode, for example, seized the nearest negatively charged chlorine end of a sodium chloride molecule and broke it away from the sodium.

There is no space here to enter into the details of this theory, which has now been superseded: but it did attempt a picture of what happens in electrolysis and received the support of the illustrious Sir Humphry Davy.

Now if some of the applied E.M.F. is required to pull the molecules to pieces there must be a minimum E.M.F. below which electrolysis cannot take place, since there will not be enough force to break the molecules. It was found, however, that electrolysis took place with the very weakest of E.M.F.S.* Moreover, the products of electrolysis appeared immediately. The breaking-up process seemed to require neither force nor time.

The ionic theory.

In 1887 Svante Arrhenius, a young Danish student of the University of Copenhagen, submitted a paper for a doctor's degree on a theory which was neither solely physics nor chemistry. And, to speak truth, neither department was anxious to sponsor it. Eventually the physicists accepted it, and Arrhenius obtained his degree.

The paper, based on the work of Clausius and van 't Hoff, was epoch-making. It set forth the *Ionic Theory* on which the conduction of electricity through electrolytes is now explained, and which, moreover, marked the birth of physical chemistry.

* Except when there is a back E.M.F. In the electrolysis of dil. H_2SO_4 films of H_2 and O_2 at the platinum electrodes act like the plates of an opposing cell and set up a back E.M.F. of 1·47 volts. But in the electrolysis of $CuSO_4$, using Cu electrodes, there is no back E.M.F.

The novel idea in the theory is that *an electrolyte such as sodium chloride splits up, or dissociates, into charged* Na^+ *and* Cl^- *atoms, immediately on entering solution.* Some of the molecules are undissociated and there is a constant breaking up and re-uniting, expressed by the following equation:

$$NaCl \rightleftarrows Na^+ + Cl^-.$$

We have chosen sodium chloride as an example because it was around this molecule that, later, the fiercest controversy raged. Nowadays, we believe that the sodium atom loses an electron, and the chlorine atom gains it, giving them their respective positive and negative charges. Faraday named these charged atoms *ions* (Greek, wanderers).

When a P.D. is applied between two electrodes dipping into the solution there is a steady drift of ions to the electrodes owing to the attraction of opposite charges. The positively charged Na^+ ions are attracted to the cathode and the negatively charged Cl^- ions to the anode. These ions carry the current (see Fig. 101). The Cl^- ions take electrons to the anode and the Na^+ ions receive electrons from the cathode. When they have given up or neutralised their charges they become ordinary atoms and react with the water or the electrode. The P.D. between the electrodes, therefore, is not destructive but directive.

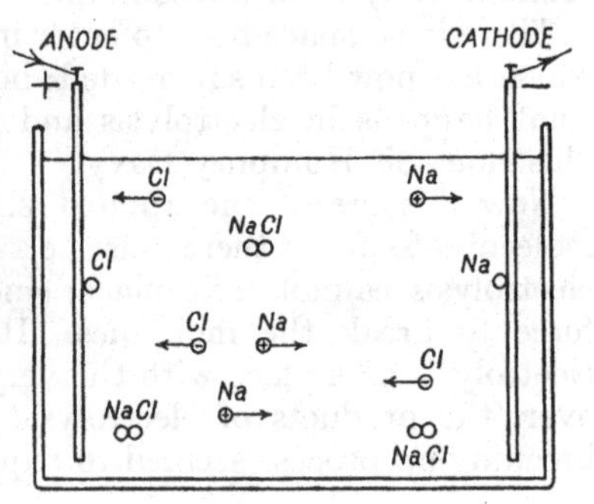

Fig. 101

The ionic theory gave rise to a long and violent controversy. Critics asked how and why the molecules dissociate. This is the strongest criticism: no satisfactory answer has been given and the process is still a mysterious one. Again, it was said that such intensely reactive atoms as those of sodium and chlorine could never remain free and uncombined in water. Every student of chemistry knows that when a pellet of sodium is thrown on to water there is a flash and a bang as the sodium reacts with the water to form caustic soda. But it was pointed out that a charged ion of sodium is not the same as an uncharged atom. The loss of an electron must somehow weaken its chemical affinity.

It was also asked how the ions retained their charges. The supporters of the theory retorted that pure water is a good insulator.

We may add that there is strong chemical evidence in favour of the theory. Electrolytes possess abnormal freezing points, boiling points and osmotic pressures, and these can be accounted for very satisfactorily in terms of ions.

To sum up, then, the passage of a current through an electrolyte is markedly different from that in a wire. A current in a wire is a flow of electrons. A current in an electrolyte is the movement of oppositely charged ions in opposite directions, the ions being already present in the solution before the P.D. is applied. The negative ion gives its electron to the anode and the positive ion receives an electron from the cathode, thereby neutralising its own positive charge. In this way electrons are continually removed from the cathode and supplied at the anode: thus the flow of electrons in the rest of the circuit is maintained.

Explanation of Faraday's laws.

Faraday's first law follows from the fact that the current in the electrolyte is carried by the ions. The weight of the ions reaching, and being liberated at, the electrodes is clearly proportional to the charge they carry.

Faraday's second law is explained by assuming that every monovalent ion carries the same charge, $\pm e$, the charge of the electron. Divalent ions carry a charge of $\pm 2e$, and so on (SO_4^{--} is an example of a divalent ion):

$$H_2SO_4 \rightleftarrows 2H^+ + SO_4^{--}.$$

Electrolysis provided the first evidence that electricity is atomic in nature, since each ion carries the "atom" of electricity, the electron. Faraday, with true scientific caution, hesitated to take this bold step. The hypothesis was put forward by Helmholtz in 1880 and was confirmed by Sir J. J. Thomson when he isolated the electron in 1895.

We can, indeed, actually calculate the charge of the electron from the fact that 96,500 coulombs liberate 1 gm. of hydrogen in electrolysis, knowing that 1 gm. of hydrogen contains $6{\cdot}06 \times 10^{23}$ atoms. Each atom carries a charge of 1 electron.

$$\text{Hence charge of the electron} = \frac{96{,}500}{6{\cdot}06 \times 10^{23}}$$

$$= 1{\cdot}59 \times 10^{-19} \text{ coulombs.}$$

By courtesy of Messrs W. Canning and Co. Ltd.

Fig. 102. A sectiona view of a nickel plating vat. The anodes and cathode suspenders are marked. The plating solution is warmed by heating coils and agitated by air which is blown through it.

The commercial applications of electrolysis

Electroplating.

Electrolysis provides an ideal method of covering one metal with a thin layer of another. A cheap metal may be coated with silver or gold, for example, or a metal like iron, which is prone to rust, protected with a fine layer of non-corrosible and brilliant chromium.

The surface to be electroplated must first be freed entirely of grease and rust: it must be made chemically clean. This is usually done by means of wire brushes or a sand blast, followed by "pickling" in solutions such as warm sulphuric acid.

The plating vats are made of wood lined with chemically pure lead, or iron lined with enamel or cement.

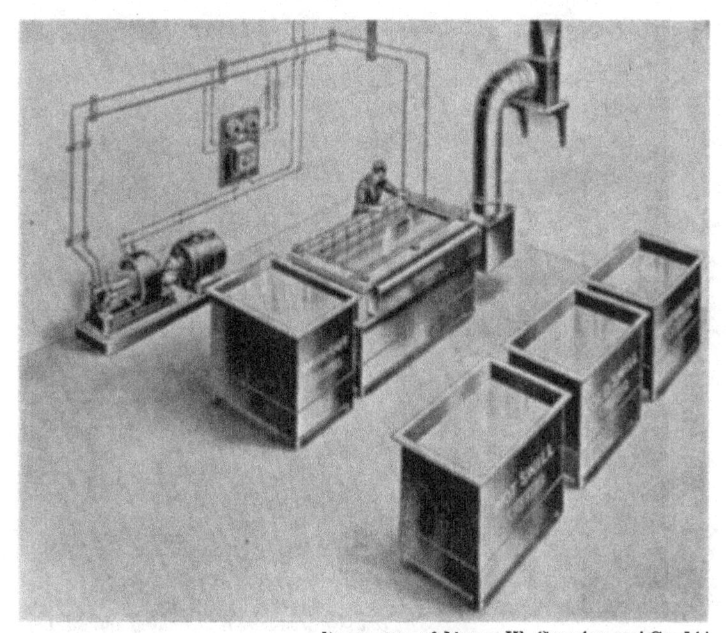

By courtesy of Messrs W. Canning and Co. Ltd.

Fig. 103. A complete chrome plating installation, with electrical equipments. Note the motor-car radiator suspended in the plating bath. After plating the radiator is immersed in the four washing baths to remove all trace of the plating solution prior to drying.

Across the vats are placed anode and cathode rods (see Figs. 102, 103 and 104), and from these rods are hung, by hooks of nickel or brass, the anodes and objects to be plated. In the case of silver plating, the anodes consist of silver plates and the cathodes, perhaps, of spoons and forks. The solution is a silver salt (silver cyanide, AgCN, dissolved in an excess of potassium cyanide, KCN), which has been found by experience to give a smooth and dense deposit. The time in plating is often shortened by agitating the solution with compressed air or by revolving the articles forming the cathodes. Time cannot be saved by using a large current: for if the current exceeds a certain value the articles may be "burned"—covered with a dark-coloured deposit.

By courtesy of Messrs W. Canning and Co. Ltd.

Fig. 104. A shop for nickel plating motor and cycle accessories. Note the cathode rods above the plating bath from which the articles to be plated are suspended.

Objects are often copper plated before they are silver plated, since copper serves as an excellent base for the silver. Similarly, nickel is used as a base for chromium, which is always made into as thin a skin as possible owing to its expense. Chromium plating was introduced into America in 1925 and then into this country in 1927. It is now widely used for the radiators and fittings of motor cars and also for household fittings.

Electrotypes.

Copies of the type of the pages of a book or engravings are often made by electrolysis to save the wear of the originals: such copies are called electrotypes.

First a mould or negative of the type is made by forcing upon it, by hydraulic pressure, a sheet of wax or guttapercha. The mould is dusted with graphite to make it conducting, and it is then placed as the cathode in a bath of copper sulphate. A current is passed until a deposit of copper, about the thickness of a stout sheet of paper, is formed inside the mould. The wax mould is stripped off, leaving a "positive" copper replica of the type. The replica is strengthened by filling and backing it with molten metal (until it is about $\frac{1}{4}$ in. to $\frac{3}{8}$ in. thick). Sometimes the copper plates are nickel or steel plated, giving them a harder surface and enabling sharper copies to be taken from them.

Plaster casts are often given a metal surface by coating them with graphite to render them conducting and then depositing the metal by electrolysis.

The manufacture of aluminium.

Before the discovery of the electrolytic method of separating it from its ores, aluminium was much too expensive for practical use. To-day a quarter of a million tons of it are manufactured annually.

An electric current is passed through a solution of molten alumina (Al_2O_3) in molten cryolite (a double fluoride of aluminium and sodium), and pure aluminium is liberated at the cathode.

The furnaces used (see Fig. 105) are comparatively shallow. The cathode consists of a bed of iron rods covered with a carbon mixture, which is carefully prepared. The anode consists of a number of carbon rods immersed in the molten alumina and cryolite: these rods burn away and have to be continually replaced. The aluminium gradually forms in a molten layer on the bed of the furnace and is run out at intervals. Fresh alumina and cryolite are constantly fed in.

The latest furnaces take a current of 40,000 amperes and the voltage per furnace is about 6 volts. Each furnace may operate without stopping for months or even years until eventually iron is found in the aluminium, showing that the carbon of the cathode has been eaten away. The furnace has then to be taken right down and rebuilt.

The British Aluminium Company have built their factories in the Highlands of Scotland, where electricity can be generated very cheaply by water power.

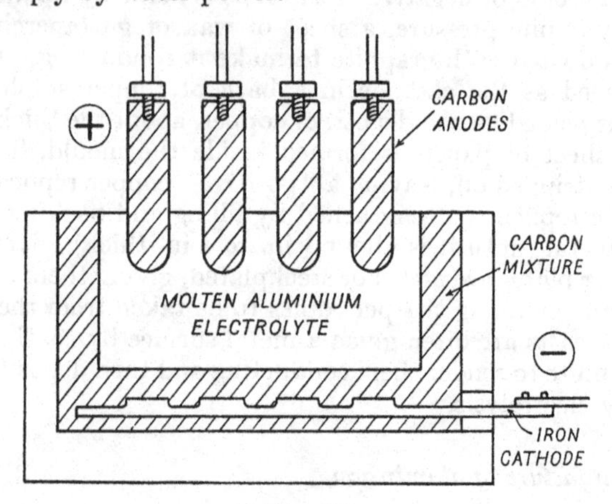

Fig. 105

Summary

Certain liquids called electrolytes conduct electricity and are split up in the process. **Hydrogen (in acids) and the metals are liberated at the cathode, and oxygen and the non-metals at the anode.**

Faraday's laws. (1) The weight of a substance liberated in electrolysis is proportional to the quantity of electricity passed.

(2) When the same quantity of electricity is passed through different electrolytes the weights of substances liberated are in the ratio of their equivalent weights.

Quantity of electricity is measured in *coulombs*, the product of the current in amperes and the time in seconds.

The electrochemical equivalent of a substance is the weight liberated in electrolysis by 1 coulomb.

Current may be measured by determining the weight of copper deposited in a measured time in a copper *voltameter*.

The *ionic theory* assumes that an electrolyte dissociates into charged *ions* as soon as it enters solution. These charged ions are attracted to the electrodes when a P.D. is maintained between the latter. The ions carry the current through the liquid.

Electrolysis is utilised commercially for (1) electroplating, (2) electrotyping, (3) refining metals such as copper, (4) the manufacture of aluminium.

QUESTIONS

1. Explain fully what you would expect to happen when a current of electricity is passed through a solution of nickel chloride, $NiCl_2$, in water, using carbon electrodes.

2. Explain clearly:

(*a*) When a solution of copper sulphate is electrolysed, platinum electrodes being used, the blue colour of the solution gradually disappears. No such effect is observed if copper electrodes are used.

(*b*) Sodium chloride does not contain hydrogen and is not an alkali, but when an aqueous solution of it is electrolysed, hydrogen is evolved at one of the electrodes, and the solution around this electrode becomes alkaline. (O. and C.)

3. Explain fully what happens when a current of electricity is passed through a dilute solution of sulphuric acid in water, using platinum electrodes.

4. What would you expect to happen when a solution of silver nitrate, $AgNO_3$, is electrolysed, using silver electrodes?

5. Salt is said to be an electrolyte and sugar a non-electrolyte. Explain what is meant by these statements. How would you demonstrate their truth in the laboratory?

A certain quantity of electricity is passed through a solution of sodium chloride. Explain the state of affairs in the solution and round the two platinum electrodes (*a*) before, and (*b*) during the passage of the current. Account for any changes in the acidity or alkalinity of the solution near the electrodes. (O.)

6. What would you expect to happen if a solution of sodium sulphate were electrolysed between copper electrodes? (O.)

7. Write a short account of the industrial applications of electrolysis.

8. How many coulombs are there in an ampere-hour, i.e. the quantity of electricity which passes when a current of 1 ampere flows for 1 hour?

9. State the laws of electrolysis, and explain how they can be verified experimentally. (N.)

10. What is meant by saying that the electrochemical equivalent of copper is 0·0003296 gm. per coulomb? An electric current is passed through a solution of copper sulphate, using platinum electrodes. Explain how you can determine from this experiment (*a*) the direction of the current, (*b*) the magnitude of the current. In the latter case point out the precautions taken to procure a good result. (L.)

11. State the laws of electrolysis.

A copper voltameter is connected in series with a source of electric supply and a current is passed for 1 hour. It is found that one electrode gains 1·1 gm. of copper in this time. Calculate the current and draw a circuit diagram which shows which electrode increases in weight. (Electrochemical equivalent of copper = 0·00033 gm. per coulomb.) (L.)

12. A current was passed for 20 min. 29 sec. through a silver voltameter and an ammeter: the mean reading of the latter was 0·126 ampere; the weight of silver deposited was found to be 0·174 gm. Taking the electrochemical equivalent of silver as 0·00112 gm. per sec. per ampere, calculate the mean current passing, and the percentage error of the ammeter reading. (C.)

13. Explain the meaning of *electrochemical equivalent.*

The value for that of hydrogen is 0·00001045. What weight of hydrogen would be liberated in one hour by a current of 100 amperes? How many litres at standard temperature and pressure would this occupy, the density of the gas being 0·09 gm. per litre? (N.)

14. A metal plate of total surface area 250 sq. cm. is to be coated with copper by electrolysis. How long will it take to make a deposit of copper 0·01 cm. in thickness, if a current of 1·5 amperes is used? (Electrochemical equivalent of copper = 0·000329 gm. per coulomb. Density of copper = 8·93 gm. per c.c.) (C.)

15. Give a short account of the electrolysis of dilute sulphuric acid, when platinum electrodes are used. A current of 50 amperes is used to generate hydrogen by electrolysis. How many hours will it

take to produce one cubic metre measured at 0° C. and 760 mm. pressure? (Electrochemical equivalent of hydrogen = 0·0000104 gm. per coulomb. Density of hydrogen at N.T.P. = 0·09 gm. per litre.)

(O. and C.)

16. A silver and a gold electroplating vat are arranged in series. What weight of gold is deposited while 1 gm. of silver is being deposited? (Equivalent weights of silver and gold are 108 and 65·7, respectively.)

17. An electric current is passing through three cells in series containing respectively solutions of the following salts: copper sulphate, silver nitrate, a lead salt.

What weights of silver and lead will be deposited while 1 gm. of copper is deposited?

Atomic weight of copper = 63;	Valency = 2.	
„ silver = 108;	„ = 1.	
„ lead = 208;	„ = 2.	(O.)

18. Given that 1 gm. equivalent weight of copper is liberated by 96,500 coulombs, calculate the electrochemical equivalent of copper. (Atomic weight = 63·6; Valency = 2.)

19. A constant current which deposits 0·85 gm. of copper in 25 minutes passes through a standard coil of resistance 2 ohms. A voltmeter attached to the ends of the coil reads 3·50 volts. Calculate the error in this reading. (The current taken by the voltmeter is to be considered negligible. 1 coulomb of electricity deposits 0·000328 gm. of copper.)

(C.)

20. Write a short account of the ionic theory.

21. A cell of constant E.M.F. is connected to copper plates immersed in pure water. Copper sulphate is then added to the water a little at a time. How would you expect the current flowing through the cell to change during the experiment?

How does the ionic theory account for these changes? How will the current change (*a*) if the solution is warmed, (*b*) if the plates are brought nearer together?

(O.)

Chapter VIII

CELLS

The simple voltaic cell.

The simple voltaic cell consists of plates of copper and zinc immersed in dilute sulphuric acid. The copper is the positive plate and the zinc the negative: when the plates are connected by a wire a current of positive electricity is said to flow through the wire from the copper to the zinc.

A voltmeter connected across the two plates reads about 1 volt. Now the P.D. between the plates of the cell (when it is on open circuit) is a measure of the total "driving force" inside the cell, called the Electro-Motive-Force, E.M.F. Thus the E.M.F. of the simple cell is 1 volt.

The E.M.F. of the cell is unaffected by its size, and depends only on the chemical nature of the two plates and of the exciting liquid. An increase in the size of a cell merely decreases its internal resistance and increases its capacity, i.e. the total output of electricity in coulombs which it can produce before it is worn out.

The source of the energy of the cell is the action of the sulphuric acid on the zinc, which is gradually worn away:

$$Zn + H_2SO_4 = ZnSO_4 + H_2.$$

The zinc may be regarded as the fuel of the cell, just as coal and petrol are consumed in steam and internal combustion engines.

Polarisation.

After a few minutes a simple cell stops giving a current: a lamp connected to its plates will go out.

The reason is that bubbles of hydrogen, liberated by the action of the sulphuric acid on the zinc, collect on the copper plate and form a cell with copper as the negative plate and hydrogen as the positive plate. This hydrogen-copper cell works in opposition to the copper-zinc cell, setting up what is called a back E.M.F. The defect is called *polarisation.*

The collection of hydrogen on the *copper* plate is very odd and needs explaining, as we shall endeavour to do on p. 130. For the hydrogen is formed by the action of the sulphuric acid on the *zinc*, which wears out: the copper does not wear out and is unattacked by the sulphuric acid.

The cure for the defect of polarisation is to remove the hydrogen as fast as it is formed. This could be done by some form of mechanical wiper or scraper. Smee invented a cell in which the copper plate was corrugated, as this helps the hydrogen to bubble off the copper to the surface of the sulphuric acid. Such methods are, however, unsatisfactory, and a chemical method is generally employed. An oxidising agent, called a depolarising agent, is introduced into the cell to remove the hydrogen chemically by changing it to water.

Thus if some potassium bichromate ($K_2Cr_2O_7$) solution is poured into a polarised simple cell to which a lamp is connected, the lamp will relight. We shall see that various chemicals are used in different cells as depolarising agents.

Local action.

Pure zinc is not acted upon by dilute sulphuric acid. It begins to dissolve only when it is connected to another metal, such as copper, immersed also in the sulphuric acid, i.e. when it forms part of a cell.

Pure zinc is much too expensive for use in a cell, and ordinary commercial zinc contains impurities such as iron, lead and carbon. Thus the zinc plate in a simple cell is worn away steadily even while the cell is not in use, owing to the fact that it forms cells with the impurities in it. This second defect of a simple cell is known as *local action.*

It may be remedied either by taking the zinc out of the acid when the cell is not in use, or by rubbing the zinc plate over with mercury, forming an amalgam of zinc and mercury on the surface of the plate. The impurities in the zinc do not form an amalgam with mercury and hence cannot permeate it and reach the acid. Thus the mercury acts as a kind of filter, allowing only the zinc to pass through it and act upon the acid.

The Daniell Cell.

In 1836 John Frederic Daniell, professor at King's College, London, invented a cell which did not polarise, and which was the most satisfactory and efficient generator of electricity of its time.

It is a simple cell with the addition of a solution of copper sulphate to act as a depolarising agent. Since zinc, when placed in copper sulphate solution, becomes covered with a deposit of copper, Daniell had to devise a means of keeping them apart but at the same time not insulated electrically. He separated the zinc and sulphuric acid from the copper sulphate solution by means of a semi-permeable animal membrane—the windpipe of an ox. We now use an unglazed porous pot.

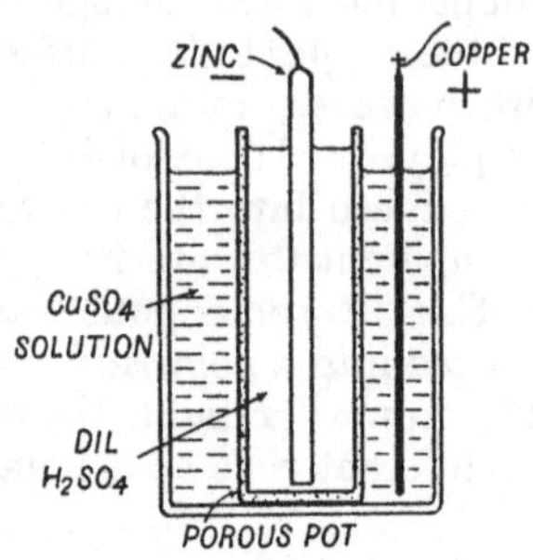

Fig. 106. Daniell cell.

The cell is shown in Fig. 106. The positive and negative plates are copper and zinc respectively, the exciting liquid is sulphuric acid, and the depolarising agent is copper-sulphate solution. The chemical action inside the cell may be represented by the following equations:

Source of the energy:

$$Zn + H_2SO_4 = ZnSO_4 + 2H.$$

Depolarising action:

$$2H + CuSO_4 = H_2SO_4 + Cu.$$

Hydrogen does not act on copper sulphate normally but only in electric cells. Actually it is the H^+ ion which acts and not the H atom.

The copper is deposited on the copper plate. Sometimes the outer container itself is made of copper and acts as the positive plate.

It can be seen from the second equation that the copper sulphate is being used up while the cell is in use. Consequently the copper sulphate solution should be concentrated, and often a ledge is provided just below the surface of the solution on which crystals of copper sulphate are placed to be dissolved and maintain the concentration of the solution.

The E.M.F. of the cell is about 1·1 volts.

Different metals used in cells.

Quite a number of pairs of substances, metals chiefly, when placed in a solution of an acid, base or salt, will form a practicable cell. The current through the external circuit is always from the uneaten to the eaten plate.

The following are suitable for use as plates of a cell and are arranged in an electrochemical series:

−ve plate end +ve plate end

Mg Al Zn Fe Pb Cu Pt C

The further apart in the list the two chosen substances, the greater the E.M.F. of the cell.

Many patents for different types of cells have been taken out: a considerable number of these, however, will not work.

Since magnesium is expensive and aluminium is apt to become passive and unable to react with the exciting liquid, the negative plate most commonly used is zinc.

For the positive plate, copper was used by Volta and Daniell, platinum by Grove (in 1839 before its price had become prohibitive), and carbon by Bunsen and Leclanché. The cells of Grove and Bunsen are seldom used to-day and we shall describe only the Leclanché cell.

The Leclanché cell.

The positive pole is carbon, the negative pole zinc, the exciting liquid salammoniac, i.e. ammonium chloride, NH_4Cl, and the depolarising agent is manganese dioxide, MnO_2 (see Fig. 107).

The manganese dioxide is in powder form and, mixed with powdered carbon to reduce the internal resistance of the cell, it is packed in a porous pot round the carbon plate.

The chemical action in the cell is as follows:

Source of energy:

$$Zn + 2NH_4Cl + 2H_2O = ZnCl_2 + 2NH_4OH + 2H.$$

Depolarising action:

$$2H + 2MnO_2 = Mn_2O_3 + H_2O.$$

The E.M.F. of the cell is about 1·45 volts.

The depolarising agent does not act rapidly, and hence the cell is suitable only for intermittent use—for electric bells and telephones. However, it has the great advantage that it can be left for years without any attention except for filling it up with water occasionally to replace that lost by evaporation.

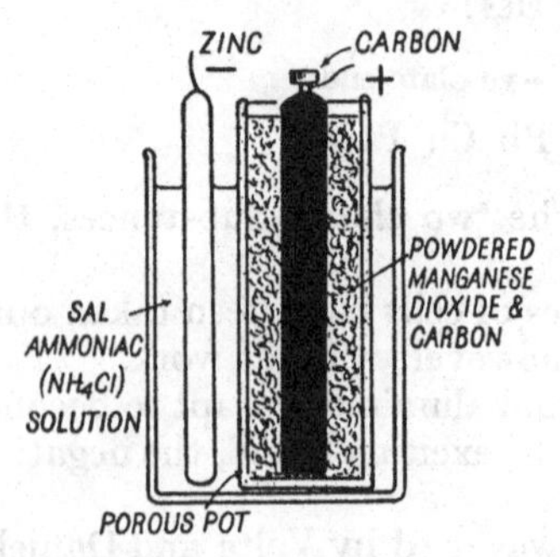

Fig. 107. Leclanché cell.

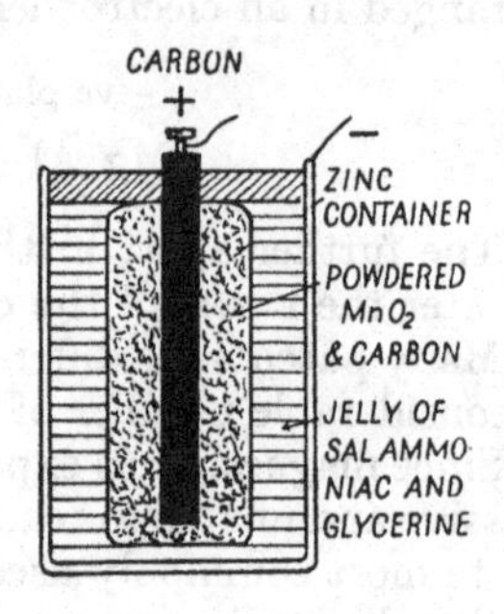

Fig. 108. Dry cell.

The dry cell.

The dry cell, such as is used in torches, and for wireless high-tension batteries, is a form of dry Leclanché cell (see Fig. 108). The positive plate is a carbon rod held in the centre of a zinc container, which acts as the negative plate; inside the container there is a jelly of salammoniac and glycerine: a paste of finely powdered manganese dioxide mixed with carbon is held round the carbon plate in a canvas bag.

The internal resistance of the dry cell is much less than that of the wet Leclanché cell for two reasons, (1) there is no porous pot, (2) the negative plate (the zinc container) completely surrounds the positive plate, from which the current flows out radially in all directions. The internal resistance increases with the age of the cell, owing to the drying up of the jelly and paste and also the incrustation of the zinc.

Owing to the low internal resistance the drop in P.D. when the cell delivers a current (the "lost volts") is quite small (see p. 135).

The E.M.F. of the cell is that of the wet Leclanché cell, 1·45 volts. The chief advantage of the cell is that it is portable: its disadvantage is that it deteriorates when left unused "on the shelf".

The Gordon magnesium cell.

A cell which has recently been put on the market has a carbon cylindrical container as its positive pole, a magnesium rod in the centre as its negative pole, and the exciting liquid is tap water. Potassium bromide is dissolved in the water to reduce the internal resistance of the cell: it is not essential to the working.

A three-cell battery designed to give a small current of 100 milliamperes is used to work a deaf-aid outfit. Its E.M.F. increases, as the cell warms up, from 2·5 to 3·1 volts in about half an hour. The chemical reaction is

$$Mg + 2H_2O = Mg(OH)_2 + H_2.$$

No depolarising agent is necessary with the small currents for which the cell is designed.

The action inside a voltaic cell.

We have stated (see p. 32) that inside every metal there is a large number of free electrons. The full complement of planetary electrons belonging to a neutral zinc atom is 29, but the two outermost of these are loosely held: they tend to break away and wander through the metal as free electrons.

A plate of zinc 12 cm. × 5 cm. × $\frac{1}{4}$ cm., i.e. of volume 15 c.c., contains rather more than 3×10^{15}, i.e. three thousand billion, of these free electrons. If such a plate were used up completely in a cell, all of these free electrons would be drained out of it and pass round the circuit, forming the current. The positively charged zinc atoms, each minus two electrons, would, as the free electrons drained away, slip gradually into the solution as ions, until finally all the zinc was completely dissolved.

But why do the free electrons drain out of the zinc into the copper plate when the two are connected by a wire? The problem is a complicated one and we can give only a simplified explanation.

The fact is, that for some reason on which we cannot speculate here, free electrons prefer to be inside a copper plate than a zinc plate. If a plate of copper and zinc (which are not immersed in sulphuric acid) are placed in contact, some free electrons at once rush out of the zinc into the copper. The migration ceases when the copper acquires such an excess of electrons that its negative charge repels the rest of the free electrons inside the zinc sufficiently strongly to stop the flow. The potential difference set up is called a contact potential difference and is of the order of 1 volt. This is an experimental fact and the contact P.D.s have been measured.

Why then cannot a cell be devised comprising copper and zinc plates without the use of a liquid such as sulphuric acid? The answer is that the electrons would not circulate round an all-metal circuit. In Fig. 109 there would be a quick rush of electrons in both wires from the zinc to the copper: for a continuous current to flow the electrons must be removed from the copper as fast as they flow from the zinc.

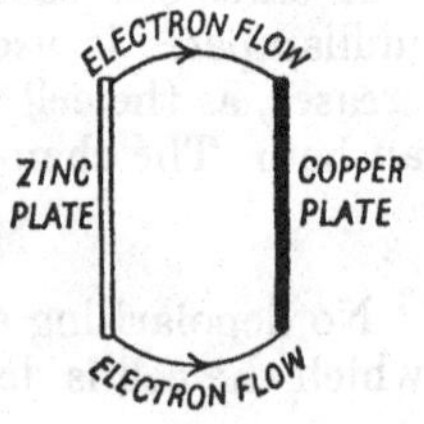

Fig. 109

Now the sulphuric acid, in which the zinc and copper plates of a simple cell are immersed, contains hydrogen and sulphate ions (see p. 114):

$$H_2SO_4 \rightleftarrows 2H^+ + SO_4^{--}.$$

When the zinc and copper plates are connected by a wire there is a rush of electrons through the wire from the zinc to the copper. These newly arrived electrons in the copper plate, with their negative charges, attract the hydrogen ions, H^+, from the liquid, and are taken up by the H^+ ions (which owe their positive charge to the fact that they have a deficiency of an electron). The hydrogen ions when they have gained an electron become ordinary hydrogen atoms and in this way the hydrogen bubbles, which we have discussed as being the cause of polarisation, collect on the copper plate.

The solution of sulphuric acid has now a number of SO_4^{--} ions which are not paired off with $2H^+$ ions. Consequently Zn^{++} ions, consisting of atoms of zinc minus two electrons, enter the solution from the zinc plate, to take the places of the escaped hydrogen ions.

This process, the dissolving of positively charged zinc ions from the zinc plate and the collection of positively charged H^+ ions at the copper plate (which subtract electrons from the copper plate), is clearly equivalent to a flow of positive electricity through the solution from the zinc to the copper. Thus the circuit is completed, since we may regard the current in the wire as a flow of positive electricity from the copper to the zinc.

Volta knew that if plates of copper and zinc are placed in contact there is a difference of potential between them. He maintained that the seat of the E.M.F. of a cell lay in the contact between the two metals, but he had very little understanding of the essential electrolytic action inside the sulphuric acid. Consequently he believed that a cell would give a current for ever: that it would never wear out. He spoke of "Cette circulation *sans fin* du fluide électrique, ce mouvement *perpétuel*...."

Faraday, however, perceived that the acid in a cell is an electrolyte and that it should be subject to the laws of electrolysis. Accordingly, he connected a cell and a voltameter in series and showed that the zinc was dissolved in the cell as fast as it was deposited in the voltameter.

Example. A cell gives a current of 0·5 *ampere for* 12 *hours. Calculate the weight of zinc consumed.*

The electrochemical equivalent of zinc is 0·000338 *gm. per coulomb.*

Quantity of electricity passed $= 0{\cdot}5 \times 12 \times 60 \times 60$

$= 21600$ coulombs.

When 1 coulomb is passed weight of zinc consumed

$= 0{\cdot}000338$ gm.

$\therefore$ When 21600 coulombs are passed weight of zinc consumed

$= 21600 \times 0{\cdot}000338$

$= 7{\cdot}30$ gm.

The accumulator.

Put two lead plates in dilute sulphuric acid and pass a current of 2 or 3 amperes for a few minutes (see Fig. 110). A brown deposit of lead peroxide, PbO_2, forms on the anode, and hydrogen is given off at the cathode.

When the current is switched off, connect a voltmeter across the plates. It will be found to read about 2 volts. The lead plates

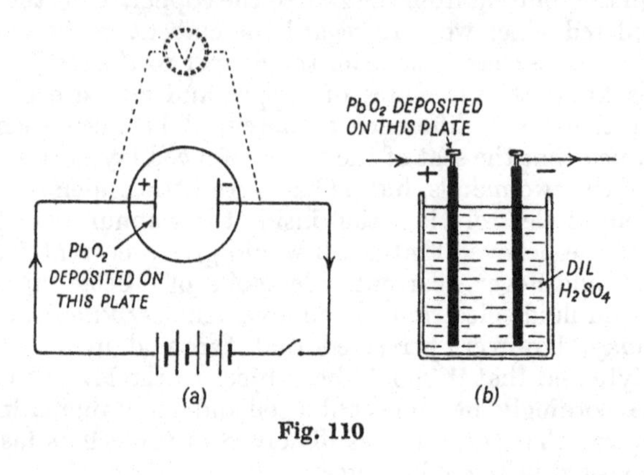

Fig. 110

and the sulphuric acid now act as a cell, and will light a lamp or ring a bell for a short time.

The cell is called a secondary cell or an accumulator, since when it is discharged, it can be charged again.

When the charging current is passed through it, the sulphuric acid is electrolysed, hydrogen ions going to the cathode and SO_4 ions to the anode. The SO_4 acts on the lead plate and the water, forming PbO_2:

$$Pb + 2H_2O + 2SO_4 = PbO_2 + 2H_2SO_4.$$

The accumulator behaves exactly like an ordinary cell, with a positive plate of PbO_2 and a negative plate of Pb, the exciting liquid being H_2SO_4. Where it differs from an ordinary cell is that it can be recharged: when the PbO_2 is all used up it can be regenerated by passing a charging current through the cell in a direction opposite to the discharge current.

Charging and discharging.

Fig. 111 represents what happens when an accumulator is charged or discharged. During charge the sulphuric acid is electrolysed and hydrogen ions go to the cathode (the negative plate of the cell), while SO_4 ions go to the anode. When charged the positive plate consists of PbO_2 and the negative plate of lead. During discharge, the current through the cell is in the opposite direction: the hydrogen ions go to the positive plate and the SO_4 ions to the negative plate. Both positive and negative plates are converted into a sulphate of lead, $PbSO_4$.

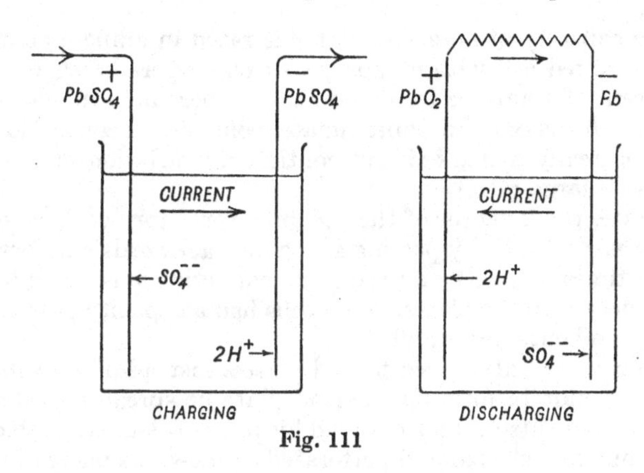

Fig. 111

We will give the chemical equations.

Discharging:

$$\underset{}{H_2SO_4} \rightarrow \underset{+\text{ve plate}}{H_2} + \underset{-\text{ve plate}}{SO_4}.$$

At −ve plate:

$$Pb + SO_4 = PbSO_4.$$

At +ve plate:

$$PbO_2 + H_2 = H_2O + PbO,$$
$$PbO + H_2SO_4 = PbSO_4 + H_2O.$$

Charging:

$$\underset{}{H_2SO_4} \rightarrow \underset{-\text{ve plate}}{H_2} + \underset{+\text{ve plate}}{SO_4}.$$

At −ve plate:

$$PbSO_4 + H_2 = Pb + H_2SO_4.$$

At +ve plate:

$$PbSO_4 + SO_4 = Pb\,(SO_4)_2.$$

$$Pb\,(SO_4)_2 + 2H_2O = PbO_2 + 2H_2SO_4.$$

Practical details.

Plain lead plates are not suitable for a commercial cell, since the chemical action is confined to the surface of the plates. Such a cell would not give a current for long: its capacity is small.

The capacity of an accumulator is rated in ampere-hours. A fully charged cell with a capacity of 60 ampere-hours will give a current of 1 ampere for 60 hours, ½ ampere for 120 hours, and so on. (This is only an approximate, commercial conception, for there is really a marked and continuous variation of capacity with discharge rate.)

Planté, the inventor of the accumulator, "formed" the plates of the earliest cells by passing a current backwards and forwards many times. In this way the chemical action was made to bite quite deep into the plates. Such cells had a capacity of from 6 to 9 ampere-hours per lb. wt.

Then Fauré introduced pasted plates: instead of allowing the lead peroxide to form on the lead plate he spread a paste of it there at the outset. Fauré spread his paste on smooth surfaces at first: but now slotted and perforated frameworks made of a lead antimony alloy (which is stronger than pure lead), filled with paste, are used (see Fig. 112). The introduction of pasted plates raised the capacity of lead accumulators to 10–15 ampere-hours per lb. wt.

It will be seen from the chemical equations that during discharge the sulphuric acid is used to form lead sulphate at the plates: hence its density decreases. Many accumulators have density indicators (see Fig. 113).

Care must be taken not to discharge an accumulator too much (the voltage should not fall below 1·85 volts), or to leave it for a long period discharged. For in these circumstances the plates become covered with a white insoluble sulphate of lead and the accumulator is ruined. The phenomenon is called "sulphating".

The industry for manufacturing secondary cells grew rapidly when accumulators began to be used in motor cars.

Many inventors have tried to reduce the weight of accumulators. Edison abandoned the use of lead and invented a cell, now used for driving vehicles, with plates of nickel oxide and iron in caustic potash. If only a cheap, light method of storing electricity in really large amounts could be found it would revolutionise industry. If we could store the electricity we could make from tides, say, for even twelve hours, the price of electricity would be enormously reduced.

Chloride Elec. Storage Co. Ltd.

Fig. 112. The plates or "elements" of an accumulator.

Chloride Elec. Storage Co. Ltd.

Fig. 113. An accumulator.

The drop in potential difference when a cell delivers a current.

When a cell (or dynamo) delivers a current, the P.D. between its terminals falls. The cell is then said to be on *closed circuit*, and the drop in P.D. is sometimes called the *lost volts*. The drop in the voltage is not fixed but increases with the current.

In a similar way the reading of the pressure gauge of a boiler, with steam up, drops as soon as the steam is made to do something, such as drive an engine. And the more steam that is used, the bigger the drop in the pressure gauge.

The P.D. maintained between the cell's terminals on closed circuit drives the current through the external resistance (to which the terminals are connected): the lost volts drive the current through the internal resistance of the cell. If the internal resistance is negligible, the lost volts are negligible: if the internal resistance is large, the lost volts are large.

The reader should study carefully the following example, which demonstrates the mathematical theory.

Example. A cell of E.M.F. *1·5 volts and internal resistance 1 ohm is connected to a wire of resistance 4 ohms (see Fig. 114). Find the current flowing and the* P.D. *across the 4-ohm resistance.*

Applying Ohm's law to the complete circuit:

$$\frac{E}{i}=R+r.$$

$$\therefore\ i=\frac{E}{R+r}=\frac{1\cdot5}{4+1}=0\cdot3 \text{ ampere.}$$

Applying Ohm's law to the 4-ohm resistance:

$$\frac{V}{i}=R,$$

$V=$ P.D. between ends of 4-ohm resistance.

$$\therefore\ \frac{V}{0\cdot3}=4,$$

$$\therefore\ V=1\cdot2 \text{ volts.}$$

This is also the P.D. between the terminals of the cell, since the ends of the 4-ohm resistance are connected to them.

Hence when the cell delivers a current of 0·3 ampere the P.D. *between its terminals falls from 1·5 volts to 1·2 volts.*

The general circuit.

Applying Ohm's law to a complete circuit, we have

$$\frac{E}{i}=R+r,$$

i.e.

E	=	iR	+	ir
E.M.F.		voltage to drive current hrough external resistance		"lost volts" driving current through internal resistance of cell

It is clear from this equation that the E.M.F. has two functions, to drive the current through both the external and internal resistances: also that the "lost volts", ir, are greater the greater the internal resistance, r, and the greater the current, i.

Example. The E.M.F. of a cell as read by a voltmeter is 1·08 volts. When a current of 0·5 ampere (read by an ammeter) is taken from the cell the reading of the voltmeter drops to 0·93 volt. Find the internal resistance of the cell (see Fig. 115).

$$\text{"Lost volts"} = 1{\cdot}08 - 0{\cdot}93 = 0{\cdot}15 \text{ volt.}$$

This voltage is required to drive the current through the cell.

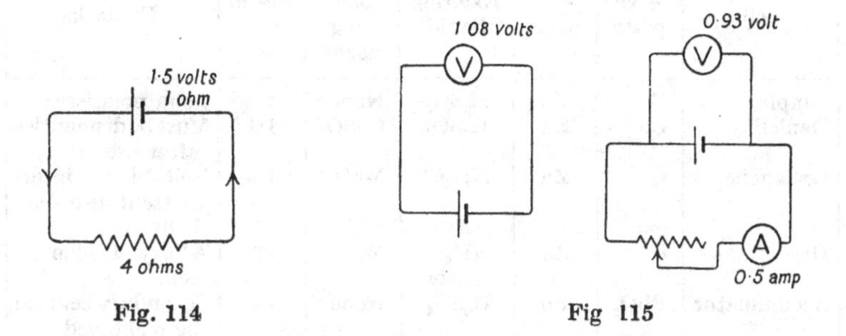

Fig. 114 Fig 115

Applying Ohm's law to the cell:

$$\frac{V}{i} = r,$$

V = P.D. required to drive the current through the cell.

$$\therefore \frac{0{\cdot}15}{0{\cdot}5} = r.$$

$$\therefore r = 0{\cdot}3 \text{ ohm.}$$

This method may be used experimentally to determine the internal resistance of a cell.

SUMMARY

The defects of a simple cell are

(1) *polarisation*, the setting up of a back E.M.F. due to the collection of hydrogen bubbles on the copper plate, remedied by an oxidising agent;

(2) *local action*, the wearing away of impure zinc while the cell is not in action, remedied by amalgamating the zinc with mercury.

Cell	+ve plate	−ve plate	Exciting liquid	De-polari-sing agent	E.M.F. volts	Remarks
Simple	Cu	Zn	H_2SO_4	None	1	Soon polarises
Daniell	Cu	Zn	H_2SO_4	$CuSO_4$	1·1	Must be dismantled after use
Leclanché	C	Zn	NH_4Cl	MnO_2	1·5	Suitable for intermittent use e.g. bells
Dry	C	Zn	NH_4Cl paste	MnO_2	1·5	A "dry" Leclanché cell
Accumulator	PbO_2	Pb	H_2SO_4	None	2·0	Secondary cell; can be recharged

The weight of zinc consumed in a cell is equal to the weight of zinc which would be deposited in electrolysis by a quantity of electricity equal to that generated by the cell.

When a cell delivers a current the P.D. between its terminals falls. The drop in P.D. is the voltage required to drive the current through the internal resistance of the cell:

$$E = iR + ir$$

E.M.F. = P.D. to drive the current through the external circuit + "lost volts," P.D. to drive the current through the cell

QUESTIONS

1. Describe the construction and action of a simple voltaic cell. How would you demonstrate the presence of local action and polarisation? Indicate briefly how these defects are remedied in a practical form of cell. (N.)

2. Describe the construction of a Leclanché cell and give a brief account of its action. Mention two purposes for which this cell is suitable.

Show how the Leclanché cell may be modified for use as a dry cell. (N.)

3. Describe fully the construction and mode of action of the different parts of an electric torch (including the battery), illustrating your answer by means of diagrams. (O. and C.)

4. State three properties which should be possessed by a good cell. Describe briefly how these properties are illustrated in each of the following: (*a*) Daniell cell, (*b*) Leclanché cell, (*c*) Storage cell (accumulator).

Distinguish between primary and secondary cells. (L.)

5. What factors determine (*a*) the electromotive force, (*b*) the resistance of a voltaic cell? (L.)

6. *X* is a small Daniell cell, *Y* is a large Daniell cell, and *Z* is a bichromate cell. If each is connected separately with a low-resistance galvanometer, *X* gives the smallest deflection and *Y* the largest: but when separately connected with a high-resistance galvanometer, *X* and *Y* give the same deflection, which is less than that of *Z*. How can these observations be explained? (O.)

7. A Leclanché cell is connected to a high-resistance galvanometer, the needle of which is deflected. The poles of the cell are then bridged across for a short time by a piece of thick copper wire. After the removal of the thick wire the galvanometer deflection is less than before but gradually rises to its former value. Explain this. (O. and C.)

8. Describe the construction of a Leclanché cell. A Leclanché cell is connected to an ammeter through a contact key. The poles of the cell are connected permanently to a voltmeter. Describe and explain the readings of the ammeter and voltmeter when the key is (*a*) pressed down, (*b*) kept down for a considerable time, (*c*) released. (O. and C.)

9. A Leclanché cell supplies a current of 0·5 ampere for 10 minutes. Find the mass of zinc used up. (Electrochemical equivalent of zinc = 0·000338 gm. per coulomb.) (O. and C.)

10. How much zinc will be consumed in a battery the current from which deposits 60 gm. of silver from a bath of silver nitrate if 20 per cent of the zinc is wasted through local action. (Atomic weights: Zn = 65, Ag = 108.) (O. and C.)

11. The zinc plate of a Daniell cell weighs 50 gm. What is the maximum amount of electricity obtainable from the cell (which is kept saturated with copper sulphate)? Atomic weight of zinc = 65·4, valency of zinc = 2, 1 Faraday = 96,540 coulombs. (1 Faraday is the quantity of electricity flowing when 1 gm. equivalent weight of zinc is consumed.) (O. and C.)

12. Explain clearly: In the electrolysis of acidulated water, hydrogen is liberated at the negative electrode, but in the Leclanché cell it is the positive electrode which polarises. (O. and C.)

13. What changes take place in (*a*) a Leclanché cell, (*b*) a storage cell (or accumulator) during discharge? Why cannot a Leclanché cell be recharged in the same way as a storage cell? (C.)

14. Explain the meaning of *electromotive force, current strength, quantity of electricity.*

Five accumulators, each having an E.M.F. of 2 volts and a resistance of $\frac{1}{10}$ ohm, are joined up in series to a 100-volt main. What additional resistance must be put into the circuit if a charging current of 5 amperes is required? (O. and C.)

15. Describe a lead accumulator and outline the changes that take place in it during charging and discharging.

The 100-volt mains are to be used to charge a 6-volt accumulator. If a charging current of 4 amperes is to be used, and the internal resistance of the accumulators is negligible, what series resistance will be necessary? Show on a diagram how you would connect up the circuit. (O. and C.)

16. A 2-volt 10-ampere-hour accumulator having an internal resistance of 0·1 ohm is charged by means of three Daniell cells, each of E.M.F. 1·1 volt and internal resistance 4·3 ohms. How long will the charging take to complete? Show on a diagram how to connect up the apparatus and the direction in which the current will flow. (L.)

17. Explain what would happen if an attempt were made to "charge" a simple cell, i.e. if a current were passed through the cell from the copper to the zinc.

18. What is meant by the difference of potential between two points in a circuit? What is the relation between the current and the potential difference?

Two Daniell cells and a wire of resistance 2·5 ohms are connected in series. If each cell has an E.M.F. of 1·1 volts and an internal resistance of 0·5 ohm, what is the difference of potential between the terminals of the battery? (C.)

19. What do you understand by the E.M.F. of a cell? Why is it not always identical with the difference of potential between the poles of the cell? When the terminals of a cell of E.M.F. 2 volts are connected by a wire of 10 ohms' resistance the potential difference between the cell terminals drops to 1·9 volts. Find the current round the circuit and the resistance of the cell. (O. and C.)

20. Describe the Daniell cell and explain its action. If the difference of potential between the poles of a cell when no current is flowing is 1·1 volts and this becomes 0·80 volt when the poles are joined by a wire of 1·5 ohms' resistance, find the internal resistance of the cell. (N.)

21. A cell of E.M.F. 1·50 volts is connected in series with an ammeter of resistance 0·20 ohm and a coil of resistance R. The reading of the ammeter is 0·32 ampere and the difference of potential between the terminals of the cell is found to be 1·35 volts. Find the internal resistance of the cell and the value of R. (C.)

22. State Ohm's law and define the resistance of a circuit.

A cell of E.M.F. 1·45 volts is connected in series with a resistance coil and an ammeter, and the current is found to be 0·56 ampere. A voltmeter connected across the terminals of the cell reads 1·12 volts. Find (*a*) the total resistance of the circuit, (*b*) the internal resistance of the cell. (O. and C.)

23. A battery of E.M.F. 105 volts and internal resistance 1 ohm sends current through three lamps in parallel. The resistance of each lamp is 207 ohms. Find (*a*) the current through the circuit, (*b*) the difference of potential between the terminals of the battery. (O. and C.)

24. When the terminals of a cell are connected by a wire of resistance 1·3 ohms the current passing through the circuit is $\frac{1}{3}$ ampere. If a coil of 3·5 ohms' resistance is used instead of the 1·3-ohm coil, the current is $\frac{1}{5}$ ampere. Calculate the E.M.F. and the internal resistance of the cell. (L.)

Chapter IX

MEASUREMENT OF RESISTANCE AND POTENTIAL DIFFERENCE

The pioneers of electric telegraphy, men like Charles Wheatstone, chief post office engineer, were particularly interested in the resistances of metals. A slight difference in the conducting powers of two metals makes a very considerable difference in the resistance of a telegraph wire several miles long: moreover, a fault (or earth connection) in a line can be located by finding the resistance of the line from one end to the fault (see p. 147).

Accordingly, methods were devised for measuring resistance accurately. There were no ammeters or voltmeters in the early days. The very units, ampere, volt and ohm, had not been defined. Wheatstone took as his unit of resistance that of a uniform copper wire 1 foot long and weighing 100 grains; in France 1 kilometre of iron wire was chosen as a standard.

Wheatstone's method was to compare the currents, i_1 and i_2, passed through the unknown and the standard resistances, R_1 and R_2, by the same cell. Using Ohm's law (and Wheatstone was one of the first to realise the importance of Ohm's law),

$$R_1=\frac{V}{i_1} \text{ and } R_2=\frac{V}{i_2}.$$

Assuming that V remains constant,

$$\frac{R_1}{R_2}=\frac{i_2}{i_1}.$$

Wheatstone measured his currents by the deflections of a galvanometer: he found that a serious source of error was the lack of constancy of the potential difference, V, supplied by his battery.

Methods of measuring resistance.

We have not the space to describe all the methods used by the early experimenters with the primitive apparatus at their disposal. We shall confine ourselves to methods utilising the modern apparatus which is to be found in every school laboratory.

1. *Method of substitution.*

Fig. 116 represents a resistance box. Inside the box are coils of resistances 1, 2, 2, 5, 10, etc. ohms: by using the appropriate coils in series, it is possible to obtain any resistance up to say 100 or 1000 or 10,000 ohms. In order to put a coil into the circuit a plug is removed from the box (see Fig. 117). When the plug is in its socket it short-circuits the coil: when it is removed,

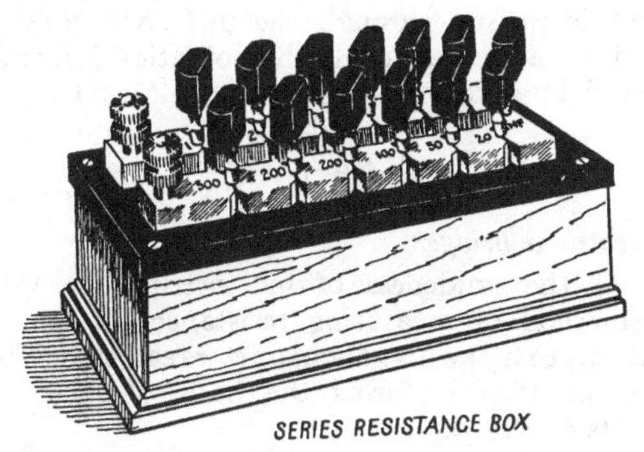

Fig. 116

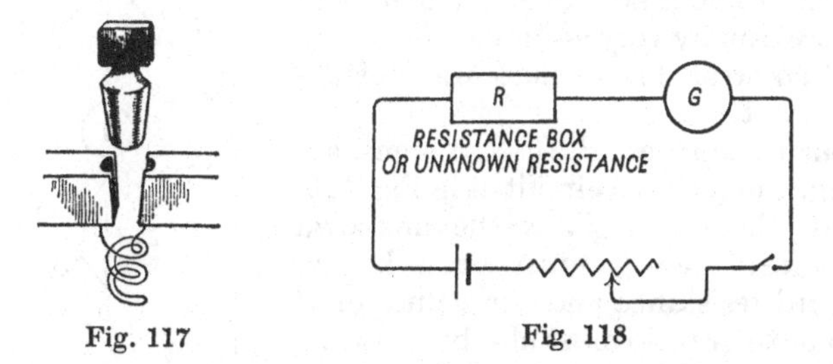

Fig. 117

Fig. 118

the current, instead of passing through the thick brass plug of negligible resistance, has to pass through the coil.

To determine an unknown resistance by the method of substitution, the unknown resistance is placed in series with a cell, and a galvanometer and a variable resistance or rheostat (see Fig. 118). The rheostat is adjusted until there is a suitable deflection in the galvanometer. The unknown resistance is now replaced by a resistance box, the rest of the circuit remaining unchanged. Plugs are taken out of the box until the deflection of the galvanometer is the same as before. Then the resistance of the box is equal to the unknown resistance.

This method is suitable only for high resistances of 100 ohms or more, since with an ordinary resistance box, the value can be obtained only to the nearest ohm.

2. *Ammeter and voltmeter method.*

A current is passed through the unknown resistance and measured by an ammeter, while the potential difference across it is measured by a voltmeter (see pp. 42–43). Then, by Ohm's law,

$$R=\frac{V}{i}.$$

3. *The Wheatstone bridge.*

Because of the crudeness of his instruments Wheatstone devised a method of measuring resistance wherein the final adjustment was to make his galvanometer read zero. The method is so ingenious that it forms the basis of most resistance measurements to this day.

Methods depending on the zero reading of an instrument are called *null* methods: they are usually very accurate, since they entail no error due to inaccuracy of the instrument.

Four resistances, P, Q, R, S ohms, are arranged in a closed circuit as in Fig. 119. One of these, say P, is the unknown resistance: then Q must be a known standard resistance and the values of R and S or their ratio must also be known.

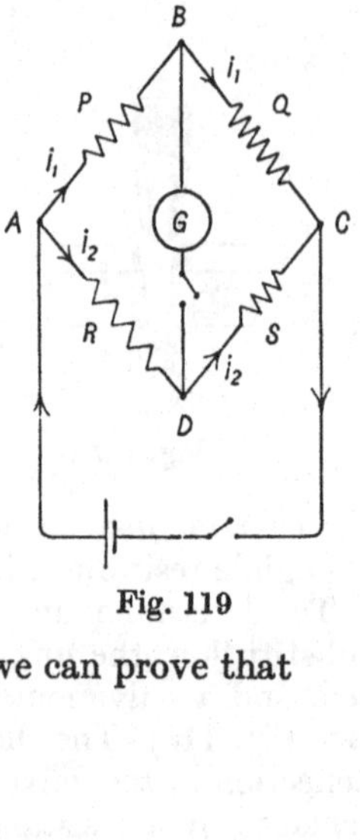

A sensitive galvanometer and a cell are connected as in Fig. 119. Then if the resistances R and S are adjusted so that no current flows in the galvanometer (when the bridge is said to be balanced), we can prove that

$$\frac{P}{Q}=\frac{R}{S},$$

whence P can be calculated.

Fig. 119

Proof. Suppose the current from the cell on reaching A divides in such a way that i_1 amperes flow through P, and i_2 amperes through R. Since no current flows through the galvanometer, i_1 amperes flow on through Q and i_2 amperes through S.

Again, just as water will not flow between two points at the same level, so electricity will not flow between two points at the same potential. Since no current flows through the galvanometer,

Potential at B = Potential at D.

$\therefore$ P.D. between A and B = P.D. between A and D.

By Ohm's law ($V = iR$),

$$i_1 P = i_2 R.$$

Similarly,

P.D. between B and C = P.D. between D and C,

$$\text{i.e. } i_1 Q = i_2 S.$$

Dividing,

$$\frac{i_1 P}{i_1 Q} = \frac{i_2 R}{i_2 S},$$

$$\text{i.e. } \frac{P}{Q} = \frac{R}{S}.$$

The metre bridge.

The commonest experimental arrangement of the Wheatstone bridge, used in the school laboratory, is the metre bridge (see Fig. 120).

A wire AC of uniform cross-section and 1 metre long, made of some alloy such as German silver, so that its resistance is of the order of 1 ohm, lies stretched between two thick brass or copper strips bearing terminals, above a metre ruler.

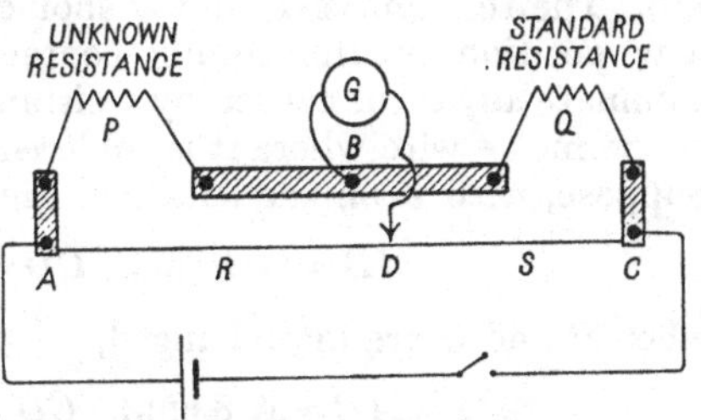

Fig. 120. Metre bridge.

There is another thick brass strip bearing three terminals to facilitate connections, and also a sliding contact, D, which can move along the metre wire.

Fig. 120 has been lettered identically with Fig. 119. The two diagrams should be examined carefully until the reader is convinced that they are identical circuits.

The position of the sliding contact is adjusted until no current flows in the galvanometer.

Then
$$\frac{P}{Q}=\frac{R}{S}.$$

But
$$\frac{R}{S}=\frac{\text{Length } AD}{\text{Length } CD},$$

since the wire AC is of uniform cross-section.

$$\therefore \frac{P}{Q}=\frac{\text{Length } AD}{\text{Length } CD}.$$

There are several experimental precautions to be noted:

1. The battery switch should be depressed before the galvanometer switch (for reasons which cannot be understood until the phenomenon of electromagnetic induction has been studied).

2. The value of the standard resistance, Q, should be chosen so that the sliding contact is near the middle of the metre wire when the bridge is balanced. This ensures the maximum accuracy. For suppose $AD=50$ cm.: an error of 1 mm. in the final adjustment is an error of 1 in 500 or $\frac{1}{5}$ per cent. Suppose now that $AD=5$ cm., an error of 1 mm. in the final adjustment is an error of 1 in 50, i.e. 2 per cent.

3. The resistances P and Q should be interchanged and the new position of the sliding contact determined. This is to eliminate any error caused by resistances introduced at the ends of the metre wire where it is soldered to the thick metal strips. Suppose, with P on the left as in Fig. 120,

$$AD=47{\cdot}4 \text{ cm.,} \quad CD=52{\cdot}6 \text{ cm.};$$

when P and Q are interchanged,

$$AD=52{\cdot}8 \text{ cm.,} \quad CD=47{\cdot}2 \text{ cm.}$$

We must average the two sets of values, when

$$\frac{P}{Q}=\frac{47{\cdot}3}{52{\cdot}7}.$$

It is advisable to take especial care to ensure that this fraction is "the correct way up". It is obvious that if we had written our fraction $\frac{P}{Q}=\frac{52 \cdot 7}{47 \cdot 3}$ we should have obtained an entirely wrong value for P.

The metre bridge is unsuitable for measuring resistances below 0·1 ohm, since the resistance of the connecting wires would then have an appreciable effect on the result: nor is it suitable for measuring very high resistances (above 100,000 ohms), when a substitution or direct deflection method would be employed.

Locating a fault in a line.

The Wheatstone bridge is used for locating a fault in a telegraph line or submarine cable. The resistance of the line to the breakage is found, and knowing the resistance of the line per yard or per mile, the position of the fault can be located. It is essential, for the method to be applicable, that the end of the line, at the break, should touch the earth. In the case of a submarine cable the water ensures a good earth connection.

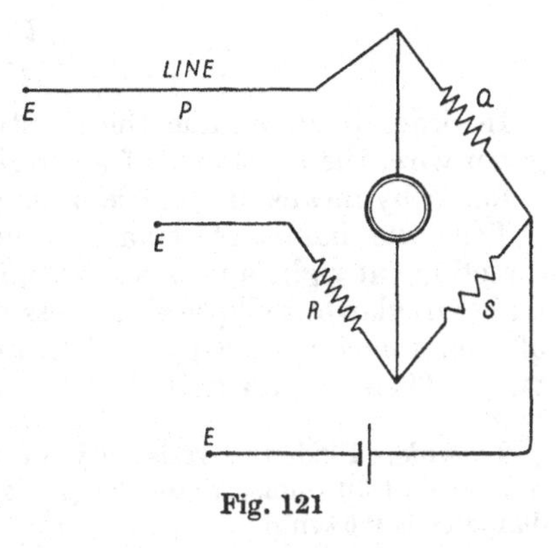

Fig. 121

The connections are made as in Fig. 121, E standing for earth:

$$\frac{P}{Q}=\frac{R}{S}.$$

Hence P can be calculated.

Specific resistance.

In order to compare the resisting powers of different substances a term **resistivity** or **specific resistance** is used.

The resistivity of a substance is the resistance of a wire of the substance 1 cm. long and 1 sq. cm. in cross-section.

Now if the length of a wire is doubled this is equivalent to two of the original lengths in series: hence the resistance is doubled. Thus the resistance R of a wire is proportional to its length, l, i.e. $R \propto l$.

On the other hand, if the area of cross-section of a wire is doubled, this is equivalent to two wires in parallel: hence the resistance is halved. Thus the resistance is inversely proportional to the area of cross-section, A, i.e.

$$R \propto \frac{l}{A}.$$

If S = resistivity of the substance of which the wire is made,

$$R = S\frac{l}{A}.$$

In order to determine the resistivity of the substance of a given wire, the resistance of a certain length of it, say 100 cm., is found by means of a metre bridge.

Then the diameters at various places along the wire, in two directions at right angles at each place, in case the wire is not truly circular but elliptical in cross-section, are found by means of a micrometer screw-gauge. The average of these diameters is taken. Then the resistivity can be calculated.

Example. Find the resistivity of the substance of a wire of resistance 1·52 ohms, whose length is 100 cm. and whose average diameter is 0·64 mm.

Radius of wire $= 0{\cdot}32$ mm. $= 0{\cdot}032$ cm.

∴ Area of cross-section of wire

$$= \pi\,(0{\cdot}032)^2 \text{ sq. cm.}$$

$$\text{Resistivity} = \frac{RA}{l}$$

$$= \frac{1{\cdot}52 \times \pi\,(0{\cdot}032)^2}{100}$$

$$= 0{\cdot}000049 \text{ ohm-cm.}$$

Since its value is so small, resistivity is often expressed in microhms, millionths of an ohm. Thus the above value may be expressed 49 microhm-cm. It may also be written 49×10^{-6} ohm-cm.

Table of Resistivities

(at normal temperatures)

Conductors	Microhm-cm.	Use
Silver	1·6	Contacts
Copper (drawn)	1·8	Connecting wires and cables
Aluminium	2·9	Grid cables carried by pylons
Tungsten	5·0	Filaments of lamps: contacts
Iron	9–15	Grid cables: resistances
Platinum	11	Contacts now replaced by tungsten
Mercury	95	International standard
Nichrome (75% Ni, 12% Fe, 11% Cr, 2% Mn)	110	Heating elements
German silver (62 % Cu, 15 % Ni, 22 % Zn)	16–40	Resistances
Manganin (84 % Cu, 12 % Mn, 4 % Ni)	44	Resistances
Eureka (or constantan) (60 % Cu, 40 % Ni)	49	Resistances
Platinoid	34	Resistances
Carbon	4000–7000	Resistances
Insulators	Megohm-cm. (approx. values at normal temperatures)	
Porcelain	500	Insulators on grid pylons
Gutta percha	1000	Insulation of wires and cables
Micanite (mica flakes cemented with shellac)	3000	Slot insulation on high voltage generators
Paraffin wax	24000	

The effect of temperature on resistance.

The resistance of most substances increases when the temperature rises. For purposes of comparison and calculation, since the resistance increases uniformly with rise in temperature, we use the concept *temperature coefficient of resistance which is the increase in resistance per* 1° *C. rise in temperature, of a resistance of* 1 *ohm of the substance at* 0° *C.*

The phenomenon is utilised, in the platinum-resistance thermometer, for measuring temperature. The resistance of a coil of platinum wire is found at certain known temperatures, called fixed points, by means of a specially sensitive Wheatstone bridge. Hence a graph of temperature and resistance may be drawn. Then to determine an unknown temperature, of an oven say, the platinum coil is placed in the oven and its resistance found. From the graph the temperature corresponding to this resistance may be determined.

The advantage of the thermometer is its great range. It can be used to find the temperature of a furnace or of liquid air. Furthermore, it is particularly suited to distance thermometry, since the platinum coil can be connected by long wires to the reading instruments and observer.

Again, the phenomenon is utilised in the Tucker hot-wire microphone, used for sound ranging in the Great War. A platinum wire, heated by an electric current, is cooled at the arrival of the compression sound wave set up in the air by the firing of an enemy gun. The cooling of the wire results in a slight decrease in resistance and causes a current to flow through a string galvanometer (see p. 81). Hence the time of arrival of the sound wave is recorded.

The temperature coefficients of resistance of most pure metals do not differ widely. That of platinum is 0·0037 per ° C., or $\frac{1}{273}$ per ° C.—the value of the coefficient of expansion of gases. This remarkable fact suggests that the cloud of electrons in a wire may behave in a manner similar to a gas.

Now the volume of a gas, when cooled to the temperature of absolute zero, −273° C., if it continued to contract by $\frac{1}{273}$ of its volume at 0° C., would become zero. We are prompted to ask whether the resistance is affected in a similar way.

Experiments performed in 1911, and subsequently, have

shown that when rings of mercury, lead, tin and thallium are cooled to the temperature of liquid helium (about $-269°$ C.) their resistances disappear almost entirely. A current once started in the ring will continue for days: the electrons go round and round without experiencing any appreciable obstruction. The phenomenon is called *supraconductivity.*

As the temperature of a pure metal is raised from 0° C. its resistance increases by approximately $\frac{1}{273}$rd of its resistance at 0° C. for each 1° C. rise. Hence at 273° C. its resistance has approximately doubled. Since the temperature of the filament of a gas-filled tungsten lamp when lit is over 2730° C., it follows that the resistance of the lamp when lit is more than ten times its resistance when cold.

The temperature coefficient of resistance of most alloys is considerably less than that of pure metals. Instead of $\frac{1}{273}$ or 0·0036, the value for manganin is of the order of 0·00005 and for eureka 0·00001. These two alloys are used for making resistance coils, since a slight change of temperature has no appreciable effect on their resistance.

Carbon is of special interest here because its resistance decreases as its temperature rises. This was particularly unfortunate in the days when the filaments of electric lamps were made of carbon. For as the filament got hotter its resistance decreased, and it took more current: the increase in the current made the filament hotter still and the resistance decreased further, and so on. Thus fluctuations in the voltage applied to the lamp were apt to cause the filament to become so hot that it burnt out. To-day carbon has been replaced by tungsten and an alloy of tungsten and osmium called osram.

Boron, glass, porcelain, and non-metallic liquid conductors, in addition to carbon, show a decrease in resistance when the temperature rises, i.e. have a negative temperature coefficient of resistance.

The potentiometer.

The potentiometer is a very sensitive instrument for measuring potential differences. It consists of a uniform wire *AB* (see Fig. 122), through which a steady current is passed, by means of a battery *D.* This wire is usually made of some alloy, such as German silver, and its resistance is of the order of several ohms.

Since a current flows through the wire from A to B there must be a difference of potential between A and B. Further, since the wire is uniform in cross-section, the potential must drop in a steady and regular way along the wire from A to B.

The idea of a steady potential drop from A to B is so important that we will describe a simple way in which it can be demonstrated.

Connect a voltmeter between A and a sliding contact K: this will, of course, measure the P.D. between A and K. Suppose the P.D. between A and B is found to be 2 volts. Then when K is half-way from A to B the voltmeter will read $\frac{1}{2} \times 2 = 1$ volt: when K is one-third of the way from A to B it will read $\frac{1}{3} \times 2$ volts, and so on. Thus the P.D. between A and K is proportional to the length AK.

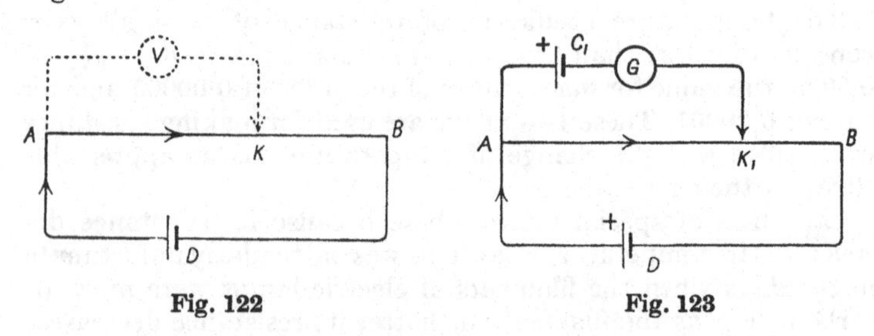

Fig. 122 Fig. 123

The potentiometer method of measuring an unknown P.D. is to obtain an equal P.D. from the wire, by tapping off a suitable length AK: this is done, as we shall see, by counterbalancing the two P.D.S.

Comparing the electromotive forces of two cells.

To compare the E.M.F.S of two cells, connect one of the cells, C_1, between A and the sliding contact through a sensitive galvanometer, as shown in Fig. 123. It is essential that the same poles (say the positive) of C_1 and D should be connected to A.

Suppose the E.M.F. (E_1) of the cell C_1 is, for the sake of concreteness, 1·5 volts. Then if the sliding contact is moved along AB to K_1, where the potential is 1·5 volts below that of A, the potential of K_1 will be the same as that of the negative pole of the cell C_1. Hence no current flows when they are connected, and there will be no deflection in the galvanometer.

Thus when the position of K_1 is adjusted so that no current flows in the galvanometer, the E.M.F., E_1, of the cell C_1 is proportional to the length AK_1, i.e. $E_1 \propto AK_1$.

The cell C_1 is now replaced by the other cell, C_2 (of E.M.F. E_2), with which it is to be compared. A new adjustment of the sliding contact K_2 is made.

Then $$E_2 \propto AK_2.$$

Hence $$\frac{E_1}{E_2}=\frac{AK_1}{AK_2}.$$

Example. **The E.M.F.s of a Leclanché cell, E_1, and of a Daniell cell, E_2, are compared by means of a potentiometer and the readings are**

$$AK_1=80{\cdot}4 \text{ cm. and } AK_2=60{\cdot}0 \text{ cm.}$$

Then $$\frac{\text{E.M.F. of Leclanché cell}}{\text{E.M.F. of Daniell cell}}=\frac{80{\cdot}4}{60}=1{\cdot}34.$$

Thus the E.M.F. of the Leclanché cell is 1·34 times as great as that of the Daniell cell.

The standard cell.

The potentiometer enables us to compare the E.M.F.S of two cells: but we cannot deduce from this result the actual E.M.F.S, in volts, of the cells. We can only say that one cell has an E.M.F. so many times as great as the other.

If, however, the E.M.F. of one of the cells is known accurately, the E.M.F. of the other can be calculated. A special cell, known as a Weston standard cadmium cell, has been devised, which, when constructed according to certain specifications, yields very accurately a known E.M.F. This E.M.F., 1·0183 volts at 20° C., has been accepted by the National Physical Laboratory, the Washington Bureau of Standards and the Berlin Reichsanstalt as the practical standard of E.M.F.

The cell and its constituents are shown in Fig. 124.

Thus by the use of a Weston standard cell with a potentiometer the E.M.F. of any cell may be determined accurately.

One of the chief advantages of the potentiometer method is that it measures the true E.M.F. of a cell, i.e. its P.D. on open circuit: for at the final adjustment, no current is supplied by the cell. When a voltmeter is used, the cell must supply a small current to operate the voltmeter, and hence the P.D. between its terminals drops slightly (see p. 135). The voltmeter, therefore, records a value slightly lower than the true E.M.F.

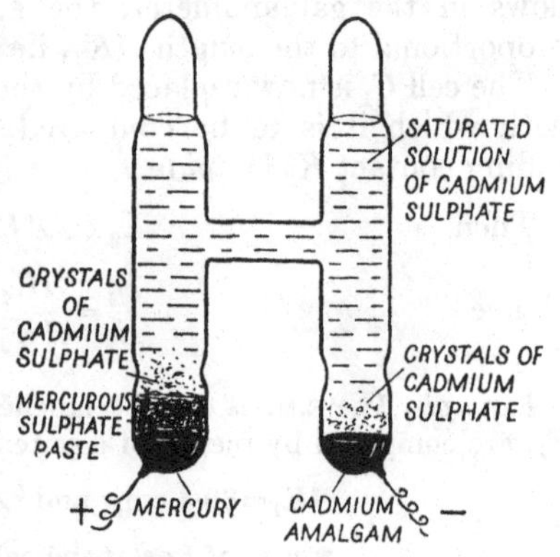

Fig. 124. Weston standard cell.

Example. In Fig. 122 the cell D has an E.M.F. of 2·1 volts and an internal resistance of 0·1 ohm. The resistance of the wire AB is 2 ohms. Find the current through AB and also the P.D. between A and B. If the voltmeter (which has a very high resistance) reads 1·4 volts, find the length AK, taking the length of AB as 100 cm.

Applying Ohm's law to the circuit ABD:

$$\frac{E}{i} = R + r,$$

$$\text{i.e. } \frac{2 \cdot 1}{i} = 2 + 0 \cdot 1.$$

$$\therefore i = 1 \text{ ampere}.$$

Hence current through AB is 1 ampere.

Applying Ohm's law to the wire AB;

$$\frac{V}{i} = R,$$

$$\text{i.e. } \frac{V}{1} = 2, \text{ i.e. } V = 2 \text{ volts}.$$

Hence P.D. between A and B is 2 volts.

$$\text{Now} \quad \frac{\text{P.D. between } A \text{ and } K}{\text{P.D. between } A \text{ and } B} = \frac{\text{Length } AK}{\text{Length } AB},$$

$$\text{i.e. } \frac{1 \cdot 4}{2} = \frac{\text{Length } AK}{100}.$$

$$\therefore \text{Length } AK = 70 \text{ cm}.$$

SUMMARY

Resistance may be measured by

(1) *the method of substitution;*

(2) *an ammeter and voltmeter,*

$$\frac{V}{i}=R;$$

(3) *the Wheatstone bridge,*

$$\frac{\mathbf{P}}{\mathbf{Q}}=\frac{\mathbf{R}}{\mathbf{S}}.$$

The resistivity of a substance is the resistance of a wire of the substance 1 cm. long and 1 sq. cm. in cross-section:

$$\mathbf{R}=\mathbf{S}\frac{\mathbf{L}}{\mathbf{A}}.$$

The resistance of most substances (with the notable exception of carbon) increases when the temperature rises.

The *potentiometer* is an instrument for measuring potential difference. It consists of a uniform wire through which a steady current is passed giving a uniform drop of potential down the wire.

QUESTIONS

1. (*a*) The resistance of 3 metres of a certain kind of wire is 6 ohms. Find the resistance of 540 cm. of it.

(*b*) The resistance of a wire is 5 ohms. What will be the resistance of wires of the same material and length and of (i) twice the cross-section, (ii) twice the diameter?

(*c*) What length of German silver wire will have the same resistance as 200 cm. of manganin wire of the same diameter? (Specific resistances of German silver and manganin are 16 and 44 microhm-cm. respectively.)

(*d*) What length of copper wire will have the same resistance as 50 feet of iron wire of three times the diameter? (Specific resistances of copper and iron are 1·8 and 9·0 microhm-cm. respectively.)

2. The specific resistance of aluminium is about twice that of copper, but its density is only one-third of copper. If the price (by

weight) of aluminium is twice that of copper, which is the cheaper metal to use for a certain length of cable of given resistance? Show your working fully.

3. A wire of length 2 metres and cross-section 1 sq. mm. has a resistance of 0·1 ohm. What is the specific resistance of its material? (C.)

4. If the resistance of a wire of length 80 cm. and diameter 0·2 cm. is found to be 0·12 ohm, what is the specific resistance of the material? (N.)

5. A strip of tinfoil of length 60 cm., width 0·2 cm., and thickness 0·0019 cm., has a resistance of 1·8 ohms. What is the specific resistance of tin? (O. and C.)

6. Explain fully: A wire is stretched uniformly until its length is doubled. Its electrical resistance is now found to be four times as big as it was before it was stretched. (O. and C.)

7. Explain upon what factors the resistance of a metal wire depends.

A 10-ohm coil is constructed of manganin wire, 0·71 mm. diameter. If the specific resistance of manganin is 0·000048 ohm per cm., calculate the length of wire used. (L.)

8. A metal wire of length 1 metre and uniform diameter has a resistance of 1·05 ohms: calculate the resistance of a coil made from the wire of the same material 50 metres long but having twice the diameter. (L.)

9. The resistances of two wires of the same material and the same length are 2·1 and 1·4 ohms. What is the ratio of their diameters?

10. What do you understand by the statement that the specific resistance of copper is $1·59 \times 10^{-6}$ at 18° C.? Why is it necessary to specify the temperature?

A centimetre cube of copper is drawn out into a wire of diameter 0·5 mm. What is the resistance of this wire, and how is it altered if the wire is cut up into three equal strips which are then joined in parallel? (O. and C.)

11. What is meant by the specific resistance of a substance? How would you measure the specific resistance of a substance in the form of a wire? (C.)

12. A cell of electromotive force 1·07 volts and one of 1·40 volts are connected in turn to a galvanometer, and the one of smaller electromotive force gives the greater deflection. What explanation can you give of this? What modification would you make to ensure that the deflections, although differing from the previous readings, are now proportional to the electromotive forces of the cells? (L.)

13. Describe the Wheatstone bridge method of comparing resistances.

Draw a diagram of the metre bridge marking the balance point, if the resistances to be compared are 3 and 2 ohms respectively. If the 2-ohm coil is shunted with a 10-ohm coil, where is the new balance point? (L.)

14. Describe the use of the Wheatstone bridge. If the bridge wire tapered slightly along its length, what error would be introduced and how could it be minimized?

In one arm of the bridge is placed a 5-ohm coil. In the other arm a 3-ohm coil and a 6-ohm coil are joined in parallel. If the bridge wire is a metre long, where must the jockey make contact for the bridge to be balanced? (O. and C.)

15. Why, when using a metre bridge, should the known and unknown resistances be (*a*) interchanged, (*b*) as nearly equal as possible?

16. Explain how a Wheatstone bridge is used to compare resistances.

Two resistance coils, A and B, are connected to a slide-wire Wheatstone bridge in the usual way, and a balance is obtained when the sliding contact is 30 cm. from the left-hand end of the wire. The resistances A and B are now interchanged, and a balance is obtained with the slider 120 cm. from the left-hand end of the wire. Deduce the length of the bridge wire (assumed uniform), and the ratio of the resistances A and B. (C.)

17. Without going into experimental details, explain the principle connected with Wheatstone's bridge method of measuring resistances.

A P.D. of 12 volts is maintained between two points A and B which are connected by two parallel branches ACB and ADB. If the respective resistances in ohms are $AC=2$, $CB=4$, $AD=3$, $DB=3$, what is the P.D. between C and D? (L.)

18. The metre bridge in Fig. 120 is balanced. If the sliding contact is moved to the right towards C, will the current in the galvanometer flow from B to D or in the reverse direction? Give the reason fully.

19. (*a*) In Fig. 125 find the resistance of the network *ABCD* between the points *A* and *C*.

(*b*) Find the total current in the circuit, *i*.

(*c*) Find the currents i_1 and i_2.

(*d*) Find the P.D. between *A* and *B* and between *A* and *D*.

(*e*) Find the P.D. between *B* and *D*.

(*f*) If a galvanometer were connected between *B* and *D*, would it register a current?

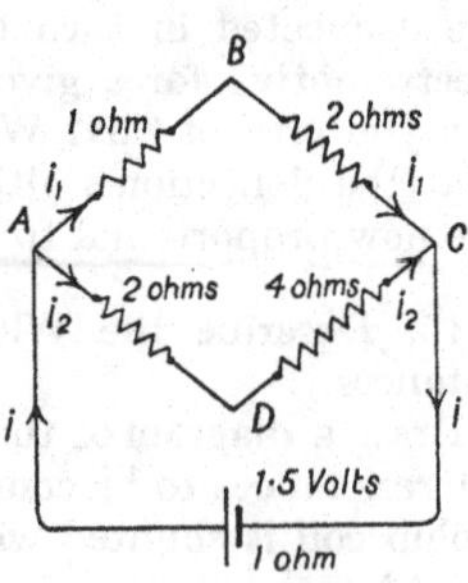

Fig. 125

20. How does the difference of potential between two points in a wire vary the current flowing along the wire? What factors determine the resistance of a conductor? The resistance of a 100-volt metal filament lamp is measured by a Wheatstone bridge method using a Leclanché cell for the bridge and found to be 20 ohms. When the lamp is running on the 100-volt lighting circuit, however, a current of only 0·6 ampere flows through it. How do you reconcile these two results? (C.)

21. How would you measure the resistance of the filament of an electric lamp (*a*) when it is glowing, owing to the passage of its normal current, and (*b*) when it is cold? (N.)

22. An electric lamp is connected through a variable resistance and an ammeter to an accumulator, and a voltmeter is available for use. When the filament is just glowing the current flowing is 0·92 ampere and the P.D. between the lamp terminals is 3·91 volts. The variable resistance is then increased and the current falls to 0·82 ampere, and the P.D. at the lamp terminals becomes 2·46 volts. Calculate the resistance of the lamp filament in each case, and decide whether the filament is made of carbon or of a metal.

Draw a diagram showing how all the apparatus is connected. (Durham.)

23. A piece of platinum wire has a resistance of 6 ohms at 0° C. and of 34 ohms at the temperature of a furnace. If the temperature coefficient of platinum is 0·004 per ° C., find the temperature of the furnace.

24. What is the drop in potential per cm. length along a potentiometer wire 2 metres long of resistance 7 ohms when connected to a cell of E.M.F. 2 volts and internal resistance 1 ohm?

25. In a potentiometer experiment, the galvanometer is found to give deflections in the same direction when connection is made at any point on the potentiometer wire. State and explain any cause which could produce this effect.

26. How could you test whether the wire of a metre bridge is of uniform resistance?

27. What is meant by electrical resistance? Five cells, each of E.M.F. 1·1 volts and internal resistance 1 ohm, are connected in series, and the terminals of the battery so formed are joined by a uniform wire 8 metres long having a resistance of 0·5 ohm per metre. Find the distance between two points on the wire such that the potential difference between them is 1 volt. (C.)

28. How would you use a potentiometer to compare the E.M.F.s of two cells?

The wire of a potentiometer is 200 cm. long and has a resistance of 4·4 ohms. When a Daniell's cell (E.M.F. 1·1 volts) in series with a galvanometer is being tested on the potentiometer wire, no current flows through the galvanometer when the points tapped are 125 cm. apart. Find the current flowing in the potentiometer wire. (C.)

29. Explain carefully why the insertion of a resistance of 1000 ohms between the galvanometer and K_1 in Fig. 123 will not affect the position of the balance point K_1, whereas 1000 ohms inserted between the cell D and the potentiometer wire would have a very great effect.

30. Four cells, each of E.M.F. 1·1 volts and internal resistance 0·5 ohm, are connected in series with a wire AB of length 1 metre and resistance 3 ohms, as shown in Fig. 126. A high-resistance voltmeter V connecting A with a point C in the wire reads 1·0 volt. Calculate (*a*) the current along the wire AB, (*b*) the length of the part AC of the wire. (O. and C.)

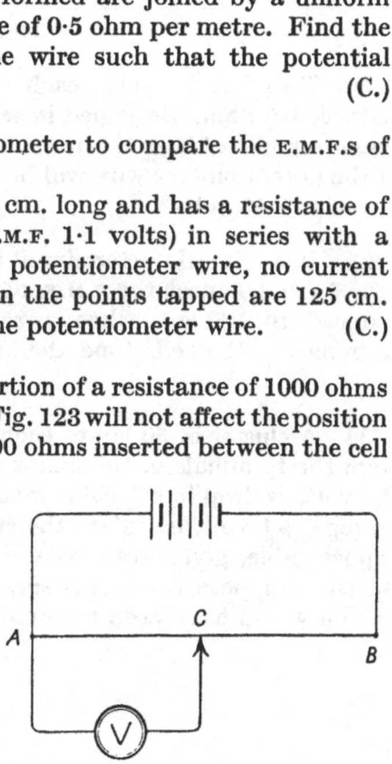

A C B V

Fig. 126

31. A 2-volt accumulator A of negligible internal resistance, a Daniell cell D (E.M.F. 1 volt, internal resistance 2 ohms), a 10-ohm resistance R, a sensitive galvanometer G and a 100-cm. potentiometer wire BC are connected as in Fig. 127, and the sliding contact K is adjusted until the galvanometer shows no deflection. Find (*a*) the current through R, (*b*) the potential drop through R, and (*c*) the distance BK. (O. and C.)

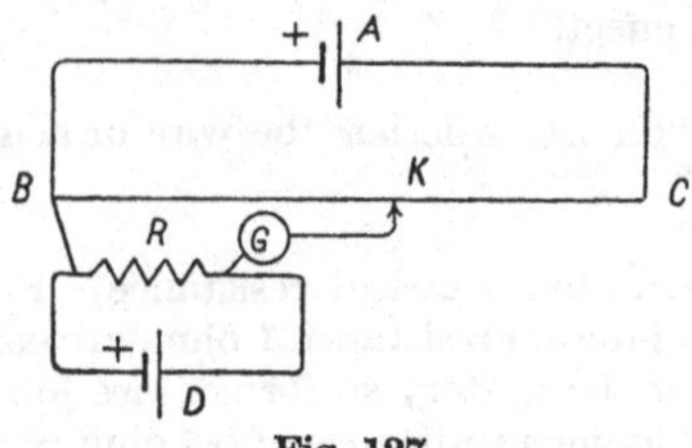

Fig. 127

32. Two Daniell cells, each of E.M.F. 1·1 volts and internal resistance 0·5 ohm, are joined in series, and connected to the ends of a potentiometer of length 1 metre and resistance 2·5 ohms. What length of the potentiometer wire will be required to balance a Leclanché cell of E.M.F. 1·45 volts? (O. and C.)

33. In a potentiometer circuit the length of the potentiometer wire of 200 cm. balances the E.M.F. of a cell in open circuit. This length is reduced to 160 cm. when a resistance of 10 ohms connects the terminals of the cell. Find the internal resistance of the cell. (O. and C.)

34. A cluster of 50 lamps each of 60 watts are to be lit 300 metres from the terminals of the mains supply at 200 volts. Assuming that the voltage drop in the cable must not exceed 3 per cent of the mains voltage +1 volt, calculate the smallest possible cross-section of the copper cable, given resistivity of copper 1·6 microhms per cm. per sq. cm. Suppose the mains supply had been 100 volts, what cross-section would have been required in this case?

Chapter X

ELECTRICAL ENERGY

The work done by an electric current.

A current of electricity in a wire is a flow of millions of free electrons in the spaces between the atoms of the wire. We have compared it to a flow of water: we will now pursue the analogy a stage further.

Flowing water possesses energy: it can do work. When it flows over a waterfall it will drive a mill wheel. The power available depends on two factors:

(1) the flow or weight of water passing per second,

(2) the head or height through which the water falls.

Suppose the flow is 500 lb. of water per sec., and the head is 20 feet. The total work that the water could do is

$$500 \times 20 = 10{,}000 \text{ ft. lb. per sec.}$$

Even if the mill wheel is not there the water still "uses up" energy at this rate. The energy is converted into heat.

In a similar way flowing electricity can do work. It will ring a bell, drive a motor, heat a room, and so on. The power available depends again on the same two factors:

(1) the flow or current, in amperes,

(2) the "head" or P.D. through which the electricity "falls", in volts.

Suppose a current of 500 amperes flows through a P.D. of 20 volts.

$$\text{Work done} = 500 \times 20 = 10{,}000 \text{ joules per sec.}$$

The joule is the unit of work on the metric system. It is equivalent to about $\frac{2}{3}$ ft. lb. (the precise definition is given on p. 325).

Power.

The rate at which a machine can do work is known as its *power*. The unit of power on the British system of units is the horse-power: on the metric system it is the *watt*, which is defined as 1 *joule per sec.* A *kilowatt* is 1000 watts and is equal to about $1\frac{1}{3}$ H.P.

Thus, in the example above, the work done is 10,000 joules per sec., and hence the power generated is 10,000 watts, or 10 kilowatts. It is clear that

amperes × volts = watts.

Mechanical units	Unit of	Definition
Joule Watt	Work Power	(See p. 325) 1 joule per sec.
Electrical units		
Ampere Volt	Current P.D.	(See p. 322) When 1 ampere flows through a P.D. of 1 volt, power is generated at the rate of 1 watt
Relations between these units: Amperes × volts = watts Amperes × volts × seconds = joules		

The reader should study carefully the table given above. The important relation, amperes × volts = watts, really follows from the definition of the volt: the size of the volt is actually adjusted to fit this equation.

The power in an electric lamp.

An electric lamp is always marked in some such way as the following: 210 V., 60 W.

This means that the lamp is designed to work on a 210-volt circuit, when it will absorb electrical energy at the rate of 60 watts, converting the electrical energy to heat and light energy.

We will calculate the current taken by this lamp.

$$\text{Amperes} \times \text{volts} = \text{watts}.$$

$$\therefore \text{Amperes} = \frac{\text{watts}}{\text{volts}},$$

$$\text{i.e. Current} = \frac{60}{210} = \tfrac{2}{7} \text{ ampere.}$$

We can now calculate the resistance of the lamp.

By Ohm's law: $$\frac{V}{i} = R.$$

$$\therefore R = \frac{210}{\frac{2}{7}} = 735 \text{ ohms.}$$

A 100-watt lamp designed for the same voltage will give a brighter light, since it is absorbing electrical energy at a greater rate. Hence it must take a larger current. The reader should work out the current taken by, and the resistance of, this lamp.

Buying electrical energy.

When we buy electricity, we are concerned not only with the current we are taking and the time for which it flows, but also the voltage at which it is supplied: for we buy electricity in terms of the work it can do.

The commercial Board of Trade Unit (B.T.U.) is the *kilowatt-hour*, which is the electrical energy supplied at the rate of 1000 watts, in 1 hour. Since there are 3600 seconds in an hour, and 1 watt is 1 joule per sec.,

$$1 \text{ kilowatt hour} = 3{,}600{,}000 \text{ joules.}$$

It is of interest to note that, taking 1 joule as $\frac{3}{4}$ ft. lb., 1 kilowatt-hour is approximately 2,700,000 ft. lb. In other words, for 6*d.*, or whatever is the local price of the B.T.U., we obtain sufficient energy to raise more than 1000 tons through 1 foot.

Example. Find the cost of burning seven 60-watt and four 100-watt lamps for 3 hours if the cost of electrical energy is 5*d.* per unit.

$$\text{Total wattage of lamps} = 7 \times 60 + 4 \times 100$$

$$= 820 \text{ watts.}$$

$\therefore$ Electrical energy used in 3 hours

$$= \frac{820 \times 3}{1000}$$

$$= 2{\cdot}46 \text{ kilowatt-hours.}$$

$$\therefore \text{Cost} = 2{\cdot}46 \times 5$$

$$= 12{\cdot}3 \text{ pence.}$$

The heating effect of an electric current.

We have already mentioned that when water falls, although it may not have done useful work in driving a mill wheel, it has nevertheless "used up" energy, and that this energy is transformed into heat.

The number of ft. lb. of energy absorbed in a certain time is equal to the product of the total number of lb. which have flowed and the distance in feet through which they have fallen.

In the same way when an electric current passes through a wire, between the ends of which there is a P.D., energy is absorbed and appears in the form of heat. The number of joules of energy absorbed in a certain time is equal to the product of the total number of coulombs (amperes $\times$ seconds) which have flowed and the P.D. in volts through which they have fallen:

$$\text{i.e. Energy absorbed in the wire} = Vit \text{ joules,}$$

where $V =$ P.D. in volts, $i =$ current in amperes, $t =$ time in seconds.

Now when mechanical energy is converted into heat there is always a fixed rate of exchange, known as the *mechanical equivalent of heat—the amount of work equivalent to* 1 *unit of heat.*

$$1 \textit{ calorie} = 4{\cdot}2 \textit{ joules.}$$

(The unit of heat, 1 calorie, is the heat required to raise the temperature of 1 gm. of water through 1° C.)

$$\therefore \text{ Heat generated in the wire} = \frac{Vit}{4{\cdot}2} \text{ cals.}$$

Using Ohm's law, $V = iR$, we can write this expression in an alternative form:

$$\textbf{Heat generated} = \frac{\mathbf{Vit}}{\mathbf{4{\cdot}2}} \textbf{ cals.}$$

$$= \frac{\mathbf{i^2Rt}}{\mathbf{4{\cdot}2}} \textbf{ cals.}$$

Heat is the kinetic energy or energy of vibration of the atoms or molecules. When the electrons constituting a current pass through a wire they keep hitting and jostling the atoms in their path, causing the latter to vibrate more energetically and in this way generate heat in the wire.

Example. An electric heater working on a 235-volt supply takes a current of 2 amperes. Find the heat it generates in 1 minute and also how long it will take to raise 1000 gm. of water, originally at 15° C., to its boiling point.

$$\text{Heat generated in 1 minute} = \frac{Vit}{4{\cdot}2} \text{ cals.}$$

$$= \frac{235 \times 2 \times 60}{4{\cdot}2}$$

$$= \frac{47000}{7}$$

$$= 6710 \text{ cals.}$$

Let x sec. = time required to heat the water.

$$\text{Heat generated in } x \text{ sec.} = \frac{235 \times 2 \times x}{4{\cdot}2} \text{ cals.}$$

$$\text{Heat required by the water} = 1000\,(100 - 15)$$

$$= 85000 \text{ cals.}$$

$$\therefore \frac{235 \times 2 \times x}{4{\cdot}2} = 85000.$$

$$\therefore x = 760 \text{ sec.} = 12\tfrac{2}{3} \text{ min.}$$

The work of Joule.

The experimental work on which the foregoing theory is based was first performed by Joule of Manchester, a rich man who had his own private laboratory, and who devoted the proceeds from his brewery to his all-absorbing scientific pursuits.

Joule showed by experiment that the heat generated in a wire is proportional to

(1) the resistance of the wire,
(2) the square of the current,
(3) the time for which the current passes.

This is sometimes known as *Joule's law*, and may be summarised in the expression we have already derived:

$$\text{Heat generated} = \frac{i^2 Rt}{4{\cdot}2} \text{ cals.}$$

Joule's experiments took the form of passing currents through wires immersed in weighed quantities of water and finding the rate of rise of temperature.

We shall describe the method he employed to determine the mechanical equivalent of heat, which can be performed quite simply in the school laboratory.

Experimental determination of the mechanical equivalent of heat.

In this experiment electrical energy is converted into heat and the two are measured. It may be regarded as a determination of the electrical equivalent of heat. But since electrical energy is measured in mechanical units, joules, it is, in effect, a determination of the mechanical equivalent of heat, i.e. the number of joules of work equivalent to one calorie of heat.

Weigh a copper calorimeter empty and three-quarters full of water. Immerse in the water a coil of suitable wire of known resistance, and connect it in series with a battery and an ammeter (see Fig. 128). Take the temperature of the water, and then pass a current for a measured time, reading the ammeter at regular intervals of, say, $\frac{1}{2}$ minute in case the current is not constant. When the current is switched off, determine the final temperature, which should be between 10° C. and 20° C. higher than the initial temperature. Calculate the average ammeter reading.

Heat developed $=(W+w)(T_2-T_1)$,

where W = weight of water, w = water equivalent of calorimeter, T_1 and T_2 = initial and final temperatures.

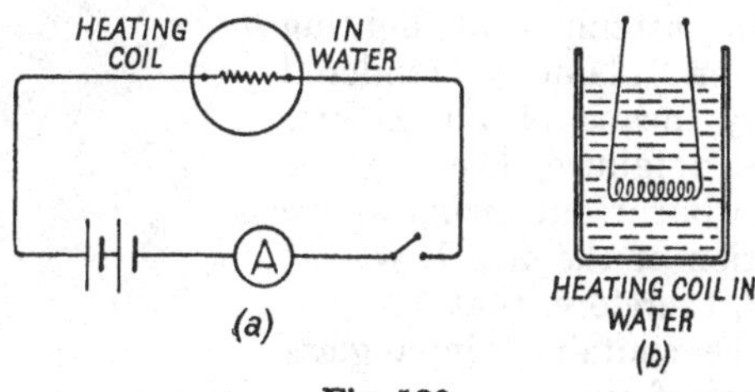

Fig. 128

Work done by the current in forcing its way through the resistance of the coil, thereby generating heat,

$$=i^2Rt \text{ joules,}$$

where i = current, R = resistance of coil, t = time (sec.).

∴ Mechanical equivalent of heat

$$=\frac{i^2Rt}{(W+w)(T_2-T_1)} \text{ joules per calorie.}$$

Electric heating.

Electricity is used for heating purposes in electric radiators, kettles, irons, hotplates, and so on. The heating element commonly used is a coil or length (see Fig. 129) of nichrome wire, made of an alloy of nickel, chromium and iron. This alloy has the advantages of a high resistance and also of not oxidising when heated.

The length and cross-section of the nichrome wire depend on the power required and the available voltage. Suppose a radiator is required on a 200-volt supply with a power of 1 kilowatt. Then the current it takes is $\frac{1000}{200}\left(\frac{\text{watts}}{\text{volts}}\right)=5$ amperes: its resistance must be $\frac{200}{5}=40$ ohms. Hence a length of nichrome wire is required of such cross-section that its resistance is

40 ohms, and that it glows to red heat when a current of 5 amperes passes through it.

Electric lamps.

The earliest attempts at lighting with a wire or "filament" rendered white hot by means of an electric current were a failure, because the filament soon burnt out owing to the oxidising action of the air. It was at once realised, therefore, that the filament must be contained in a glass bulb exhausted of air.

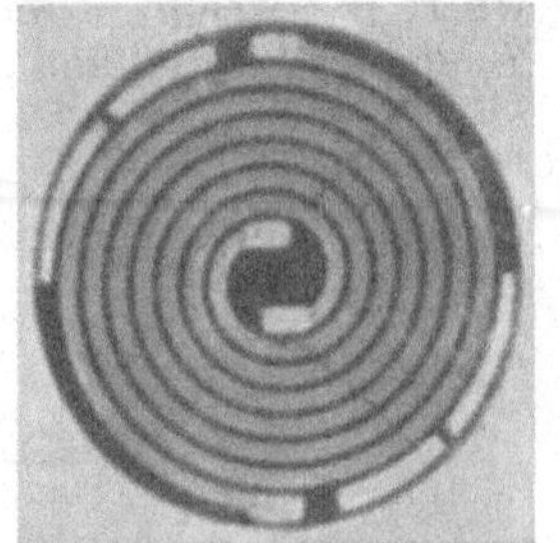

General Electric Co. Ltd.

Fig. 129. An X-ray photograph of a hot plate. The heating element consists of a coiled wire bent in the double spiral shown.

The considerable technical difficulties in doing this were overcome, independently, by Edison and Swan about 1880, when they produced the carbon-filament lamp. Carbon possesses the advantage of not melting when raised to very high temperatures, but it proved to have the disadvantage of disintegrating or "evaporating" slowly, when white hot in a vacuum. Thus the bulbs of carbon-filament lamps slowly blackened inside owing to a fine deposit of carbon, and the life of a filament was not long.

Carbon was eventually superseded by various metals with a high melting point, the one which is now most commonly used being tungsten.

The light given out by a hot filament is greater the higher the temperature; and even metals with a high melting point tend to disintegrate rapidly in a vacuum when at too high a temperature. Langmuir, an American physicist, conceived the idea, some twenty years ago, of preventing this disintegration by filling the lamp bulb with an inert gas (which will not cause oxidation of the filament), such as argon or nitrogen. The pressure of the gas tends to prevent disintegration, and in a modern gas-filled lamp the filament is run at a temperature of about 2700° C. as compared with 1600° C. in the carbon-filament lamp. The fraction of the electrical energy converted into light in a gas-filled lamp is about 10 per cent (the rest is wasted as heat), while the corresponding fraction for the carbon-filament lamp is about 3 per cent: the light given out by the former is whiter and less yellow than that of the latter.

The filament of a vacuum lamp (which is still made, but with a metal filament) is in the form of a large zigzag, but in a gas-filled lamp the filament is close-coiled (see Fig. 131). In the latter

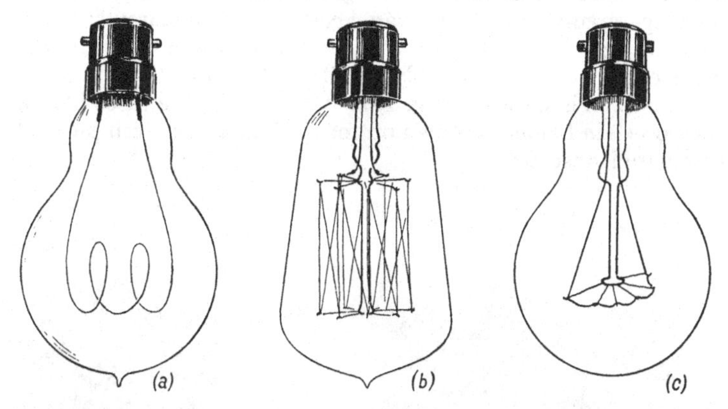

Fig. 130. (*a*) and (*b*) are vacuum lamps with carbon and tungsten filaments, respectively. (*c*) is a gas-filled lamp whose filament is shown magnified in Fig. 131.

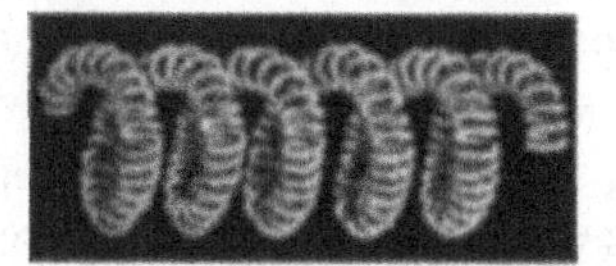

General Electric Co. Ltd.

Fig. 131. A coiled coil lamp filament.

case the coils, being very close together, tend, as it were, to keep each other warm and do not lose heat so rapidly by convection in the gas

The efficiency of a lamp is rated by the number of watts per candle-power. The efficiency of the carbon-filament lamp was about 4 watts per candle-power; that of the gas-filled tungsten-filament lamp is about 1 watt per candle-power.

Fuses.

A fuse is a safety valve in an electric circuit. It consists of a wire made of a tin-lead alloy (with a low melting point), or of a thin copper wire, which melts and breaks the circuit when the current exceeds a certain value. If there were no fuse in, say, a lighting circuit of a house, an accidental short-circuit might cause a very large current to flow which would heat up the connecting wires and possibly set fire to the house.

A fuse box is shown in Fig. 132. When one of the fuses blows, the porcelain holder is drawn out of the box and a fresh piece of fuse wire is inserted

General Elec. Co. Ltd.

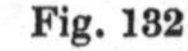

Fig. 132

General Elec. Co. Ltd.

Fig. 133

Fig. 132. A fuse-board for lighting installations. Two fuses have been removed and are lying on the ground. The fuse wire connects the two metal terminals, in each case.

Fig. 133. A domestic meter. Currents are induced in the copper disc (in the middle), due to the alternating field of the electromagnet, with the result that the disc slowly rotates—see Chapter XI.

The wiring of a house.

The electricity supply of a house is drawn, usually, from a cable under the road. This cable, composed of a conductor of copper strands insulated with paper or rubber enclosed in a moisture-proof sheath of lead wound round with steel tape armouring, is tapped at what is called a joint box.

From the joint box a service cable enters the house, where it is connected to the company's fuses or cut-outs, which should be opened only by officials of the company. If the house is wired both for lighting and heating two separate meters are required, since power for heating is supplied at a cheaper rate than for lighting. (This is to encourage the use of electric radiators and cookers, the cost of which would be prohibitive at the price of power for lighting.)

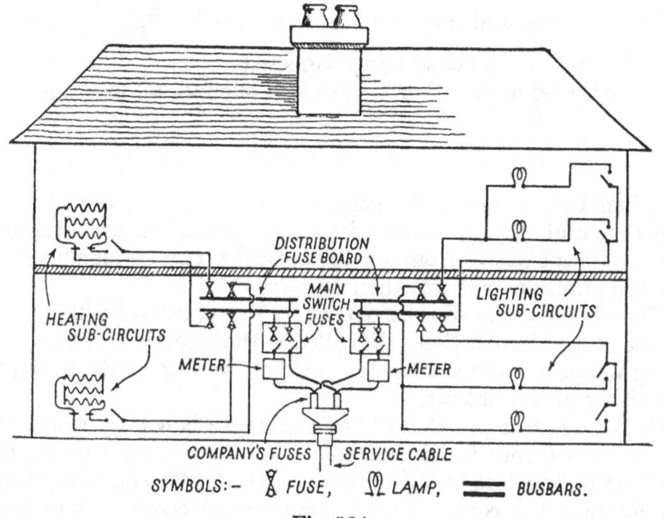

Fig. 134

Beyond the meters, the circuits proceed to the main switches, main-switch fuses, the distribution fuse board and so to the heating and lighting subcircuits (see Fig. 134). The lighting fuses on the distribution fuse board are usually of about 4 amperes' capacity, and the heating fuses of 15 amperes' capacity.

In the early days, electricians, taking their cue from nature, evolved the "tree" system of distribution. The sap of a tree rises through the trunk to its branches, and thence to smaller branches, and finally to twigs. We may regard the distribution system of Fig. 134 as being (inevitably) of this type, with the important modification that the main branch, where it enters the house, divides into four equal sub-branches (each with its own fuses), rather than proceeding throughout the house and shooting out small twigs here and there.

It is very important to note that all lamps and heaters are connected in parallel across the full voltage of the mains. A lamp is so constructed that it will give its correct illumination only at the specified voltage, say 230 volts. If two such lamps were connected in series, the voltage across each of them would be only $\frac{230}{2} = 115$ volts, and they would give a very dim light. Moreover, if all the lamps in a house were connected in series, all would have to be switched on together, and should one fail, all would go out.

The electric arc.

When two carbon rods, connected in a circuit, are brought together and then separated for a short distance, a continuous sheet of flame passes from one to the other. The voltage between the rods must be not less than 40.

The phenomenon was discovered by Davy, using the 2000 cells at the Royal Institution. He called it the electric arc, since the flame between horizontal rods tended to arch upwards, owing to the rising of the hot air.

On starting or "striking" the arc, there is a high resistance where the carbons touch and much heat is evolved, causing the surfaces to be raised to incandescence. On separating the carbons the air becomes conducting: a stream of electrons passes from the negative to the positive rod. The flame consists of white-hot particles of carbon torn from the positive rod, in which a crater is formed. Both carbons do, in fact, burn away gradually, but the positive carbon wears away twice as fast as the negative; hence the positive rods are often made twice as thick as the negative rods. To maintain the arc the carbons must be moved together, either at intervals by hand or by means of an automatic electromagnetic device. If the carbons are enclosed in a globe which is nearly air-tight, they burn away very slowly, about 1 mm. per hour, owing to lack of oxygen.

The electric arc provides the highest continuous temperatures attainable by man: the temperature in the crater of the positive carbon is about 3500° C., and that of the negative carbon is about

3000° C. Carbon rods are used, since all metals melt at such high temperatures.

The electric arc gives an intensely bright and dazzling light and used to be employed extensively for street lighting. It has now been replaced, for this purpose, by high-candle-power metal-filament lamps and gas-discharge lamps. It is still used for projectors and searchlights, however, since, being a source of small area, its light can be easily focused: a parabolic reflector (like that of a motor-car headlamp) will produce a powerful, parallel beam when the arc is placed at its principal focus.

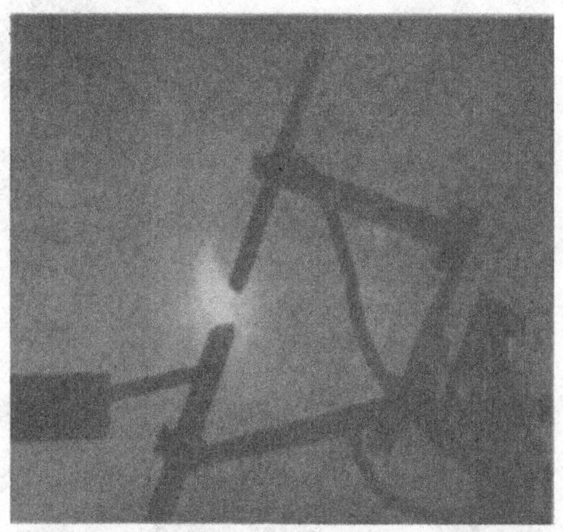

B. F. Brown

Fig. 135. An electric arc being deflected by a magnet (bottom left in the clamp). The arc consists of a stream of electrons which are deflected by a magnet (see also Fig. 288).

Electric welding.

Boiler plates, the plates forming the sides of a ship, a cracked cylinder of a motor-car engine, and so on, can be welded by moving an electric arc along the seam. The positive rod is made of steel and is held near to the plates, which themselves form the negative electrode. The intense heat melts the plates at the seam, and molten metal from the positive rod effects a join. The light emitted is so dazzling that the welder must wear special goggles (see Fig. 136).

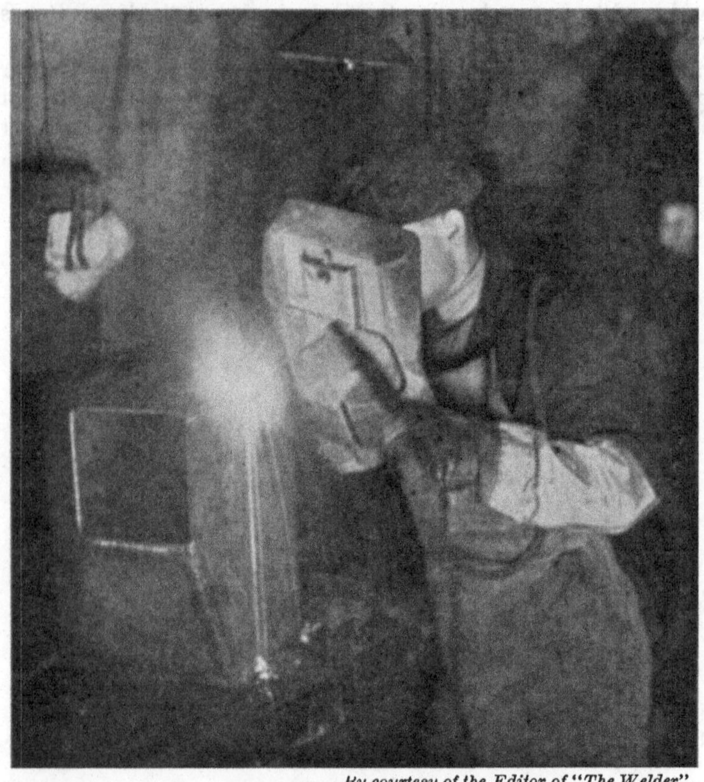

By courtesy of the Editor of "The Welder".

Fig. 136. Arc welding the seams of a steel box. The light is so dazzling that the welder must use a shield.

The electric furnace.

Electric furnaces are now commonly used for melting steel and other metallurgical processes on account of their convenience, cleanliness and economy. Fig. 137 shows a direct-arc furnace for melting steel.

By courtesy of the General Electric Co. Ltd.

Fig. 137. An electric-arc furnace for melting steel. It is a G.E.C. Brown-Boverie, 3-phase, 3-electrode Heroult type.

SUMMARY

The rate at which electrical energy is absorbed, in a lamp for example, is measured in **watts. A watt is 1 joule per sec.**

Amperes × volts = watts.

The B.T.U., or unit in which electrical energy is bought and sold, is **1 kilowatt–hour,**

i.e. **1000 watts for 1 hour** = 3,600,000 joules.

The mechanical equivalent of heat is 4·2 joules per calorie.

Joule's law. Heat produced by an electric current

$$= \frac{i^2Rt}{4{\cdot}2} \text{ calories,}$$

also

$$= \frac{iVt}{4{\cdot}2} \quad ,,$$

The lamps in a house are wired in parallel. A fuse is a thin wire which melts when the current exceeds a certain safe value.

The electric arc is a stream of electrons from the negative to the positive carbon. It provides an intense light and very high temperature—about 3500° C.

Electric welding and the electric-arc furnace are commercial applications of the electric arc.

QUESTIONS

1. (*a*) What current is taken by a 100-watt lamp from 230-volt mains? What is the resistance of the lamp?

(*b*) A lamp has a resistance (when lit) of 800 ohms, and takes a current of ¼ ampere. What is the wattage of the lamp?

(*c*) A vacuum cleaner is rated at 160 watts on 240-volt mains. What current will it take? How much will it cost to run per hour, assuming that 1 kilowatt hour costs 1*d.*?

2. A motor-car battery can deliver a current of 100 amperes (for starting the engine) at a P.D. of 12 volts. What is its power (*a*) in kilowatts, (*b*) horse-power? (760 watts = 1 horse-power.)

3. (*a*) Explain why the filament of a lamp becomes white hot while the leads remain quite cool.

(*b*) Will the filament of a 100-watt lamp have a larger or smaller resistance than that of a 60-watt lamp working at the same voltage? Explain fully.

4. Explain briefly a method of wiring a house for electric lighting. A house is lighted by five 20-watt lamps, four 40-watt lamps, and three 60-watt lamps. If the supply voltage is 200 and all the lamps are switched on, find the total current taken from the mains and the cost per hour. (Price of electric power = 7*d.* per unit, i.e. 1 kilowatt-hour.) (O. and C.)

5. Explain the action and use of a fuse used in the wiring of the electric supply of a house. A room has three points supplied from the same pair of leads from the 200-volt mains. These are used to supply electricity to (*a*) a 2-kilowatt electric fire, (*b*) a 100-watt lamp, (*c*) a motor which costs 2*d*. per hour to run.

What current is taken in each of the three cases, and how much would it cost to run the whole for 3 hours at 3*d*. per unit? What would be a suitable fuse to use in the leads supplying the current? (L.)

6. Two lamps in parallel on a 110-volt circuit take 60 watts and 50 watts, respectively. What are their resistances? If these resistances remain unaltered when the lamps are in series on a 220-volt circuit, what power will they now take together? (L.)

7. Explain the probable cause of the trouble when

(*a*) one lamp in a house goes out;

(*b*) the lights in one part of the house go out;

(*c*) all the lights in the house go out;

(*d*) all the lights in all the neighbouring houses go out.

8. Some of the coils in the filament of a lamp touch and are short-circuited. How will this affect (1) the current flowing, (2) the brightness, (3) the life of the lamp?

9. Describe, with the aid of a diagram, some form of fuse which is used in the electric-lighting circuit of a house. Explain why it is inserted and why a fuse wire must not be replaced by an ordinary wire.

A room is lighted by five 200-volt 60-watt lamps and heated by an electric fire, all connected in parallel. Fires are made which consume power in multiples of 500 watts (e.g. 500 watts, 1000 watts, and so on). Calculate the greatest permissible power of the fire so that a 15-ampere fuse, placed in the circuit leading to the room, is not burnt out when the fire and all the lamps are in use. (N.)

10. Twenty lamps connected in series on a 220-volt circuit take a current of 0·2 ampere. What is the resistance of each lamp and what power is being used?

If it is desired to light forty lamps from the same main terminals, how should they be connected? What power will now be used? (N.)

11. An incandescent bulb operated on a 110-volt circuit has a resistance of 140 ohms and a candle-power of 32. Calculate the number of watts per candle-power and the cost of operation for 500 hours if electricity is supplied at 7*d.* per kilowatt-hour. (O. and C.)

12. A shaving-water heater has a resistance, when working, of 50 ohms and takes a current of 2 amperes. How many calories of heat will it generate in 1 minute? How long will it take to raise 250 gm. of water from 10° C. to 100° C.?

13. An electric heater on a 250-volt circuit is required to raise 800 gm. of water from 15° C. to boiling point in 6 minutes. Heat losses by radiation being neglected, what current must be supplied to the heater, and how much will it cost, if electricity is supplied at 2*d.* per kilowatt hour? (1 joule = 0·238 calorie.) (O. and C.)

14. A current of 5 amperes flows through a wire of resistance 10 ohms for 2 minutes. If the heat produced is supplied to 100 gm. of water, through how many degrees will the temperature be raised? (N.)

15. Show how, by measuring the heat developed in a wire by a current of known strength, the mechanical equivalent of heat can be determined if the resistance of the wire is known. (O. and C.)

16. (*a*) Calculate the number of calories equivalent to a Board of Trade Unit (the kilowatt-hour), (*b*) Calculate the rate at which heat is generated in a 60-watt lamp.

17. What is the resistance of a 60-volt lamp which takes a current of 0·5 ampere from 60-volt mains? How much electrical energy would be converted into heat and light by such a lamp in 1 minute? What should be the resistance of the filament of a similar lamp, designed for twice the power of the former and to be used on 120-volt mains? (N.)

18. Equal lengths of platinum and silver wires of the same diameter are arranged in series. When a steadily increasing current is passed through the two wires, the platinum gets red hot first. If they are arranged in parallel, the silver wire gets red hot first. Explain this difference. (Platinum has a higher resistivity than silver.) (L.)

19. Describe an experiment to verify Joule's law connecting the heat produced by an electric current with the resistance of the conductor through which it flows, the strength of the current and the time of flow.

A glow lamp takes 30 watts at 110 volts. Calculate its resistance. (C.)

20. How does the heat generated per second by a current in a wire depend upon (*a*) the difference of potential E between the ends of the wire and (*b*) the resistance R?

Current from 100-volt mains is sent through coils of resistance 50 ohms and 20 ohms connected in parallel. How much heat is generated per second in each coil? ($J = 4{\cdot}2$ joules per calorie.) (O. and C.)

21. Explain the meaning of joule, watt, kilowatt-hour. An electric radiator takes 2 kilowatts when connected with the 250-volt mains. Calculate (*a*) the current taken from the mains, (*b*) the resistance of the radiator, (*c*) the cost of running the radiator for 5 hours, if the price of electrical energy is $1\frac{1}{2}d.$ per kilowatt-hour. (O. and C.)

22. A radiator is to be run at a temperature of 600° C., using energy at the rate of 1000 watts and at a voltage of 200. Calculate what length of No. 20 S.W.G. nickel chrome resistance wire will be required for the heating element, given that its resistance at 600° C. is 1·7 ohms per yard.

23. An electric iron used on a 110-volt mains takes a current of 3·2 amperes. What is the resistance of the iron, and how much heat is generated per second? How might this iron be adapted to generate the same amount of heat per second on 220-volt mains? (N.)

24. Derive an expression for the rate of production of heat by the passage of an electric current through a resistance.

A current of $\frac{1}{2}$ ampere flowing in an electric lamp generates 900 calories per minute. Calculate (*a*) the resistance of the lamp at its working temperature, (*b*) the power expended in lighting the lamp. (Mechanical equivalent of heat $= 4{\cdot}2$ joules per calorie.)

If you measured the resistance of the same lamp by including it in one arm of a Wheatstone bridge, would you expect to obtain the same value of the resistance? Give your reasons. (O. and C.)

25. Assuming that a supply of electricity at a pressure of 10 volts can be generated at a cost of $2d.$ per kilowatt-hour, calculate (*a*) the number of coulombs required, and (*b*) the cost in electrical power of depositing 1 kilogram of copper from an electrolytic solution. (The electrochemical equivalent of copper is 0·00033 gm. per coulomb.) (O. and C.)

26. It is desired to charge a battery of four accumulator cells from the 210-volt direct-current circuit, using an ordinary electric lamp as a resistance. Give a diagram, showing clearly the connections for the circuit. Assuming that each cell has an E.M.F. of 2·5 volts when on charge and that the charging current is to be $\frac{1}{2}$ ampere, specify in volts and watts the most suitable size of lamp to use. Calculate also the cost of giving the battery a charge of 30 ampere-hours, if the price of electrical energy is 6*d.* per Board of Trade Unit. (1 Board of Trade Unit = 1 kilowatt-hour.) (C.)

27. What factors determine the temperature finally attained by a conductor carrying a current? If the constancy of the coils of a resistance box is endangered when the energy used in them exceeds 0·0002 watt per ohm, what is the maximum voltage which may be applied when a 1000-ohm coil is used? (O. and C.)

Chapter XI

ELECTROMAGNETIC INDUCTION

An electric current flowing along a wire creates a magnetic field. Will a magnetic field applied near to a wire create an electric current in it?

The former effect was first demonstrated by Oersted in 1819 and the lines of force are concentric circles round the wire. The latter effect was eventually demonstrated by Faraday in 1831 after a number of fruitless experiments during the preceding ten years. Faraday's mind had been running on the likelihood of this phenomenon ever since Oersted's experiment, and as early as 1822 he made a memorandum in his note-book, "Convert magnetism into electricity". His biographer, Tyndall, records that he carried an iron core and a coil of copper wire in his pocket as a constant reminder of the problem.

Inducing a current.

Suppose we hold a wire, connected to a sensitive galvanometer, near to the poles of a powerful electromagnet. So long as the wire is held at rest no current flows in it. But if the wire is moved down sharply between the poles of the magnet (see Fig. 138) the needle of the galvanometer will give a kick, showing that a small, momentary current flows. The current, *which lasts only while the wire is moving*, is known as an *induced current*, and the phenomenon, *electromagnetic induction.*

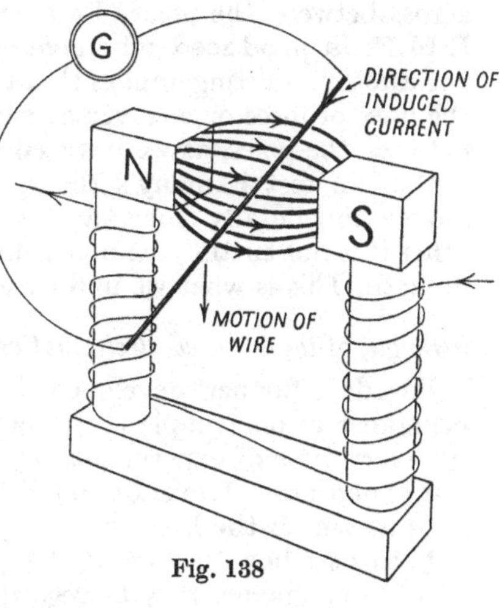

Fig. 138

When the wire is moved up between the poles there is an induced current in the opposite direction, a fact which we shall consider later.

The induced electromotive force and the induced current.

It is simpler to deal with the induced E.M.F., which causes the induced current to flow, than with the induced current: for the induced E.M.F. is independent of the rest of the circuit, whereas the induced current depends upon the resistance of the circuit. Thus, for the sake of precision, we shall often speak of the induced E.M.F. rather than the induced current.

Faraday's theory.

The production of an induced current in a wire by means of a magnet which does not even touch the wire is a mysterious process. Faraday argued that something must be happening in the open space between the wire and the magnet.

In order to form a mental picture of this action at a distance, Faraday made use of his theory of lines of magnetic force. When the wire is moved down between the poles of the magnet we can visualise that it *cuts* the lines of force of the magnet which stretch across between the poles. Faraday suggested that **an induced E.M.F. is produced whenever lines of force are cut by a conductor.** Cutting implies that the conductor is moving across the lines of force or vice versa: when the wire is held stationary between the poles, as we have said, no induced current flows.

We can test Faraday's theory very simply. If the wire is moved horizontally along the lines of force from one pole to the other it is not cutting them and no current, therefore, should be induced. This is what we find by experiment.

Strength of the induced electromotive force.

Faraday further developed his theory. He was able to generalise in one simple statement, or law, the various factors (in every type of experiment) on which the size of the induced E.M.F. depends. **The induced E.M.F. is proportional to the rate at which the lines of force are cut.**

Let us see how this law applies to our experiment.

A strong magnet may be regarded as possessing more lines of force than a weak one. Hence a strong electro-magnet (i.e. one taking a large current) should produce a larger induced current than a weak one. This may be verified experimentally.

Again, the faster the wire is moved, the faster the lines of force are cut and the greater should be the current. Experiment once more bears out our predictions.

It is strange to read that some of Faraday's contemporaries failed to appreciate the value of his theory of lines of force. The Astronomer Royal, Airy, openly scoffed at it.

But we have still to mention its crowning merit. It is capable of exact numerical development. We may cite here, without explanation, that an induced E.M.F. of 1 volt is produced when the lines of force are cut at the rate of 100,000,000 per sec., taking 1 line of force per sq. cm. to represent a field of unit strength (1 oersted).

The direction of the induced current.

The direction of the induced current may be predicted by means of the following rule, known as *Fleming's right-hand rule.*

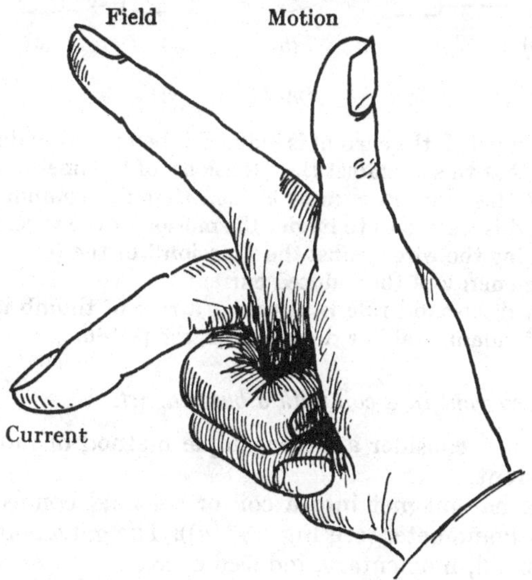

Fig. 139. Fleming's right-hand rule.

Hold the *right hand* so that the thumb and first and second fingers are mutually at right angles (see Fig. 139). *Then if the thumb indicates the direction of motion, and the first finger the direction of the magnetic field, the second finger will indicate the direction of the induced current.*

Apply this rule to Fig. 138. Assuming the field to be in the direction shown and the wire to be moved downwards, show that the induced current will be in the direction indicated. Show also that the current is in the opposite direction when the wire is moved upwards. (Remember that Fleming's *left*-hand rule gives the direction of the force exerted on a wire in which a current is already flowing. It is essential to use the correct hand in each case. The *right* hand gives the direction of the *induced* current.)

Fig. 140 represents the magnetic fields of the electromagnet and the induced current (cf. Fig. 62) and also their combined fields.

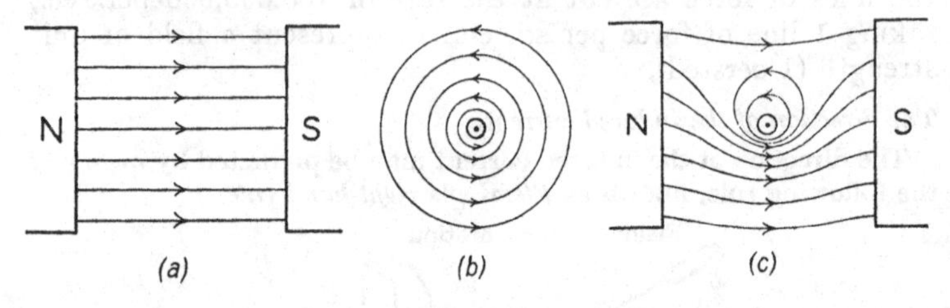

Fig. 140

The wire in which the current is induced is being moved down and it will be seen that this is against the "tension" of the lines of force: the direction of the induced current is such that the combined field is distorted in this way so as to hinder the motion of the wire. The work done in moving the wire against the "tension" of the lines of force is equal to the energy of the induced current.

Fleming's right-hand rule is, literally, a rule of thumb method of applying a fundamental law due to Lenz, see p. 186.

Inducing a current in a coil with a bar magnet.

We will now consider another simple method of inducing an electric current.

Plunge a bar magnet into a coil or solenoid connected to a sensitive galvanometer (see Fig. 141 (*a*)). The galvanometer will register a small, momentary, induced current.

Note that here, in contrast to the former experiment, we are moving the magnet while the conductor, in which the current is induced, is at rest. But there is relative motion between the magnetic field and the conductor. The current, once again, only lasts during the motion.

A coil is preferable to a straight wire since it gives a greater induced current, just as, in the reverse phenomenon of producing magnetism by electricity, it gives a stronger magnetic field.

Now in the case of a coil, as opposed to a straight wire, it is simpler to make a slight modification of Faraday's theory of the "cutting" of lines of force. When the bar magnet is plunged into the coil we can imagine that its lines of force *thread* the coil rather as cotton threads the eye of a needle. **Whenever the number of lines of force threading a coil changes, an**

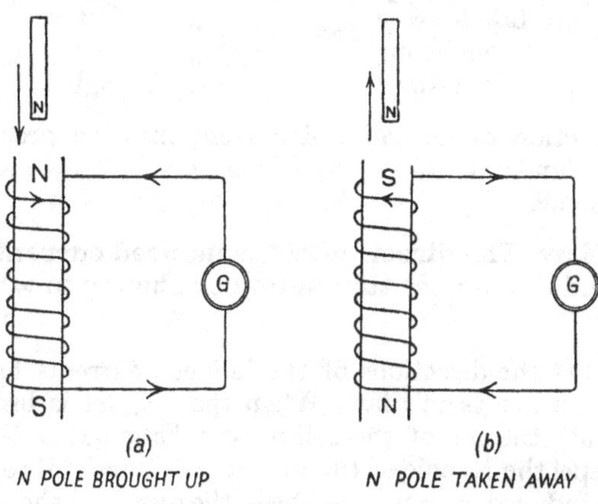

Fig. 141

E.M.F. is induced in the coil. Note that in the experiment of Fig. 138, the number of lines of force threading the circuit change when the wire is moved: the conception of threading works just as well as cutting for that experiment too.

A number of lines of force is sometimes called a *magnetic flux*. **The strength of the induced current is proportional to the rate at which the magnetic flux through the coil varies.** Thus a strong magnet, which possesses more lines of force, should produce a larger current than a weak magnet. The faster the magnet is plunged in the greater the rate of change of the magnetic flux in the coil and hence the larger should be the induced current. Again, since the lines of force thread each turn of the coil the induced E.M.F. should be proportional to the number of turns in the coil. All these deductions may be verified experimentally.

The direction of the induced current.

When the magnet is plunged into and then withdrawn from a coil two induced currents in opposite directions are obtained. If previously the N. pole of the magnet was leading, both effects are reversed by repeating with the S. pole leading. Below, these results are tabulated:

N. pole brought up	Deflection	right (say)
N. ,, taken away	,,	left
S. ,, brought up	,,	left
S. ,, taken away	,,	right

The direction of an induced current may be predicted by means of a fundamental law which was enunciated by a Russian scientist, Lenz.

Lenz's law. The direction of the induced current is such that it tends to oppose the motion or change to which it is due.

In Fig. 141 the directions of the induced currents have been marked in, using Lenz's law. When the magnet is brought up (Fig. 141 (*a*)), the top of the coil must behave like a N. pole in order to repel the N. pole of the magnet and hinder its approach. Hence the induced current must be in the direction shown. When the magnet is taken out of the coil, the top of the coil must behave like a S. pole and attract the N. pole of the magnet, thus tending to prevent its departure. Hence the induced current is in the opposite direction to that when the magnet is brought up.

The reader should draw similar diagrams for the bringing up and taking away of the S. pole of a magnet.

Lenz's law is really a special application of the Principle of the Conservation of Energy, namely that energy cannot be created from nothing. The energy of the induced current must have some source, and that source is the work done by the person who moves the magnet, say, in overcoming the opposition set up by the induced current. Thus when the N. pole of a magnet is plunged into a coil, the work done in overcoming the repulsion between it and the N. pole set up in the solenoid is transformed into the electrical energy of the induced current. We will leave it to the ingenuity of the reader to invent a perpetual-motion machine assuming the opposite of Lenz's law to be true, namely that the induced current helps the motion to which it is due. Note that it is only by inverting the laws of Nature that your magic machine will work.

Mutual induction.

In Fig. 142 a coil, connected to a cell, is placed inside another coil connected to a sensitive galvanometer. Whenever the current is switched on or off in the inside coil, an induced current flows

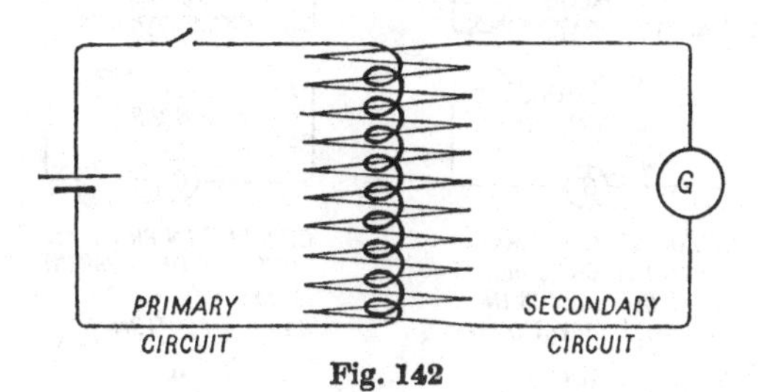

Fig. 142

in the outer one. This phenomenon is known as *Mutual Induction.* The inner coil in which the original current flows is known as the *primary,* and the outer coil in which the induced current flows is known as the *secondary.* When the current is switched on in the primary, it sets up a magnetic field, and the growing lines of force thread the secondary, thereby, while their number is increasing, causing an induced current to flow. Similarly, when the current is switched off in the primary, the magnetic field disappears and the dying lines of force, collapsing, as it were, on to the primary, cease to thread the secondary: the change gives rise to an induced current.

If a bar of soft iron is inserted into the primary, the induced current is enormously increased. This is because the iron becomes magnetised and demagnetised when the current is switched on and off and adds a large number of magnetic lines of force of its own to those of the primary.

The direction of the induced current in the secondary is in the opposite direction to that in the primary when the latter is switched on (see Fig. 143 (*a*)), since, applying Lenz's law, it tends to prevent the growth of the magnetic field which is causing it. It therefore creates a magnetic field of its own which is opposite to, and tends to nullify, the magnetic field due to the primary. On the other hand, when the current in the primary is switched off (see Fig. 143 (*b*)), the current in the secondary is in the same direction, and tends to prevent the magnetic field from dying away, by creating a field of its own in the same direction.

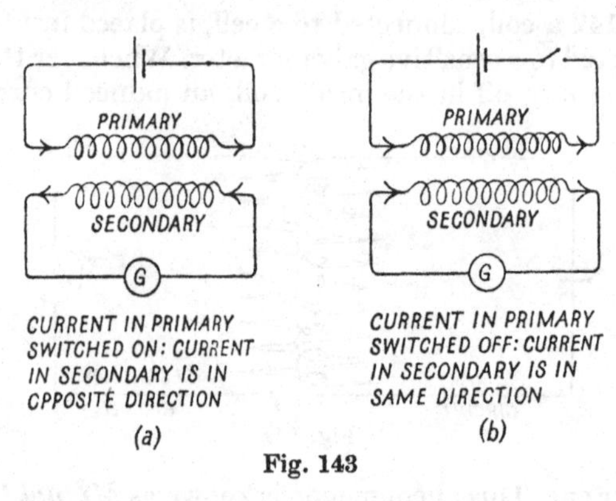
PRIMARY
SECONDARY
G
CURRENT IN PRIMARY SWITCHED ON: CURRENT IN SECONDARY IS IN CPPOSITE DIRECTION
(a)
PRIMARY
SECONDARY
G
CURRENT IN PRIMARY SWITCHED OFF: CURRENT IN SECONDARY IS IN SAME DIRECTION
(b)

Fig. 143

The induction coil.

In the experiment just described, if the current is continually switched on and off in the primary, an induced current will surge backwards and forwards in the secondary. A current which keeps changing its direction in this way is known as an *alternating current.*

This is a simple model of the induction coil, first invented in Europe by Ruhmkorff in 1851. The purpose of an induction coil is to obtain a high E.M.F. from a low one. If the secondary has a thousand times as many turns as the primary, the E.M.F. in the secondary will be roughly 1000 times as great as that in the primary. Induction coils are used for coil ignition in motor cars, in the production of X-rays, in modern research, and for medical treatment to give mild shocks in the cure of muscular complaints.

Since no machine can give out more energy than is put into it, the current in the secondary is reduced in about the same ratio as the E.M.F. is increased. For

Electrical power = current × voltage.

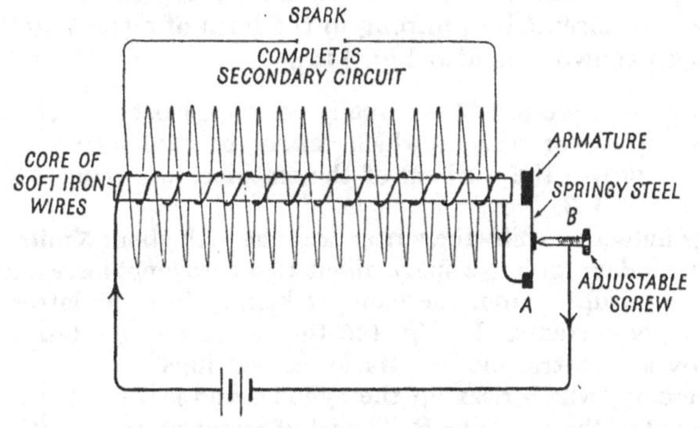

Fig. 144. Induction coil.

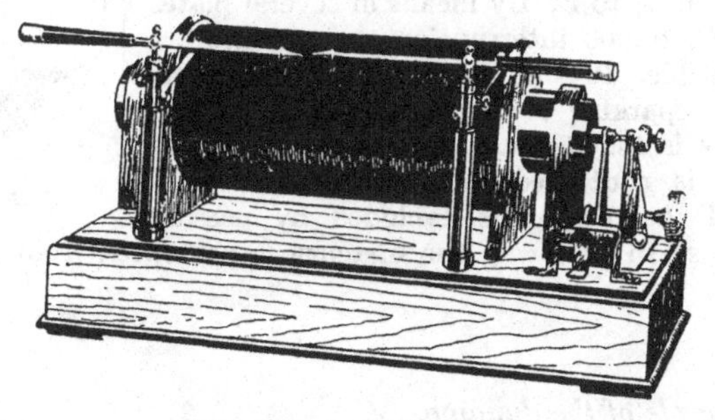

Fig. 145. Induction coil.

Thus if an induction coil were 100 per cent efficient and the E.M.F. in the secondary 1000 times that in the primary, the current in the secondary would be $\frac{1}{1000}$th of that in the primary.

The essential features of an induction coil are

(1) a soft-iron core,

(2) a primary coil consisting of a few thick turns,

(3) a secondary coil consisting of a large number of fine turns,

(4) some device for switching the current on and off in the primary as rapidly as possible.

Fig. 144 shows a diagram of an induction coil with a hammer break similar to that of an electric bell. When the machine is working the induced current in the secondary may be made to complete its circuit by jumping in the form of a spark across the spark gap shown (see also Fig. 145).

A condenser, see p. 347, is usually connected between *A* and *B* in Fig. 144. This, for reasons which cannot be explained here, considerably improves the working of the machine, and reduces sparking between *A* and *B*.

Large induction coils (the writer uses one with about 5 miles of wire in the secondary, giving a spark about 10 inches long) have a different type of interrupter from the hammer break, since the latter cannot carry heavy currents. In Fig. 146 the cylinder *A* is rotated at high speed by an electric motor. Its lower end dips into mercury, which rises up the cylinder and is thrown out of the side tube *B*. The jet of mercury hits *C* once in each revolution and carries the current from *A* to *C*. By means of several plates such as *C*, 10,000 interruptions per minute are made possible. It is obvious that a considerable amount of sparking between the mercury jet and the plates is inevitable. Consequently, the instrument is contained in a closed case from which all the oxygen is excluded by passing in coal gas. In this way the mercury is prevented from becoming oxidised.

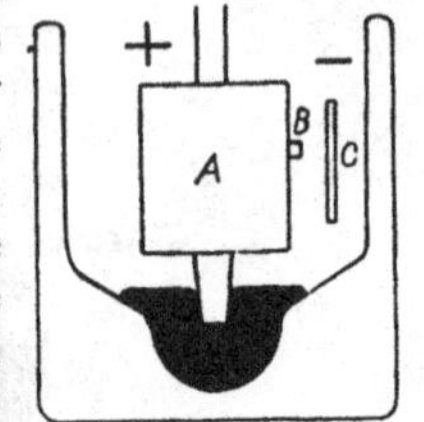

Fig. 146

The principle of the dynamo.

The dynamo is a machine for generating electricity by means of electromagnetic induction. It consists, in the simplest form, of a single rectangular coil which is rotated between the poles of a permanent magnet (see Fig. 147). The coil cuts the lines of force due to the magnet, and hence an induced current is set up in it.

Each time the plane of the coil reaches the vertical position the direction of the induced current is reversed.

For as the coil passes through the vertical position that side of it which was ascending now begins to descend and hence the current in it is reversed since it is cutting the lines of force in the opposite direction.

The actual direction of the induced current, assuming that the coil is made to rotate in the direction of the arrow, is marked in Fig. 147. This may be deduced in either of two ways: (*a*) by Lenz's law the upper face of the coil must be a S. pole since there must be repulsion between it and the S. pole of the magnet to oppose rotation; (*b*) by applying Fleming's right-hand rule to one side of the coil.

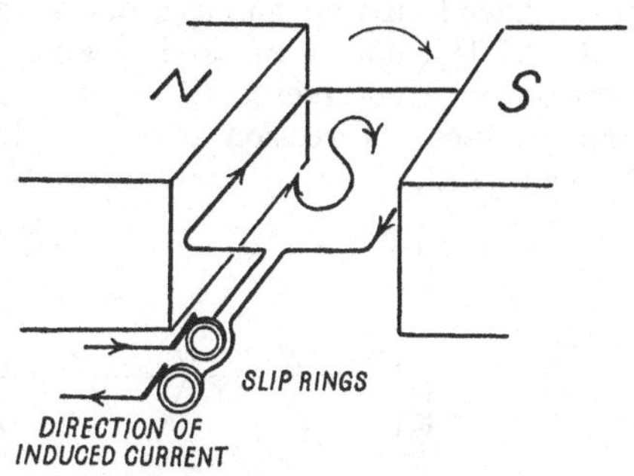

Fig. 147. The principle of the alternating current dynamo.

The way the strength of the current varies as the coil is rotated.

The induced current is a maximum when the plane of the coil is horizontal. To prove that this is true we must show that the coil cuts the magnetic lines of force most rapidly when it is moving through the horizontal position. In Fig. 148 the coil is represented as horizontal, at 45° to the horizontal, and vertical by the lines AB, $A'B'$, and $A''B''$, respectively. The lines of force are drawn, equally spaced. It will be seen that the coil cuts many more lines of force in moving from the position AB to $A'B'$ than it does in moving from $A'B'$ to $A''B''$. The sides which are most effectively cutting the lines of force are those which are perpendicular to the paper and whose sections are represented by the points A and B.

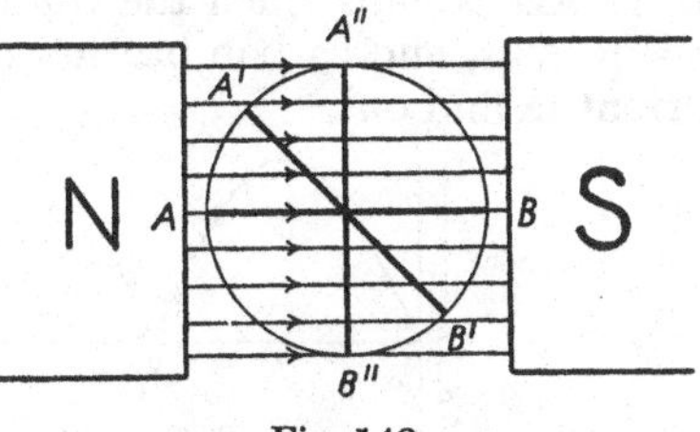

Fig. 148

As the coil moves from the horizontal position AB to the vertical position $A''B''$, the induced current decreases from its maximum value to zero. The graph in Fig. 149 represents how the induced current changes in one complete revolution of the coil. At the points marked *V* and *H* the coil is vertical and horizontal respectively. It will be seen that the current is zero and reverses in direction at the points *V*, and is a maximum at the points *H*.

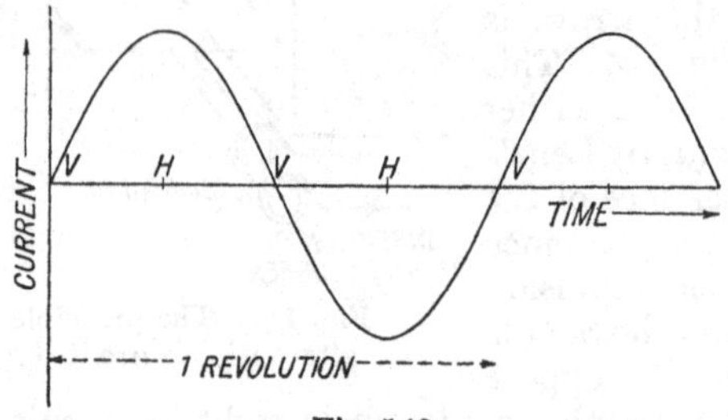

Fig. 149

Slip rings or commutator.

It is clear that since the coil is revolving, a rather special arrangement will be needed for leading the current in and out of it, otherwise the wires will become twisted and tangled. If the alternating current induced in the coil does not require to be rectified (i.e. converted into direct current), the ends of the coil are connected to two separate rings, insulated from each other, on the axle about which the coil is revolving. They are known as slip rings, and carbon brushes press against them to lead the current in and out.

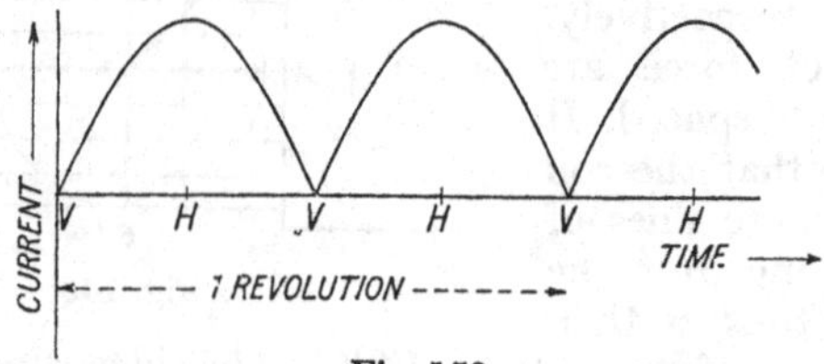

Fig. 150

If the dynamo is required to produce direct current then a commutator similar to that of an electric motor must be used. It consists of a split ring, the two halves of which are insulated from each other and connected to the ends of the coil (see Fig. 63). The gaps between the two halves of the ring must be opposite the brushes when the coil is vertical: then simultaneously the current reverses and each brush makes contact with the opposite end of the coil. Hence the current always flows out at one brush and in at the other.

The current-time graph may then be represented as in Fig. 150.

Eddy currents.

Eddy currents are the induced currents set up in a mass of metal which is cutting magnetic lines of force.

To demonstrate their existence allow a copper or aluminium pendulum to swing between the poles of a powerful electromagnet (see Fig. 151). Switch on the current through the coils of the electromagnet: the pendulum will be "braked" strongly as it cuts the lines of force and will stop swinging. The eddy currents set up in the metal pendulum are the cause of this braking action. By Lenz's law, they are in such a direction as to stop the motion to which they are due, i.e. the swinging of the pendulum.

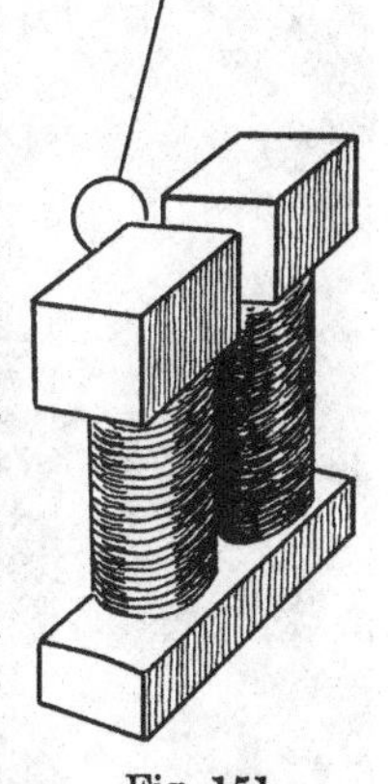

Fig. 151

Owing to eddy currents a very great effort is required to push a copper bar between the poles of the powerful electromagnet in Fig. 152, and the bar cannot be moved quickly. A hollow aluminium sphere released between the poles does not fall to the ground in a normal way but sinks slowly.

Eddy currents are set up in the core of an induction coil, and if it were one solid piece of iron they would circulate round its perimeter in the same direction as the induced currents in the secondary. The core, however, is made up of thin soft-iron wires insulated from each other by varnish, and the insulation prevents the eddy currents from circulating (see Fig. 153).

We shall see in the next two chapters that eddy currents are set up in the iron cores of transformers and of motors and dynamos: their heating effect proved to be one of the most

By courtesy of the Metropolitan-Vickers Electrical Co. Ltd.

Fig. 152. A hollow aluminium sphere falling through the intense magnetic field (12,300 oersted), between the poles of a very powerful electromagnet. Owing to the eddy currents set up in it, the sphere sinks slowly as though through a viscous liquid. The electromagnet, weighing 11 tons, was constructed for Prof. Blackett for his experiments on cosmic rays. The pipes at the top are air ducts through which passes a stream of air to cool the coils. The coils round each pole are immediately below the air ducts and the white painted, rounded parts are the soft-iron yoke. The gap between the poles can be varied by sliding one of the poles in.

serious obstacles in the early days of the development of these machines. Their effect is minimised by building up the cores of thin, insulated, iron sheets, called laminations.

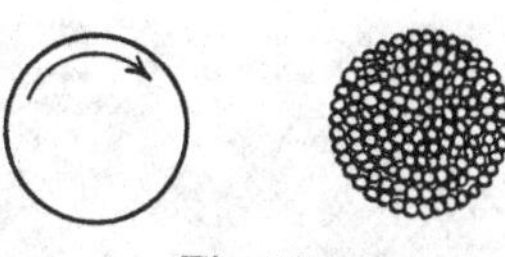

Fig. 153

Advantage is taken of eddy currents in dead-beat moving-coil galvanometers, i.e. galvanometers which return at once to their zero position without swinging. The moving coil is wound on a light metal frame and the eddy currents induced in this frame stop it from swinging.

Michael Faraday.

The discoverer of electromagnetic induction was Michael Faraday, an outstanding figure in the history of electricity and one of the greatest of experimenters.

He was born on 22 September 1791 at Newington, Surrey, the son of a blacksmith. After an elementary education at a local day school he became, at the age of thirteen, an errand boy at a bookseller's shop. He had to distribute newspapers which, being expensive in those days, were hired and not bought. His master soon accepted him, without premium, as a bookseller's apprentice, and during the next few years he made good use of his opportunity to read widely. His interest in science was aroused by an article in the *Encyclopaedia Britannica* on Electricity.

When he was twenty years old a customer gave him tickets for a course of lectures by the celebrated chemist, Sir Humphry Davy, at the Royal Institution.

Faraday's imagination was captured so completely by these lectures that he wrote to Davy, enclosing notes of the lectures beautifully bound, to ask for a post at the Royal Institution. Davy showed Faraday's letter to one of the managers of the Institution with the question, "What am I to do?" "Put him to wash bottles", was the answer. "If he is good for anything he will do it directly, if he refuses he is good for nothing."

In the upshot, Faraday was appointed as laboratory assistant, at 25*s*. per week, with the use of two living rooms at the top of the Royal Institution building in Albemarle Street. Here he was destined to continue even more gloriously the tradition

By courtesy of the Director of the National Portrait Gallery

Fig. 154. Michael Faraday.

begun by Rumford and Davy. Davy's blend of social and histrionic gifts combined with his eminence as a scientist had already made the Institution famous. Society flocked to hear him lecture. But it has been said, with some truth, that Davy's greatest discovery was Michael Faraday. Faraday, too, became a brilliant and fashionable lecturer. It was he who began the series of lectures for children which are held each Christmas at the Royal Institution.

Between October 1813 and the spring of 1815, Faraday accompanied Davy in the role of secretary and servant on the Grand Tour of the continent, then in the throes of the Napoleonic wars. They visited Paris, Geneva, Genoa, Florence, Rome and Naples, and met the greatest scientists of the day, among them, Ampère, Volta, Arago and de la Rive. When, in later years, Davy opposed Faraday's election to a fellowship of the Royal Society and displayed jealousy of the rising fame of the younger man, Faraday could never forget the debt he owed to Davy. The story is told that once, at dinner, long after Davy's death, a guest spoke slightingly of him to Faraday. Faraday made no reply, but after dinner he took the guest to the library and, pausing in front of Davy's portrait, he said: "He was a great man. It was at this spot that he first spoke to me."

On his return from the continent Faraday eagerly devoured all the papers and scientific publications of his contemporaries, and repeated their experiments. In 1821 he became superintendent of the house and laboratory of the Royal Institution. He married, and he and his bride lived above the laboratory in happy contentment for forty-six years.

By 1830 his reputation was such that he was making over £1000 a year as a consultant chemist to industrial firms. His salary from the directors of the Royal Institution was a mere £100 per annum. At this stage in his career he was faced with two alternatives, to make money or to devote himself to scientific research. He chose the latter and his income dwindled. It is perhaps significant that the following passage was marked by Faraday in his Bible: "The love of money is the root of all evil."

Faraday was a deeply religious man. He belonged to a strict sect of dissenters, known as the Sandemanians. So strict, indeed, were the elders of this sect that they excommunicated Faraday (temporarily) for failing to attend divine service when he obeyed a command to lunch with Queen Victoria.

Although he was born and died poor, Faraday was always proud and independent. Towards the end of his life he was offered a pension of £300 a year by Lord Peel, not only for his work in pure science, but also for his public services in connection with the lighthouses of the country. Before the matter was completed a new minister, Lord Melbourne, came into office, and he used language which so offended Faraday that the latter refused to have anything more to do with the offer. It was not until Lord Melbourne had been induced by friends to make a full and frank apology that Faraday would accept the pension.

His closing years were spent in retirement imposed by his age and failing memory. He lived at this time in a house at Hampton Court that had been placed at his disposal by Queen Victoria. On 25 August 1867 he died, seated in a chair in his study.

"Not half his greatness was incorporate in his science, for science could not reveal the bravery and delicacy of his heart."

Faraday's discovery of electromagnetic induction.

Faraday kept a diary, giving full details of his day to day experiments, during the period 1820–1862. It consists of 4000 closely written pages, and is an invaluable record of the method, patience and persistence required in scientific research. Faraday is said to have been satisfied if $\frac{1}{10}$ per cent of his experiments, that is 1 in 1000, led to a really important discovery. And Faraday was gifted with a flair for the significant: "He smells the truth", said one of his contemporaries.

We have not the space here to record all the experiments with negative results which Faraday performed before he discovered the phenomenon of electromagnetic induction. He tried repeatedly to produce what we now call mutual induction. In the year 1825, for example, he placed a wire, connected to a sensitive galvanometer, alongside another wire carrying a current: he placed it inside a coil carrying a current: he coiled the wire and placed it near a straight wire carrying a current. But in no case could he detect an induced current. He was looking for a phenomenon which never occurs: he expected that a steady current in one wire would induce a current in another: he had no notion that mutual induction can only occur when the original current is changing.

In 1831 he turned once again to the investigation of mutual induction and in this year obtained a small clue which led to complete success. He could not know that his discovery was the germ of a method of producing electrical power in vast quantities and hence of almost the whole of electrical engineering. Indeed, he wrote to a friend, "I am busy just now again on electro-magnetism, and think I have got hold of a good thing but cannot say. It may be a weed instead of a fish that, after all my labour, I may at last pull up."

He wound two coils of insulated copper wire, each about 200 feet, round a wooden cylinder. He connected a battery of 120 cells to one of the coils. The current was so large that it rapidly made the coil hot. The other coil was connected to a galvanometer. No effect occurred in the galvanometer while the current was flowing steadily in the first coil, but Faraday noted slight kicks of the needle when the current was switched on and off. Here at last was an effect. What was its significance?

Then followed the famous experiment of 29 August 1831. Two coils were wrapped on an iron ring: one coil was connected to a battery of ten cells and the other to a galvanometer (see Fig. 155). On making and breaking the battery circuit a large induced current flowed in the galvanometer: "the impulse at the galvanometer... was so great as to make the needle spin round rapidly four or five times...." The ring, constructed by Faraday (who had learnt from his father some of the blacksmith's art) from a round iron bar 20 inches long, is preserved at the Royal Institution (see Fig. 156).

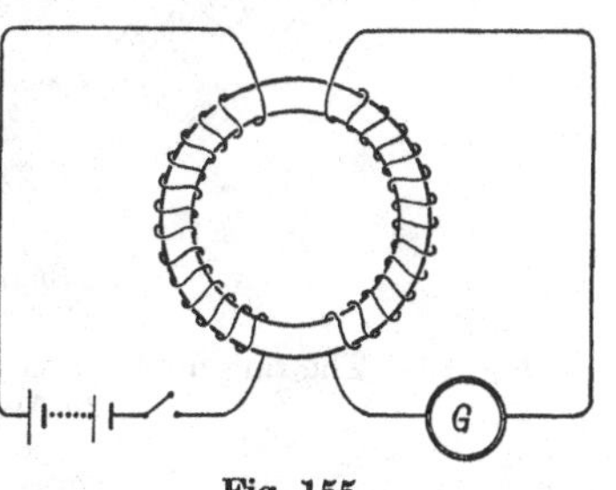

Fig. 155

Faraday successfully repeated the experiment, using two coils wound on a straight iron bar instead of an iron ring. This straight bar became a magnet when the current flowed in the primary coil. The final step, the production of electricity solely by means of a magnet, was now in sight, and a day or two later

Faraday induced a current in a coil by plunging in a bar magnet.

Finally, Faraday made a machine which generated a continuous induced current—the first dynamo. He rotated a copper disc, with rubbing contacts on the axis and circumference, between the poles of a permanent magnet so that it cut continuously the magnetic lines of force (see Fig. 157). The permanent magnet used

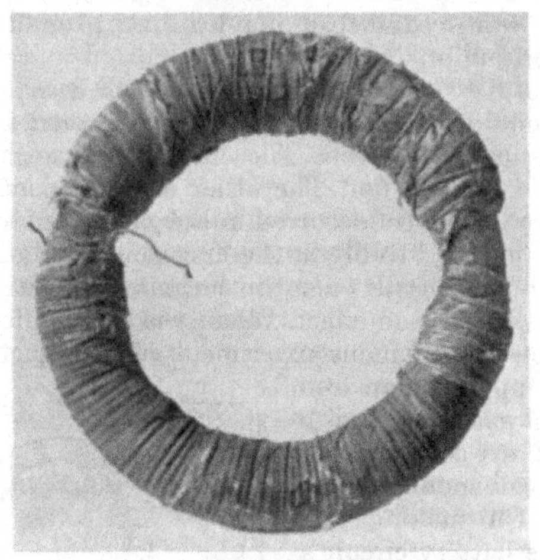

By courtesy of the Royal Institution

Fig. 156. The ring used by Faraday in his famous experiment of 29 August 1831.

was a large one consisting of 400 bars each 15 in. long, 1 in. wide and $\frac{1}{2}$ in. thick, weighing 1000 lb. It had been constructed in the middle of the eighteenth century and was the property of the Royal Society.

With Faraday's name should be coupled that of Henry, who also discovered the phenomenon of electromagnetic induction independently, in America.

Faraday's other discoveries.

Faraday had no mathematical training, but nevertheless invented the symbolism of magnetic lines of force, which may be regarded as a kind of geometry. His most fundamental ideas were put into mathematical form by Clerk Maxwell.

His discoveries in electrolysis we have already mentioned, and we shall consider in Chapter XVIII his experiments in electrostatics.

By courtesy of the Science Museum

Fig. 157. The first dynamo, Faraday's magnet and disc.

He had a profound conviction of the unity and simplicity of nature. This led him to look for, and eventually discover, a relation between electricity and magnetism and light. He passed light through uranium glass between the poles of a powerful electromagnet and found that the light suffered a subtle change (which we cannot explain here). Later, Maxwell put forward the theory that light is really an electromagnetic phenomenon, a form of wave motion, and predicted the possibility of similar waves, now known as wireless waves.

Faraday also set himself to find a relation between electricity and gravity, and dropped coils from the high roof of the Royal Institution to the floor of the lecture room below in the hope that, by some means, an induced current would be produced by gravity (not to be confused with the effect due to the earth's field). He was unsuccessful. But in recent years the prediction of Einstein, that light waves would be bent by the powerful gravitational field of the sun, has been confirmed by observations in Brazil of an eclipse. Since light is electromagnetic by nature, Faraday's intuition was a true one.

SUMMARY

Electromagnetic induction.

Whenever a magnetic field is changing in the neighbourhood of a conductor, i.e. when lines of magnetic force are being cut, or the number of them threading a coil is changing, an induced E.M.F. is set up. This change may be caused by

(1) moving a magnet near the conductor,

(2) moving the circuit near to a magnet, or revolving a coil in a magnetic field (as in a dynamo),

(3) starting, stopping or changing the current in a neighbouring circuit (mutual induction as in the induction coil).

Laws of electromagnetic induction.

1. Faraday's law. The induced E.M.F. in a conductor is proportional to the rate at which the conductor cuts the magnetic lines of force. Or

The induced E.M.F. in a circuit is proportional to the rate of change of the number of lines of force threading the circuit.

2. Lenz's law. The direction of the induced current is such that it tends to oppose the motion or change to which it is due.

Fleming's right-hand rule enables one to predict the direction of the induced current, and may be shown to conform with Lenz's law.

The induction coil and dynamo are practical applications of electromagnetic induction.

QUESTIONS

1. The lines of force between the poles of a powerful electromagnet are horizontal and a wire connected across the terminals of a sensitive galvanometer is

(*a*) moved vertically downwards between the poles,

(*b*) moved horizontally from one pole to the other,

(*c*) moved in a direction inclined to the horizontal between the poles,

(*d*) held at rest between the poles,

(*e*) bent back on itself and the double wire moved vertically between the poles.

Describe and explain the indication of the galvanometer in each case. Draw a diagram for case (*a*), marking in the N. and S. poles, the direction of motion of the wire and also of the induced current.

2. A straight vertical wire is moved (*a*) west, (*b*) north, in the earth's field. In what direction will the induced current flow?

3. Describe (*a*) an experiment to illustrate induction between a magnet and a coil, (*b*) an experiment to illustrate induction between two coils. Explain with diagrams in each case how you would apply Lenz's law to find the direction of the induced current. (O. and C.)

4. State the laws of electromagnetic induction and describe *one* experiment to illustrate each of them.

A coil is connected by long leads to a sensitive galvanometer. Explain how it is possible to move the coil in the earth's magnetic field (*a*) so as to produce a deflection in the galvanometer, and (*b*) so that there is no deflection.

How would you make the deflection in (*a*) as big as possible?

(O. and C.)

5. The S. pole of a bar magnet is plunged into a coil standing on a table. Does the induced current in the coil appear to circulate in a clockwise or anti-clockwise direction as viewed from above?

What would be the effect on the induced current of doubling (*a*) the strength of the S. pole, (*b*) the number of turns in the coil, (*c*) the speed of the pole?

6. Invent a perpetual-motion machine, assuming the opposite of Lenz's law to be true.

7. If you were provided with two flat circular coils, describe the arrangements you would make to show that the alteration in current strength in one coil may produce a current in the other. State and explain a rule for determining the direction of the induced current, and show how if the centres of the coils are maintained at a fixed distance apart, they should be arranged to give (*a*) the greatest possible induced current, (*b*) the least possible. (L.)

8. Two vertical coils, *ABC*, *DEF*, are connected in series, as shown in Fig. 158, the portions *BC* and *EF* being nearest to the observer. The coil *DEF* is in the magnetic meridian, and a small compass needle *ns* is placed at its centre. A bar magnet *NS* is brought up, from the left, to the coil *ABC*. State whether the north pole *n* of the compass needle *ns* will move to the left or right, and give your reasons. State the laws on which you rely for your conclusions. It may be assumed that the two coils are so far apart that *NS* produces no direct influence on *ns*. (O. and C.)

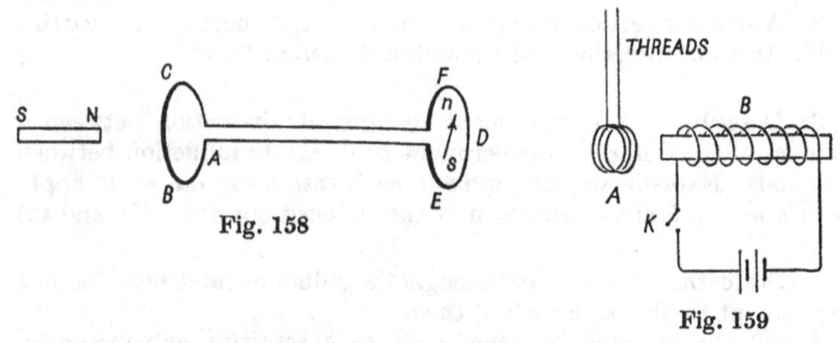

Fig. 158

Fig. 159

9. Explain fully: *A* is a *closed* coil consisting of a few turns of thick copper wire and suspended by long threads (see Fig. 159). *B* is a strong electromagnet, one pole of which faces the centre of the coil. When key *K* is pressed down, *A* is momentarily repelled. When *K* is released, *A* is momentarily attracted.

10. In Fig. 160 *A* and *B* represent two wires. A current is switched on in *A* and the lines of force are spreading outwards and up towards *B*, as shown. Copy this diagram and put in the direction of the current induced in *B*. Explain your reasoning. (This is an example of mutual induction.)

What will be the direction of the current in *B* when that in *A* is switched off. Explain fully.

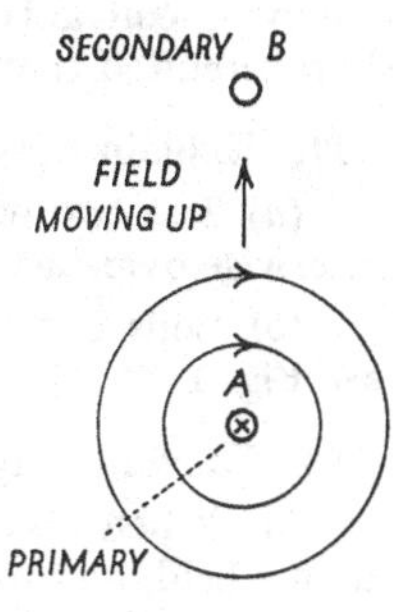

Fig. 160

11. Explain as carefully as you can why the secondary of an induction coil consists of many turns of fine wire, while the primary has only a comparatively few turns of thicker wire. Show why a core of iron in the primary produces a current of greater strength in the secondary. (N.)

12. Explain fully the effects of introducing a metal tube between the primary and secondary of an induction coil.

13. Describe, with a diagram, the construction of an induction coil. For what purpose is it used? Explain clearly (*a*) the necessity for the "make and break", (*b*) the function of the condenser, (*c*) why the secondary coil has more turns than the primary. (O.)

14. A coil of wire lies flat on a table and its ends are connected by long leads to a sensitive galvanometer. The coil is suddenly turned so that it stands perpendicular to the table and the galvanometer is found to register a current. How is the current caused and what is its direction in the coil? What effect, if any, would be produced on the current by (*a*) increasing the number of turns in the coil, (*b*) increasing the radius of the coil, all other factors remaining constant? Give reasons for your answers. (O. and C.)

15. Upon what conditions does the E.M.F. generated in a dynamo depend?

Being supplied with coils of wire of known number of turns, bar magnets, resistance boxes and a sensitive galvanometer, describe briefly the experiments you would conduct in support of your statements.

The E.M.F. of a dynamo is 110 volts and its internal resistance 0·1 ohm. When the terminals of the machine are connected by wire, a current of 20 amperes flows through the circuit. What is the resistance of the wire? (L.)

16. What are eddy currents? Why do the oscillations of the coil of a moving-coil galvanometer decay more rapidly when the coil is short-circuited than when it is on open circuit?

17. Explain fully:

(*a*) The two wires carrying A.C. must be in the same lead sheath: otherwise over-heating of the sheaths occurs.

(*b*) Coils in a resistance box are wound back along themselves (see Fig. 117).

18. Explain fully:

(*a*) When a coil of wire is rotated about a vertical axis in the earth's field, an alternating current is set up in it.

(*b*) An iron ring wrapped in two places with 10 and 100 turns of wire, respectively, can be used as a step-up or a step-down transformer.

19. A loop of wire whose ends are attached to a delicate galvanometer is carried slowly up to one end of a stationary coil of insulated wire, slipped over it without stopping and taken some distance beyond; it is then brought back over the same path to the starting-point. How will the galvanometer behave supposing a steady current has been flowing in the stationary coil all the time? How do you account for this behaviour?

20. The N. pole of a magnet is brought up to a coil connected to a sensitive galvanometer. Draw a diagram showing the direction of winding of the coil and the direction of the induced current. What will be the indications of the galvanometer needle if the magnet is passed completely through the coil. Consider the case of a magnet which is shorter than the coil. Explain fully with the aid of diagrams.

21. Why cannot the same hand be used for Fleming's two rules giving (1) the direction a conductor carrying a current in a magnetic field tends to move (motor rule), (2) the direction of the induced current when a conductor is moved in a magnetic field (dynamo rule)?

22. Explain what is meant by eddy currents.

Explain why a bar magnet suspended over a brass disc which is rapidly rotated is deflected in the same direction as the rotation of the disc. (Arago's experiment.)

23. A penny spinning (or twisting) at the end of a thread is drawn aside and made to swing also like a pendulum between the poles of an electromagnet. Explain fully why it moves in jerks.

Chapter XII

DYNAMOS AND MOTORS

A dynamo and a motor are really the same machine used in two different ways. If driven by an external source of power, such as a gas engine or a steam or water turbine, the machine generates electric current and behaves as a dynamo: if fed with electric current it will drive other machines and behave as a motor.

The reverse motor effect in a dynamo.

We have seen (pp. 190–192) that a dynamo in its simplest form consists of a coil which is made to rotate between the poles of a magnet. The coil cuts the lines of force and a current is induced in it.

As soon as the induced current begins to flow the dynamo tends to act like a motor. Since, by Lenz's law, the direction of the induced current is such that it tends to oppose the motion to which it is due, the motor endeavours to revolve in the opposite direction to that of the dynamo. The effect is known as the *reverse motor effect.* The larger the current delivered by the dynamo, the greater the current in the armature coils and hence the greater the reverse motor effect. Thus a large dynamo requires a powerful engine to turn it, and the work done against the reverse motor effect is equivalent to the electrical energy generated by the dynamo.

Electric trams and locomotives are often made to use their motors as dynamos to generate electricity in order to slow down: the reverse motor effect, which then comes into operation, provides a most efficient brake. The locomotives on the Swiss Federal Railways can slow down a 600-ton train on a down gradient of 1 in 37 from 40 m.p.h. to 3 m.p.h. by means of this "regeneration", as it is called.

The reverse dynamo effect of a motor.

We have seen (pp. 76–78) that a coil carrying a current in the magnetic field between two poles tends to turn until its plane is at right angles to the lines of force. If, each time it reaches this position, the current is reversed by means of a commutator, it continues to rotate. This is the principle of the electric motor.

While the coil is rotating it is cutting the lines of force between the poles and hence an induced E.M.F. is set up in it. According to Lenz's law this E.M.F. tends to oppose its cause: it is therefore a back E.M.F. and reduces the current in the coil. The motor is, so to speak, trying to act like a dynamo in the reverse direction: the effect is called the *reverse dynamo effect*.

Some idea of the magnitude of the effect may be gained from the following readings taken with a small motor with an armature similar to that in Fig. 67. The armature coil was connected in series with an ammeter and a voltmeter was placed across it. Readings were taken with the motor held at rest and also when running.

(*a*) *Motor not running* (*armature held at rest*).

P.D. across armature coil = 1·07 volts,

Current through armature coil = 2·67 amperes.

(*b*) *Motor running at full speed with no load.*

P.D. across armature coil = 3·24 volts,

Current through armature coil = 1·475 amperes.

Applying Ohm's law (armature at rest):

$$\text{Resistance of armature coil} = \frac{V}{i} = \frac{1{\cdot}07}{2{\cdot}67} = 0{\cdot}40 \text{ ohm.}$$

When the motor was running current was 1·475 amperes.

∴ P.D. to drive 1·475 amperes through a resistance of 0·40 ohm

$$= iR = 1{\cdot}475 \times 0{\cdot}40 = 0{\cdot}59 \text{ volt.}$$

But the voltage across the coil was 3·24 volts.

Hence **Back E.M.F. = 3·24 − 0·59 = 2·65 volts.**

Now power supplied by battery to armature coil when running

$$= \text{E.M.F. applied} \times \text{current}$$

$$= 3{\cdot}24 \times 1{\cdot}475$$

which may be written

=	$(2{\cdot}65 \times 1{\cdot}475)$	+	$(0{\cdot}59 \times 1{\cdot}475)$.
	Work done per sec. in driving current against back E.M.F.		Work done per sec. in driving current through the resistance of the coil.

The first term represents the useful power turning the motor and the second term the power wasted as heat in the coil.

Self-adjusting property of an electric motor.

When the load (or opposing torque) on an electric motor is increased, the motor tends to slow down: its back E.M.F. is therefore reduced and a bigger current flows, thereby causing a greater torque. The electric motor is thus self-adjusting: it takes, automatically, just as much current as it requires.

When a tramcar ascends a hill there is no need for a change to a lower gear as may be the case with a petrol-driven vehicle. The tram will slow down somewhat, but as it does so the current taken by its motor increases. When the tram has slowed down sufficiently to take a large enough current to produce the requisite tractive force, it will then continue at a constant speed so long as the gradient of the hill remains unaltered.

Series and shunt-wound motors.

We mentioned on p. 79 that the armature and field coils of an electric motor may be connected in series or in parallel (series-wound and shunt-wound motors, respectively).

Now the *torque* or turning force of a motor is proportional to the product of the armature current and the magnetic field, and hence to the *product of the armature and field currents.*

The chief advantage of a series-wound motor is that it produces a large torque at starting. This is due to the fact that, when the armature coil is turning slowly and there is, in consequence, a small back E.M.F., a large current flows in both the armature and field coils.

In the case of a shunt-wound motor the current in the field coil is unaffected by the back E.M.F. in the armature coil. Only the armature current, therefore, is large at starting; that in the field coil is no bigger than when the motor is running at speed. The shunt-wound motor is not, therefore, so good a starter as the series-wound motor.

On the other hand, the speed of a shunt motor fluctuates less under varying loads than that of a series motor. Suppose the load on a shunt motor decreases and the motor speeds up. The back E.M.F. in the armature increases and the current in it decreases. The current in the field coils, which are in parallel with the armature coils, is unaffected. In the case of a series motor, however, the speeding up of the motor causes a decrease not only in the armature current but also in the field current. The decrease in the field current means a still smaller back E.M.F., a still larger current in the armature and hence a further increase in speed. Thus the series-wound motor is unstable.

In order to obtain the advantages of both series- and shunt-wound motors, *compound winding* is employed: part of the field coils is in series and part in parallel with the armature coil.

Since series-wound motors are "good starters" they are used in trams and locomotives, in cranes and in motor cars. For traction and all variable speed purposes they are superior to any other form of electric motor.

Shunt-wound motors are essentially constant speed motors. They are used extensively in factories for driving such machines as lathes, steel-rolling mills, printing presses, chain-grate stokers, and rubber manufacturing machinery. They have also a domestic use in refrigerators, vacuum cleaners, and sewing machines.

Starting resistances.

If the full working voltage were put across the armature coil of a large motor when at rest, an excessive current would flow owing to the absence of back E.M.F. Such a current might cause serious damage to the insulation of the armature owing to overheating. Hence a starting resistance (see Fig. 161), which is cut out of the circuit gradually as the speed of the armature increases, is always employed.

The driver starts a tramcar by moving a handle in jerks, from stud to stud, at the top of the controller box (see Fig. 162). This box contains a starting resistance and also an arrangement for putting both of the tram's electric motors at first in series and then in parallel. When the motors are in series only half the voltage between the trolley wire and the rails (usually about 550 volts) is across each motor; the current is therefore comparatively small. When in parallel, both motors obtain the full voltage, and hence the current through each is considerably increased. This device reduces the size of the starting resistance required.

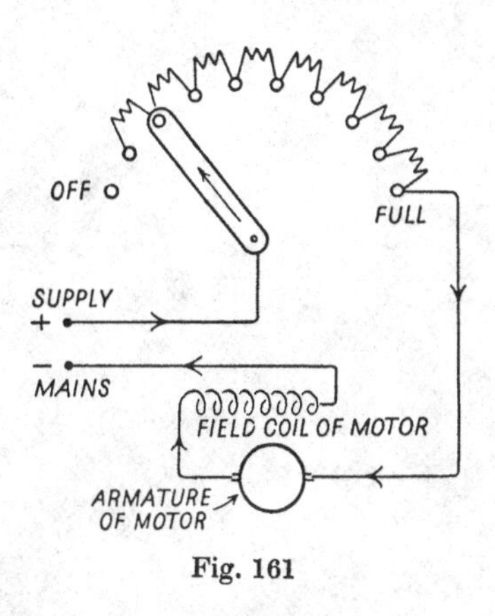

Fig. 161

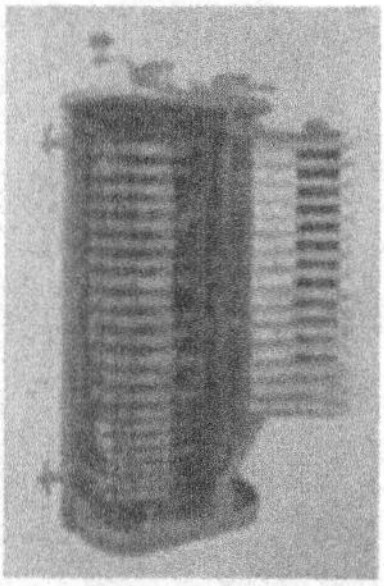

British T.-H. Co. Ltd.

Fig. 162. A tram controller, with cover removed.

The drum-wound armature.

The current produced by a single-coil direct-current dynamo is represented in Fig. 150: it is continually rising to a maximum and decreasing to zero. To eliminate this fluctuation a number of rotating coils at an angle to each other are employed in a practical dynamo: there is then always one coil passing, or about to pass, through the position in which the induced current is a maximum. With two coils only, placed at right angles, the fluctuations are very much reduced (see Fig. 164): and each additional coil helps to smoothe the resultant current.

Similarly, with a single-coil motor the torque is very uneven: it is a maximum when the coil lies along the lines of force and zero when it is at right angles to them. In order to obtain a more uniform torque a number of coils are distributed uniformly about the armature, as in the case of the dynamo.

We will describe one type of winding, the drum-wound

By courtesy of the Southern Railway

Fig. 163. The driver in the control cabin of an electric train. He is holding what is known as "the dead man's handle", which controls the current to the motors and which, should he become incapacitated, flies back causing the train to stop. The large wheel, top right, is for the operation of the hand brake. The drum below the wheel is a compressed air chamber which forms a part of the continuous air-brake system. The instruments, reading from the top, are the ammeter, speedometer and Barke air-gauge. (The vertical lever with a handle is for the operation of the window wiper.)

armature, which is suitable either for a dynamo or a motor. We will explain its action in a motor and leave the reader to perform a similar task for the dynamo.

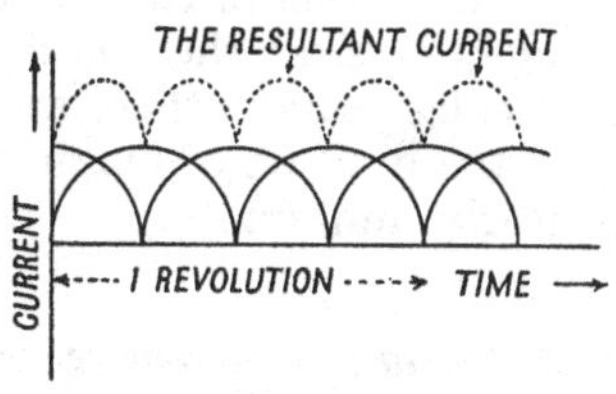

Fig. 164

The drum-wound armature consists of a soft-iron cylinder with slots, parallel to its axis round its circumference, in which copper strips are laid (see Fig. 66). In Fig. 165 there are eight copper conductors in slots, corresponding to four coils. The commutator has four segments, one to each coil.

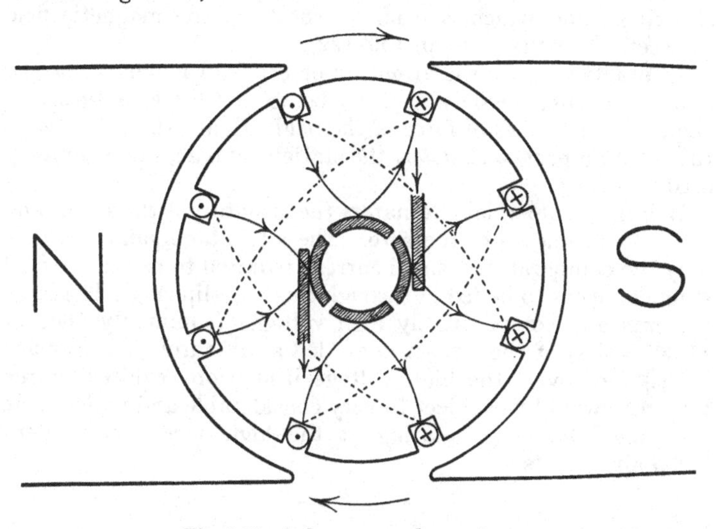

Fig. 165. A drum-wound armature.

The current, led in at one commutator brush and out at the other, divides into two paths. The reader should trace these paths. The currents in all the conductors on the left-hand side of the armature are coming out of the paper, and those on the right are going into the paper. Hence, applying Fleming's left-hand rule (see p. 75), all the conductors on the left are urged upwards and those on the right downwards: the armature therefore revolves in a clockwise direction. The forces on the conductors near the top and bottom of the armature are small and the forces on the conductors near the middle of the armature are large. At any particular instant there are always two conductors somewhere near the position of maximum force and hence the total torque is nearly uniform.

The alternating current generator.

A dynamo for generating alternating current is usually termed an A.C. generator or an alternator.

In principle it consists of a coil whose ends are connected to slip rings, and which is made to rotate in the magnetic field between two poles (see pp. 190–192).

In practice either the armature or the field magnets may be made to rotate, the only requisite being that the armature coils should cut the lines of force of the field. Hence the terms *rotor*, the rotating part, and *stator*, the stationary part, are commonly used.

In large commercial alternators the armature is the stator and the field magnets are the rotor. The advantage gained is that only the comparatively small current required to excite the field magnets needs to be fed in through rotating slip rings. The large currents and comparatively high voltages (commonly 6600 or 11,000 volts) of the armature are led away through stationary leads. Moreover, the high voltage insulation required in the armature would be subject to very considerable and undesirable centrifugal forces if rotating at the high speeds of modern turbo-alternators.

Frequency.

The *frequency* of an A.C. is the *number of cycles* or complete to and fro alternations it makes *per second*. A single coil rotated between one pair of poles generates 1 cycle per revolution. To produce A.C. at 50 cycles per sec. (the frequency in the Grid) it would be necessary to rotate the coil at 50 revs. per sec., i.e. 3000 revs. per min. Alternatively, the coil could be held stationary and the poles revolved at this speed.

If the coil were rotated between four poles arranged as in

By courtesy of the British Thomson-Houston Co. Ltd.

Fig. 166. The rotor of a 25,000 kV.A., 3000 revs. per min., turbo-alternator. Direct current is fed into the conductors lying in the slots in the iron core, causing the latter to have one pair of poles. See also Fig. 167.

By courtesy of the British Thomson-Houston Co. Ltd.

Fig. 167. The stator of a 25,000 kV.A., 3000 revs. per min., turbo-alternator. The induced currents are generated in the coils of the stator. See also Fig. 166.

Fig. 168, it is clear that the alternations in the coil would be at the rate of 2 cycles per revolution. To produce A.C. at 50 cycles per sec. the coil (or the poles) would need to rotate at 25 rev. per sec., i.e. 1500 revs. per min.

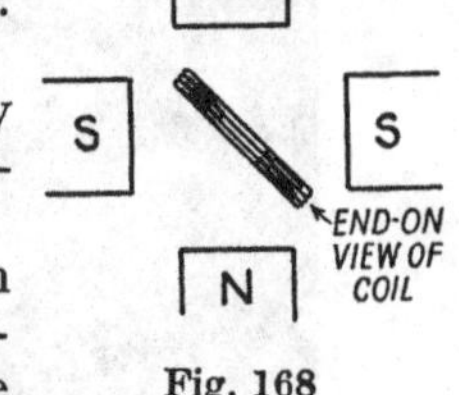

Fig. 168

Thus to generate A.C. at a given frequency there is one speed only at which an alternator can run.

The rotors of alternators driven by steam turbines, known as turbo-alternators, revolve at high speed. They may comprise two or four poles, and to produce A.C. at 50 cycles per sec. they must run at 3000 and 1500 revs. per min. respectively. Low-speed alternators driven by water turbines or reciprocating engines are multipolar. Their rotors may contain up to 100 poles.

Power stations.

In this country the bulk of electric power is derived, originally, from the chemical energy of coal. The energy of the coal is liberated in the form of heat in furnaces and raises steam in boilers: the steam drives turbines which drive the electric generators.

Fig. 169 is a view of the Battersea Power Station. The tallest part of the building, beneath the two chimneys, is the Boiler House. Adjacent and running parallel to it on its left is the Turbine House, and on the outside, also running parallel, is the Switch House. When completed there will be another Turbine and another Switch House on the right of the Boiler House, and the total capacity will be between 400,000 and 500,000 kilowatts. The present plant has only half this capacity.

The present coal consumption of the station is about 20,000 tons per week, and the ashes to be disposed of weekly amount to between 3000 and 4000 tons.

Hence the station was built with a large river frontage on the Thames and provided with a jetty to take 2000-ton colliers and also a barge berth. Two grab cranes take the coal from the colliers and discharge it into a hopper. From here it travels on belt conveyors to a control tower, where it is weighed and transmitted either to the Boiler House bunkers or to the coal store. The latter is a vast underground store between the river front and the north front of the building, with a capacity of 75,000 tons. The jetty cranes can each handle 240 tons per hour and hence unload a 2000-ton collier in one tide.

The ashes are removed by rail. It will be seen, in Fig. 169, that railway tracks encircle the site.

By courtesy of the London Power Co. Ltd.

Fig. 169. The Battersea Power Station. Under the chimneys is the Boiler House, on its left the Turbine House, and on the extreme left the Switch House. Note the river frontage for receiving coal by barges, and the railway for removing the weekly 4000 tons of ash.

The Boiler House is designed for a row of nine boilers: six have been installed (1936). Each boiler is rated at a normal evaporation of 250,000 lb. per hour of steam at 625 lb. per sq. in. and 875°–900° F. The boilers are of the water-tube type (see the author's *Heat*). They require a large amount of space, and represent a large portion of the capital cost of a station.

The flue gases or smoke are carefully washed before passing

out of the chimneys, to eliminate sulphur fumes. The chimneys are the largest in London: their gigantic size can best be appreciated when viewed from a distance against a background of other buildings.

In the Turbine House there are at present three turbo-generators, two rated at 69,000 kilowatts, and one at 105,000

By courtesy of the British Thomson-Houston Co. Ltd.

Fig. 170. A 30,000 kW. turbo-alternator in the new extension of the North Wilford Power Station, Nottingham. The steam turbine is on the right (upper storey), and the condenser is below it. The main alternator is in the middle of the picture and to the left of it are the house service generator and exciter.

kilowatts. (A 75-kilowatt machine has approximately the same power as a 100 H.P. engine.)

A turbo-generator (see Figs. 170 and 171) consists of a turbine (see the author's *Heat*), revolved by steam from the boilers, which is coupled to and drives A.C. generators or dynamos.

Each turbine consists of three cylinders, a high-pressure (H.P.), intermediate-pressure (I.P.) and low-pressure (L.P.) cylinder,

By courtesy of the British Thomson-Houston Co. Ltd.

Fig. 171. One of the three 75,000 kW. turbo-alternators in the Barking Power Station. The turbine end is in the foreground. The enormous pipes lead the steam from the intermediate-pressure cylinder to the low-pressure cylinder.

through which the steam passes in turn. The steam connections between the steam chest and the H.P. cylinder, and also between the H.P. and I.P. cylinders, are below the floor. That between the I.P. and L.P. cylinders is overhead. All are of U shape to give flexibility: the overhead connection can be seen plainly in Fig. 171.

Underneath the floor and connected to the L.P. cylinder are twin condensers. The steam condenses here and a vacuum of 29·1 inches (barometer 30 inches) is produced, that is to say, the pressure here is about 0·9 inch of mercury. The object of the vacuum is to cause the steam to rush more violently out of the turbine than it would if it were discharged into the air, and hence to increase the efficiency of the turbine. Ejector or vacuum pumps are used to remove the air, and other gases which do not condense.

At Battersea, to condense the steam 2,850,000 gallons of water are supplied per hour through bronze pipes inside the condensers. This water is drawn from, and discharged back into, the Thames by circulating pumps, through two large tunnels about 12 feet in diameter.

When a large supply of cold water, such as that afforded by the Thames, is not available at a power station, the condensing water has to be used over and over again, and must be cooled in large cooling towers (see Fig. 172).

The condensed steam, which is pure distilled water, on leaving the condensers is heated by steam "bled" from the turbines, and ultimately finds its way back to the boilers.

Each generating unit comprises (1) the main generator, (2) the house-service generator, (3) the exciter groups, which are coupled, in this order, to the turbine shaft.

The house-service generator provides power for the various feed and intake pumps and the machinery of the station.

The current passes from the generators to meters, transformers and circuit breakers in the Switch House. From here it passes along feeders, cables in an underground tunnel, into the Grid System.

To deal with the changing demand, arrangements are made for increasing or decreasing the amount of electric power generated. An engineer sits in the centre control room of the power station with telephones and automatic signalling apparatus at his elbow, like a chief of staff.

When a large extra load is required he telephones to the men in charge of the boilers to increase the pressure of the steam or to raise steam in another boiler. Each turbine can be connected to several boilers.

By courtesy of the Central Electricity Board

Fig. 172. The Hams Hall Power Station, near Birmingham. Note the vast cooling towers for cooling the water from the condensers—only necessary when there is not a plentiful supply of water, and the water must be used over and over again. The water trickles down a kind of honeycomb: some of it is lost in the form of steam which issues from the tops of the towers.

In the Boiler House there is a central panel on which there are two pointers for each boiler, a red pointer set by the control room showing the output required of the boiler and a green pointer showing the boiler's actual output. The duty of the boiler attendant is to keep the red and green pointers together.

When extra load is taken from a generator it tends to slow down. More steam is then automatically admitted into the turbine.

The speed control of a generator is very important. If the speed, and hence the frequency, rises too much the station tends to take load off other stations and vice versa. This is known as "load snatching".

Summary

A dynamo tends to act like a motor in the opposite direction. This effect acts as a kind of brake, and the work done against it is the source of the electrical energy generated by the dynamo.

A motor tends to act like a dynamo; an induced back E.M.F. is generated which opposes the current. The work done in forcing the current against this back E.M.F. is the useful work done in turning the armature of the motor. When the motor slows down the back E.M.F. is reduced and hence the current increases; thus the motor automatically takes as much current as it requires. A starting resistance is usually necessary, since the back E.M.F. at starting is small and hence the current may be excessive.

The field and armature coils of a motor may be joined in series (series-wound), in parallel (shunt-wound) or part in series and part in parallel (compound-wound). Series-wound motors are better starters than shunt-wound motors, but their speed tends to fluctuate more under varying loads.

An A.C. generator or alternator consists of a rotor comprising two, four, or more electromagnetic poles excited by D.C. and a stator in which A.C. is induced. The frequency of an A.C. is the number of cycles, or complete to and fro alternations, it makes per sec. To produce A.C. with a frequency of 50 cycles per sec. a two-pole alternator must be run at 3000 revs. per min., a four-pole generator at 1500 revs. per min., etc. Most alternators in this country are driven by steam turbines.

QUESTIONS

1. Explain what is meant by the reverse motor effect in a dynamo. What is the source of the energy generated by a dynamo?

2. Explain what is meant by the back E.M.F. of a motor. Why is a starting resistance required for a motor? What starting resistance is required by a series motor of total resistance 2 ohms, run off 100-volt mains, if the maximum safe current is 20 amperes?

3. Explain clearly how the E.M.F. generated by a dynamo depends on (1) the number of revolutions per sec., (2) the number of turns in the armature coils, (3) the strength of the magnetic field.

4. If a dynamo is called upon to deliver a greater current, where does the extra energy come from? Explain fully.

5. Describe and explain the action of a dynamo for the production of (*a*) alternating current, (*b*) direct current.

Which of these two types of current supply would be suitable for the following purposes: (*a*) lighting an electric lamp, (*b*) working an apparatus for electro-plating, (*c*) working a transformer, (*d*) charging an accumulator, (*e*) working an electric radiator? (C.)

6. Upon what conditions does the E.M.F. generated in a dynamo depend?

Being supplied with coils of known number of turns, bar magnets, resistance boxes and a sensitive galvanometer, describe briefly the experiments you would conduct in support of your statements.

The E.M.F. of a dynamo is 110 volts and its internal resistance 0·1 ohm. When the terminals of the machine are connected by a wire, a current of 20 amperes flows through the circuit. What is the resistance of the wire? (L.)

7. Describe an electric motor which you have seen and explain how it works.

The resistance of a motor used to drive a grindstone is found to be 20 ohms. When turning the stone it takes a current of 0·25 ampere at 250 volts, and when a tool is being ground the current rises to 2 amperes. Explain these facts. (L.)

8. Discuss the advantages and disadvantages of series-wound, shunt-wound and compound-wound motors.

9. Can (*a*) a series-wound, (*b*) a shunt-wound, motor be run on either D.C. or A.C.? Explain.

10. Why does the iron armature of a motor or dynamo tend to get hot? How is this effect minimised?

11. Draw a simple diagram similar to Fig. 67 representing a series-wound D.C. dynamo and put in the direction of the current. Explain why such a dynamo works only if the armature is driven in one direction.

12. Why should the field circuit of a shunt motor never be broken? (Consider what happens when the field of a shunt motor is made weaker.)

13. (*a*) Explain how the speed of a shunt motor can be varied by placing a variable resistance in series with the field coils. What happens to the speed of the motor when this resistance is increased and why?

(*b*) If a variable resistance is placed in series with the field coils of a series motor what happens to the speed of the motor and why?

(These are practical methods of varying the speeds of shunt and series motors.)

14. Make a large carefully drawn diagram similar to Fig. 165 of a drum-wound dynamo armature with 16 slotted conductors. Put in the two brushes and trace the directions of the currents by means of arrows and also by means of a · and × in the conductors (as in Fig. 165).

15. A small motor takes 10 amperes, the E.M.F. supplied being 24 volts when driven against a brake, and generates $\frac{1}{4}$ H.P. Find its efficiency. (1 H.P. = 746 watts.)

16. The starter of a motor car has an efficiency of 80 per cent. If it is worked by a 12-volt battery and takes a current of 40 amperes, calculate its horse-power. (1 H.P. = 746 watts.)

17. Explain why the current taken by an electric motor cannot be calculated by dividing the applied volts by the resistance in ohms of the motor. (L.)

18. Find the back E.M.F. of a motor working at 400 volts when the armature current is 50 amperes, the resistance of the armature circuit being 0·8 ohm.

19. Find the back E.M.F. of a series motor developing 4 H.P. and taking 30 amperes. If the resistance of the motor is 3 ohms, find the P.D. of the supply mains. (1 H.P. = 746 watts.)

20. What is the current flowing through the armature coil of a motor, resistance 0·75 ohm, if the voltage supply is 110 volts and the back E.M.F. is 105 volts?

21. Find the back E.M.F. of a shunt motor whose armature and field resistances are 0·2 ohm and 40 ohms when it is taking 20 amperes from 100-volt supply mains.

22. What current will be taken by a motor developing $\frac{1}{2}$ H.P. when connected to a 110-volt supply, if the efficiency of the motor is 75 per cent? (1 H.P. = 746 watts.)

23. A shunt motor running off 100-volt mains takes an armature current of 2 amperes. If the resistance of its armature is 0·5 ohm, calculate the back E.M.F.

The speed of the motor is 800 revs. per min. If it is now loaded so that it takes an armature current of 10 amperes, find the new speed.

24. A series motor, of resistance 2 ohms, running under load at 300 revs. per min., takes 5 amperes when supplied from 60-volt mains. At what speed does it run when unloaded and taking a current of 2 amperes? (Assume that the strength of the field is proportional to the current.)

25. How many revolutions per minute must an 8-pole generator make in order to produce A.C. at 50 cycles per sec.?

26. Give a brief general description of a power station. Explain carefully the various transformations of energy that occur in the generation of electricity from coal.

Chapter XIII

THE GRID

During the last few years Great Britain has been covered with a gridiron or network of overhead wires. These wires are connected to power stations and transmit electrical power in the form of alternating current (A.C.).

Hitherto Great Britain was divided into about 600 electrical supply areas, using current at various voltages, frequencies and prices.

In some industrial areas the "peak load" occurred during the daytime owing to the large consumption of power by factories: in others it occurred, mainly for lighting, at night.

Now the storage of electricity in bulk by accumulators at a power station is uneconomical. It must be generated as required. Hence at any period of the day or night a large proportion of the plant in some power houses up and down the country lay still and idle. In 1930–31, although the maximum demand for electrical power was less than 4,000,000 kW. (kilowatts), the plant in existence had a capacity of 7,000,000 kW.

An industrial concern works at its maximum efficiency when the whole of its plant runs at full speed day and night. The capital outlay, represented by the plant, is then earning its interest fully and unceasingly. Through the linkage of the Grid, generating stations now pool their resources so that each can work almost at full pressure continuously. In consequence, the reserve plant of the country has been reduced from over 40 per cent to 15 per cent of the total plant.

The number of generating stations has been reduced from nearly 500 to 130. Those closed down were small and comparatively inefficient: the remainder, the "selected stations", have been augmented by a number of modern super-stations, such as those of Battersea and Hams Hall near Birmingham.

The body which controls the erection and maintenance of the Grid is the Central Electricity Board. To the Board the owners of the selected stations sell their power at a price fixed by Act of

Parliament. The Board then resells the power to the 600 local authorities, many of whose power stations have been dismantled and replaced by substations, which, as we shall see, serve merely for the reception of power and its conversion to a different voltage. The Board, therefore, acts as a middleman: it distributes the power from the wholesaler to the retailer.

Although the cost of electricity still varies in different localities, its frequency and the voltage supplied to the domestic consumer are becoming standardised at 50 cycles per sec. and 230 volts.

The cost of the Grid is estimated at £30,000,000, and that of standardising the frequency at a further £16,300,000.

The full benefit of the scheme will be felt increasingly as the years pass, when the demand for electrical power increases and the price, in consequence, falls. Already purchased electrical power constitutes 50 per cent of the power used in industry. Some day, perhaps, when this figure is nearer 100 per cent, the sordid grime of our northern industrial towns may vanish like an ugly dream.

Power transmission.

The rate of generation or consumption of electrical energy in watts is equal to the product of the current in amperes and the voltage (see p. 162). Thus if the generation is at the rate of 20 kilowatts, this may be in the form of 2 amperes at 10,000 volts, or 20 amperes at 1000 volts, 80 amperes at 250 volts, and so on.

In order to transmit electric power economically along a cable, a high voltage and a small current must be used. If a small voltage and a large current are employed the loss of energy in the form of heat in the cable is prohibitive.

We will take a simple example. Suppose a generator has a capacity of 20 kilowatts, and that it generates at a voltage of 250 volts (the maximum voltage that may be supplied to a domestic consumer). The current delivered by the generator must be $\frac{20,000}{250} = 80$ amperes.

For the sake of simplicity, let us assume that the generator supplies direct current and that a single cable, of resistance 1 ohm, is connected between the positive terminal of the generator and the consumer's positive terminal (see Fig. 173). The negative terminals are earthed and the return current may be regarded as flowing through the earth. We will assume that the resistance of the earth is negligible.

Applying Ohm's law to the cable:

$$V = iR.$$

$$\text{P.D. between ends of cable} = 80 \times 1$$
$$= 80 \text{ volts.}$$

This drop in potential along the cable is required to drive the current through it.

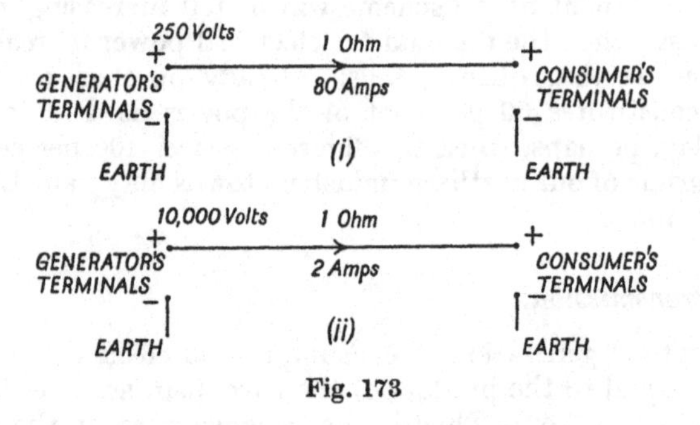

Fig. 173

Hence potential of consumer's positive terminal

$$= 250 - 80$$
$$= 170 \text{ volts.}$$

$$\text{Power reaching consumer} = 170 \times 80$$
$$= 13{,}600 \text{ watts.}$$

$$\text{Power loss in cable} = 20{,}000 - 13{,}600$$
$$= 6400 \text{ watts.}$$

Let us suppose now that the voltage at the generator end of the cable is raised from 250 volts to 10,000 volts. The current flowing in the cable is $\dfrac{20{,}000}{10{,}000} = 2$ amperes.

Applying Ohm's law to the cable:

$$V = iR.$$

P.D. between ends of cable $= 2 \times 1$

$= 2$ volts.

Hence potential of consumer's positive terminal

$= 10{,}000 - 2$

$= 9998$ volts.

Power reaching consumer $= 9998 \times 2$

$= 19{,}996$ watts.

Power loss in cable $= 4$ watts.

The enormous disparity between the losses in the cable in the two cases, 6400 watts and 4 watts, clearly emphasises the desirability of transmitting power at a high voltage and a small current. Note that the losses are i^2R watts, where R is the resistance of the cable; in the first example this was

$$80^2 \times 1 = 6400 \text{ watts},$$

and in the second, $2^2 \times 1 = 4$ watts. In the expression i^2R the factor R is invariable but i can be reduced as much as it is feasible to raise the voltage.*

In the main lines of the Grid the transmission voltage is 132,000 and in the secondary networks 66,000, 33,000 and 11,000 volts.

Advantages of alternating current over direct current.

The chief reason why A.C. is used in preference to D.C. in the Grid is that its voltage may be raised or lowered comparatively simply by means of an instrument with no moving parts, called a transformer. To perform the more difficult operation of changing the voltage of D.C. a rotary converter is required. The D.C. is made to drive a motor which turns a dynamo, and the dynamo generates D.C. at the required new voltage.

* The reader should now work out the losses using two cables. In practice the negative terminals are always connected by a cable (and the resistance of the earth is by no means negligible as we have assumed).

One of the first power stations, that of Ferranti at Deptford, was built in 1880 to supply power for the arc lighting of London. Ferranti, a man of genius and ahead of his time, designed his station to generate A.C. but, owing to inadequate financial resources, his venture was not a success. This served as an encouragement to the advocates of D.C., and during the next twenty-five years over a hundred D.C. power stations were built in this country. It was not till 1904 that the advantages of A.C. were widely appreciated and the policy was reversed. During recent years, with the building of the Grid, most of the old uneconomic D.C. power stations have been scrapped.

The transformer.

The transformer is designed for "stepping up" or "stepping down" the voltage of A.C.

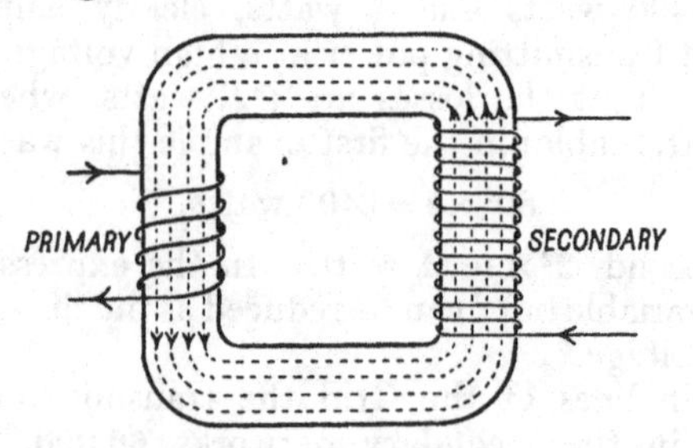

Fig. 174. A step-up transformer. The dotted lines represent the lines of magnetic force due to the primary current. The induced current in the secondary is in the direction shown only when the current in the primary is increasing, since its lines of force (not shown) are in the opposite direction to those of the primary current.

It consists of two coils wrapped on an iron ring (see Fig. 174): through one of the coils, called the *primary*, the original A.C. is passed, and in the other, the *secondary*, an A.C. is induced.

The action of the transformer is an example of mutual induction (see p. 187). As the A.C. in the primary reverses, it reverses the magnetism in the iron ring. At each reversal the lines of magnetic force in the ring die away and fresh lines in the opposite direction appear. But these magnetic lines of force thread the secondary coil: their movement induces a current in the secondary.

The induced E.M.F. or voltage in the secondary is proportional to the rate of cutting or threading of the lines of force. If the secondary has twice as many turns as the primary, since the same number of lines thread through both, the voltage in the former will be twice that in the latter. Generally,

$$\frac{\text{Voltage across secondary}}{\text{Voltage across primary}} = \frac{\text{Number of turns in secondary}}{\text{Number of turns in primary}}.$$

However, if the voltage in the secondary is double that in the primary, the secondary current will be one-half, approximately, the primary current. For, assuming the transformer to be 100 per cent efficient,

$$\text{Power in primary} = \text{Power in secondary},$$

i.e.

$$\text{Primary current} \times \text{primary voltage} = \text{Secondary current} \times \text{secondary voltage.}$$

Thus an increase in the secondary voltage (owing to an increase in the number of turns) entails a corresponding decrease in the secondary current.

Power transformers.

When two power stations are connected through a transformer, the primary and secondary coils become interchanged according to the direction of flow of power. Hence it is customary among engineers to refer to the High Tension (H.T.) and Low Tension (L.T.) windings.

A large transformer used in the Grid system is rated by its power, i.e. amperes × volts, which is approximately the same for both H.T. and L.T. windings. Suppose the voltage in the H.T. is 6000 and the current 10 amperes. Then the rating of the transformer is

$$\frac{6000 \times 10}{1000} = 60 \text{ kV.A. (kilovolt amperes).}$$

The transformer is described as follows:

60 kV.A.; 6000/400 volt transformer,

where 400 volts is the voltage in the L.T.

By courtesy of Messrs Ferranti, Ltd.

Fig. 175. The core and windings of a 60,000 kV.A., 3-phase transformer. Note the thick paper insulation at the top and the oil ducts between the coils. On each leg there are two coils, the high tension and the low tension, between which are synthetic resin paper cylinders, known as the major insulation.

Transformer design.

Transformer cores are always laminated, with light insulation between, to prevent eddy currents (see p. 193). One of the difficulties of transformer design is the reduction of noise and vibration, due to the vibration of the laminations, which is not stopped by clamping.

By courtesy of the English Electric Co. Ltd.

Fig. 176. Two 15,000 kV.A., 132,000 volt transformers, in use on the Grid. Note the stacks of tubular radiators for cooling the oil in the transformers. Much heat is produced due to eddy currents and the changing magnetisation (at 50 cycles per sec.) of the cores. Note also the overhead steel framework and porcelain insulators, for carrying the cables.

The efficiency of a transformer is about 95–98 per cent. The energy wasted appears in the form of heat. Large power transformers are usually cooled by air blowers, circulating oil, or a water-cooling system.

The overhead lines.

Since air is one of the best and certainly the cheapest insulator, the cables of the Grid are, wherever possible, carried overhead. There are nearly 3000 miles of 132,000 volt lines and over 1000 miles of secondary lines.

By courtesy of the Central Electricity Board

Fig. 177. Pylons carrying a single circuit (three-phase), in Dumfries. Note the three cables, each carrying one phase, and the neutral or earth wire connected to the tops of the pylons. Note also the arcing rings at the bottom of each insulator string.

The supporting towers, made of galvanised steel, are a familiar feature of the landscape. Since, as we shall explain, the A.C. traversing the Grid is "three phase", there are three cables to each circuit. In addition an earthed wire, called the neutral wire, connects the tops of the towers. Each 132 kV. circuit is designed to carry 50,000 kW., and each cable can carry a current of 219 amperes.

The cables, which are as thick as a man's finger, consist of

By courtesy of the Central Electricity Board

Fig. 178. One of the Thames crossing towers at Barking, 487 feet high and carrying a 3060 feet span of two, three-phase, circuits. There are six neon crosses, one on each face one-third and two-thirds of the way up except on the river side, and also a neon light at the apex, to warn aircraft at night or in fog. Two of these crosses are plainly visible in the picture about one-third of the way up. There are steel stairs to the summit.

thirty aluminium wires twisted round seven steel wires (to give them strength). They are connected to each tower by a suspension clamp supported through a string of porcelain or toughened glass insulators, usually nine in number, though sometimes more in a smoky, impure atmosphere. The cables are about 12 feet apart to prevent arcing. To eliminate the possibility of damage due to a flash-over caused, say, by a lightning surge, arcing horns and rings are provided. The rings can be seen at the bottom of the insulator strings in Fig. 177: the horns are at the top of the insulators. The arc discharges between a horn and ring (and then through the steel framework of the tower to the earth), in preference to passing over the surface of the insulators. The tower is effectively earthed if its legs are embedded in soil: if they are supported in concrete one of the legs is connected to an earthed pipe.

The tensions of the cables must be carefully adjusted so that they all swing together in a wind to prevent any chance of arcing between them. The average span between the towers on the 132 kV. lines is 900 feet, and in such a distance there is appreciable expansion and contraction due to change of temperature. The cables must be strong enough to withstand a ½-inch radial coating of ice.

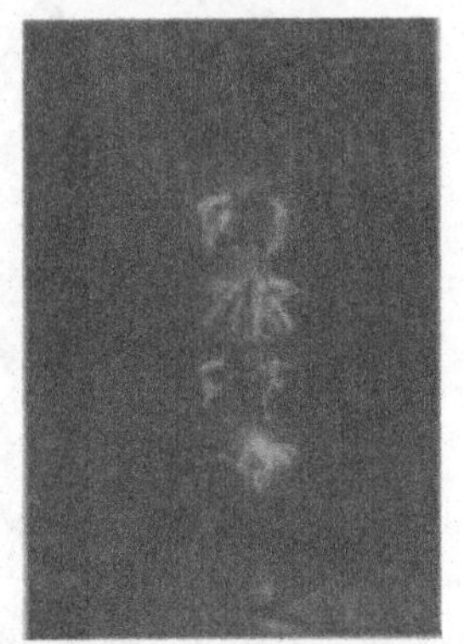

General Electric Co. Ltd.

Fig. 179. A surge flashover on a small unit post type insulator at 700 kV. with arcing horns removed.

Specimens of each type of tower are tested to destruction by winches for their resistance to snow and wind. Terminal towers, which sustain the pull of the cables in one direction only, must be especially strong. Firm anchorage is essential: an anchor is often inserted in a hole, bored in the ground with an auger, the cavity being filled with concrete.

Substations.

The Grid is a system of interconnected rings. These rings are fed, joined and tapped through switching and transforming stations, known collectively as *substations*, of which there are about 270.

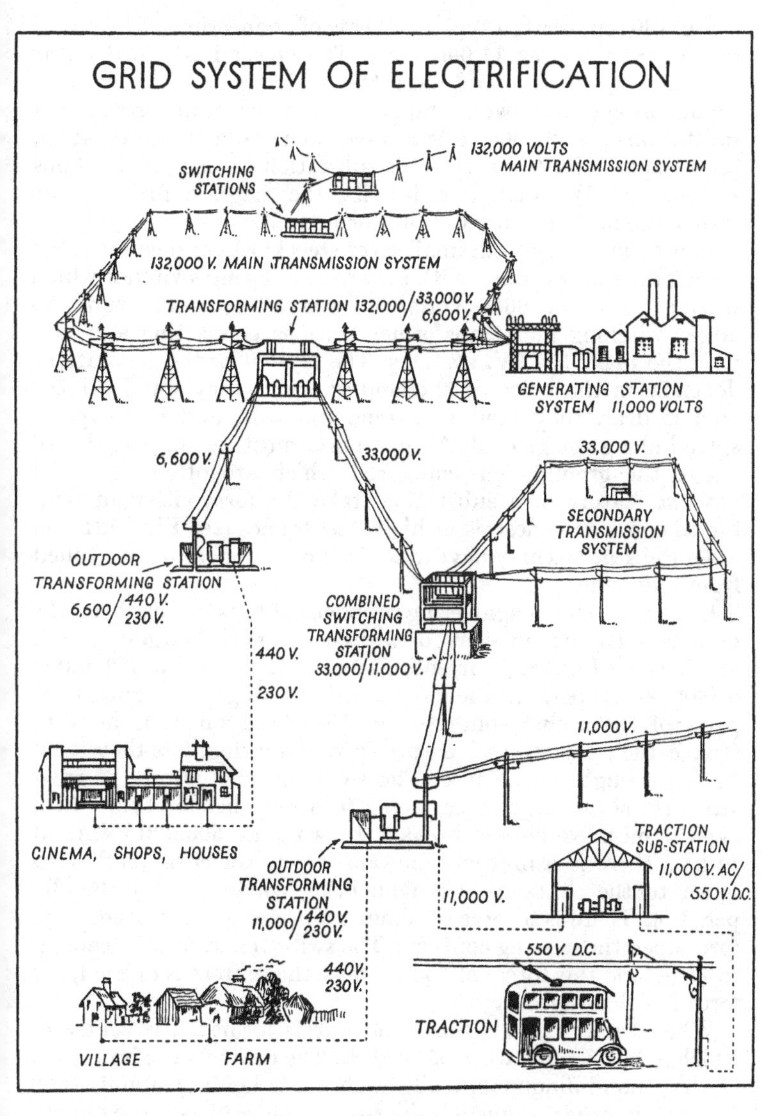
GRID SYSTEM OF ELECTRIFICATION
132,000 VOLTS
MAIN TRANSMISSION SYSTEM
SWITCHING
STATIONS
132,000 V. MAIN TRANSMISSION SYSTEM
TRANSFORMING STATION 132,000/33,000 V. 6,600 V.
GENERATING STATION
SYSTEM 11,000 VOLTS
6,600 V.
33,000 V.
33,000 V.
SECONDARY
TRANSMISSION
SYSTEM
OUTDOOR
TRANSFORMING STATION
6,600/440 V. 230 V.
COMBINED
SWITCHING
TRANSFORMING
STATION
33,000/11,000 V.
440 V.
230 V.
11,000 V.
CINEMA, SHOPS, HOUSES
OUTDOOR
TRANSFORMING
STATION
11,000/440 V. 230 V.
TRACTION
SUB-STATION
11,000 V. AC/
550 V. D.C.
11,000 V.
550 V. D.C.
440 V.
230 V.
TRACTION
VILLAGE
FARM

Fig. 180

Fig. 180 represents a typical scheme of connections. The power station generates at 11,000 volts. The power is stepped up at once to 132,000 volts and fed into a main transmission system. From this system power is tapped off through transformers in a substation (1) at 33,000 volts to a secondary transmission system, (2) at 6600 volts to an outdoor substation for supply to shops and houses. We shall not describe the diagram further: the reader should study it carefully for himself.

Substations vary from small sheet-steel kiosks or pole-mounted transformer substations with air-break isolating switches which receive power at 6600 volts and convert it to 440 volts, to large switching and transformer stations connecting and tapping the main 132 kV. systems. The lay-out of the substations depends on the price of the land on which they stand. If the land is dear, they tend to extend upwards, and if cheap, to spread along the ground. All live parts must be supported well above the ground. The supports, which are often the most obvious feature of a substation, take the form either of reinforced concrete pedestals or high steel frames (see Figs. 181 and 184). Substations may have attendant operators or be controlled from a distance.

One of the chief engineering problems of the Grid has been the design of circuit breakers or switches. In the largest circuit breakers (see Fig. 182), currents may reach a peak value of 200,000 or 300,000 amperes at the moment of breaking. The component parts of the arcing contacts (see Fig. 183), must be light to reduce their inertia and ensure speedy working, but they must be big enough not to fuse. The switches must also be of great strength, since large forces come into play inside them. In a typical oil-filled circuit breaker to work at 500,000 kV.A. at 11 kV., there is a mutual repulsion, due to the currents flowing down to the contacts in neighbouring conductors, of 1000 lb. per foot. There is a force of about 1 ton per contact tending to force open the moving contacts. The switch must be stiff enough to withstand this force: of course, once the contact is opened, the force is advantageous.

When the contact is broken an arc is formed, the centre of which is at a temperature of 2600° C. The contacts are immersed in dry, non-inflammable oil (each tank holding about 1000 gallons), in order to quench the spark. The arc heats the oil and

By courtesy of the English Electric Co. Ltd.

Fig. 181. The largest outdoor switching station on the Grid. The 132,000 volt substation at Northfleet, Kent.

By courtesy of the Central Electricity Board

Fig. 182. A line of oil circuit breakers at Hams Hall Grid substation (see also Fig. 183).

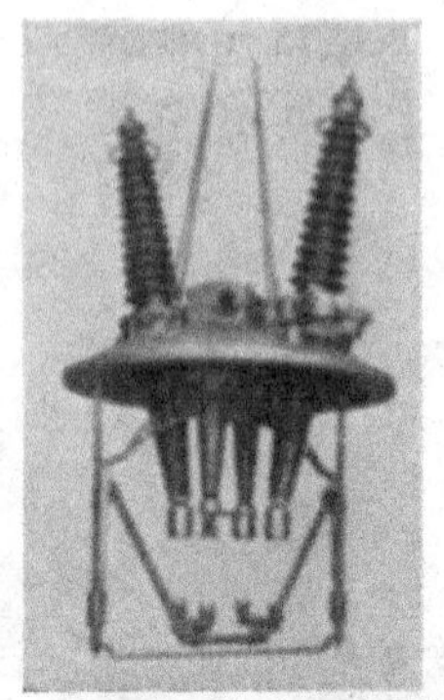

General Electric Co. Ltd.

Fig. 183. One phase of a 132 kV., 1,500,000 kV.A., oil circuit breaker removed from tank. The view shows the switch in the open position: to close it the U-shaped rod is raised. There are four contacts, all interconnected. Note the heavy, insulator bushes, and the arcing rings at the top of the two outer bushes.

By courtesy of the Metropolitan-Vickers Electrical Co. Ltd.

Fig. 184. Tandem isolating switch being operated. The rod, across the top of the three post insulators above the man, swivels about the middle post insulator and so makes or breaks a connection.

breaks it up into its constituents, mainly hydrogen, methane and acetylene gases. Since 6000 cubic inches of gas may be generated in $\frac{1}{50}$th sec., breaking has the effect of an explosion. The tanks are made of $\frac{1}{2}$-inch steel and all joints must be made very strong to withstand the pressure.

In conjunction with each switch a number of isolators operate (see Fig. 184). These consist of metal rods, supported on insulators

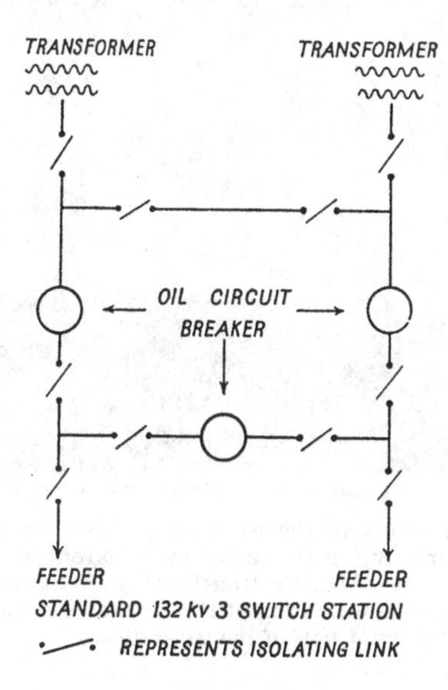

Fig. 185

in air, which can be made to rotate and break a circuit, isolating, as their name implies, part of it for purposes of repair or a change of connections. The isolators are so connected to the switch that they cannot be moved unless the switch is open, since they are not designed for making or breaking a live circuit.

Fig. 185 represents a three-switch station: note the numerous isolators. By suitable operation of the switches, the two feeder overhead circuits can be connected to either, both, or neither of the transformers. The station is so designed that should one of the oil switches or transformers be out of action the rest can continue to function.

The reader should compare Figs. 185 and 186.

By courtesy of the Metropolitan-Vickers Electrical Co. Ltd.

Fig. 186. An aerial view of the Norwich outdoor substation. Note the concrete trestles for carrying the cables and isolators, and also the lay-out (compare with Fig. 185). The two transformers are at the back; the three oil circuit breakers, each consisting of three cylindrical tanks side by side, form a triangle in the front half of the substation.

Protective gear.

In the event of the earthing of a live Grid line, or of a short-circuit between two lines, special protective gear comes into operation immediately and causes the switches controlling the defective part of the circuit to be opened. This prevents damage to the rest of the Grid, which continues to function normally. Meanwhile the unhealthy section has been isolated and can be repaired.

There are a number of different protective systems but all are too elaborate to describe in detail here. The general principle is that the unusual current caused by the fault induces a small current which operates a relay: the relay completes an auxiliary circuit controlling the large oil circuit breaker.

Metering.

Since the Central Electricity Board buys the electrical power fed into the Grid from individual power stations, and sells what

By courtesy of the Central Electricity Board

Fig. 187. Metering equipment in a Grid substation. This meter indicates the total energy taken in mega-watthours and also the maximum demand in kilowatts.

is tapped out to individual supply concerns, accurate meters are installed at points of import and export.

The instruments record the load in kW.H. in each individual circuit, the total of the loads in kW.H. on all circuits, the average demand in kW., the maximum demand in kW., and the time at which it occurs.

Control.

The Grid is controlled from seven central stations, at London, Glasgow, Birmingham, Manchester, Leeds, Newcastle and Bristol.

Each of these stations administers an area. In its central control room there is a map or chart of its area (see Fig. 188). When, by orders of the control engineer, a switch on the 132 kV.

By courtesy of the Central Electricity Board

Fig. 188. The central control room of the South-West England and South Wales region of the Grid, at Bristol.

lines is opened or closed, the change is signalled back and automatically recorded on the chart. Movements of auxiliary switches and isolators are registered manually on the chart by the engineer. The engineer can see therefore, at a glance, the connections in every part of his system. By means of instruments he can read the loads in each circuit and those imported and exported at each station. He is in telephonic communication with all the power stations and substations in his area. It is he who decides the load for each power station. Should a breakdown occur in his area he becomes instantly aware of it and he must make arrangements for a suitable transfer of the load.

General Electric Co. Ltd.

Fig. 189. A visual instruction receiver at Hams Hall generating station, with central metering panel in the background.

He can send rapid instructions to the power stations by a system of "engine room" signalling—Control—Start Up—Raise kW.—Steady—Lower kW.—Shut Down—Stand By (see Fig. 189). The engineer at the power station acknowledges by turning a pointer to the signalled instruction, whereupon a lamp in the control rooms is extinguished.

The control engineer, therefore, is the directing brain, situated at the nerve centre of each area. Not only must he, or one of his subordinates, be on duty day and night in the central control room, but with the statistics that his instruments give him, he must pre-arrange programmes to meet normal demands.

Three-phase alternating current.

We have mentioned (see p. 234) that the alternating current in the Grid is *three-phase*. We shall now explain the meaning of this term.

Suppose three separate coils, whose planes are inclined at 120° to each other, are rotated together between two magnetic poles. In Fig. 190 for clearness only half of each coil has been drawn. Three separate alternating currents will be induced in the coils, differing in "phase" by 120°. These three alternating currents are represented in Fig. 191. The curve labelled 1 represents the current induced in coil 1 starting from its position in Fig. 191. Note that one cycle is completed in 360°, i.e. one revolution. Coil 2 does not reach the present position of coil 1 until it has rotated through 120°. Hence curve 2 in Fig. 191 is 120° behind curve 1. Similarly, curve 3 is 240° behind curve 1.

Now the three coils have a total of six ends, and it might be expected that six slip rings would be necessary to lead away the currents.

However, the algebraic sum of the currents at A, B, C, or at A', B', C', is zero: in the positions of the coils of Fig. 190 there is no current in coil 1 and the currents in coils 2 and 3 are equal and opposite. The algebraic sum of the currents is indeed zero at every instant during the rotation of the coils, a fact which can most readily be appreciated by

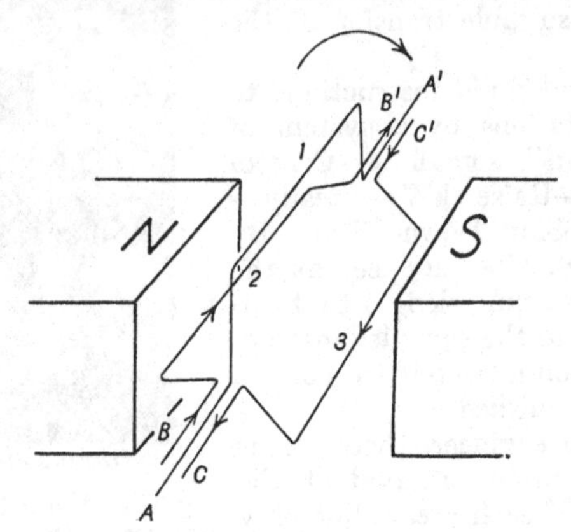

Fig. 190

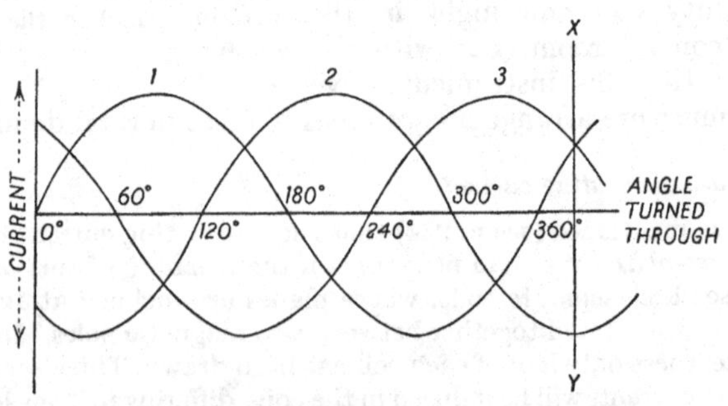

Fig. 191

drawing a line in Fig. 191, such as XY, in any position parallel to the current axis. The algebraic sum of the three values of the currents where XY cuts the three curves is always zero. Hence the three ends A, B, C, or A', B', C', can be connected together and only the other three ends require slip rings (see Fig. 192). In practice a thin lead, known as the neutral wire, is also taken from the interconnected ends.

The conventional method of representing the coils of a three-phase generator and the wires leading away the current from the slip rings is shown in Fig. 193. This method of connection is known as star- or Y-connection. The three wires represented by continuous lines in Fig. 193 are connected through a transformer to the three main cables attached to the Grid pylons. The middle of the star is connected to a comparatively thin earthed wire, the neutral wire, which is connected to the tops of the pylons. Current flows along the neutral wire only, as we shall see, when there is a lack of balance in the consumption of the three alternating currents forming the three-phase supply.

Met. Vick. Elect. Co. Ltd.

Fig. 192. Brush gear for three-phase current.

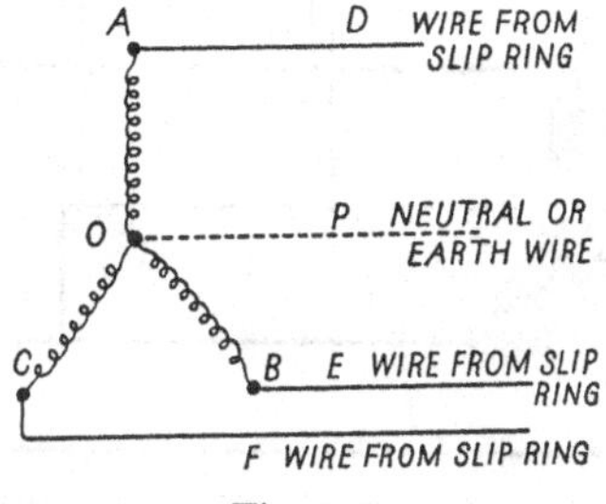

Fig. 193

Balancing.

By the addition of the neutral wire a three-phase system may be regarded as three single-phase systems. For example, in Fig. 193, if connection is made between *D* and *P* a current will flow in the single-phase circuit *ADPO*. Unless equal currents are taken in circuits *BEPO* and *CFPO* a resultant current will flow along the neutral wire *OP*: this is said to disturb the balance of the system, since it is not designed for a large current to flow along the neutral wire *OP*.

Fig. 194 represents the supply of power from a three-phase system to houses and a factory. Current is fed from the Grid to the H.T. windings of a step-down three-phase transformer. The currents from the L.T. windings of this transformer flow along underground cables. Each house taps only a single phase by means of wires connected as shown. A factory, on the other hand, takes three-phase current. The currents taken from the separate phases must be balanced: that is to say, it must be arranged that they are as nearly as possible equal in magnitude.

The usual three-phase voltage supplied to the factory is 400 volts. The power in each single-phase system is approximately $\frac{1}{3}$ of the total, and hence the voltage is $\frac{1}{\sqrt{3}}$ of the three-phase voltage, i.e.

$$\frac{400}{\sqrt{3}} = 230 \text{ volts.}$$

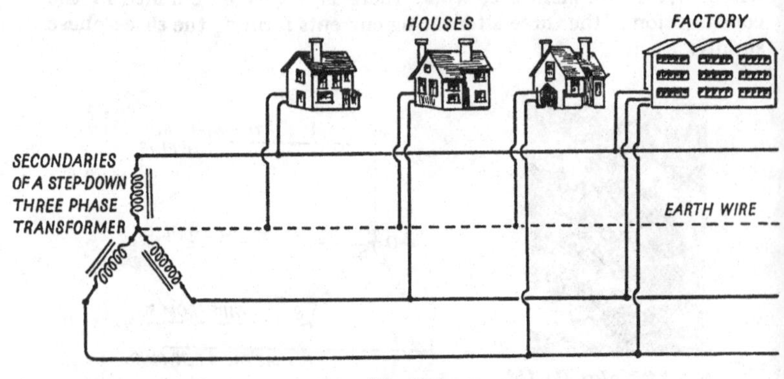

Fig. 194

Advantages of three-phase alternating current.

The advantages of three-phase A.C. are as follows:

1. Less copper or aluminium is required in the transmission lines.
2. A three-phase generator is capable of a greater output than a similar single-phase generator.
3. Three-phase motors (with which we have had no space to deal) have considerable advantages over single-phase A.C. motors.

Summary

The Grid is a system of interconnected circuits (mainly overhead wires) carrying A.C. and joining power stations and substations. Substations are transforming and switching stations.

The advantage of A.C. over D.C. is that its voltage can be stepped up or down by means of a transformer.

A transformer consists of a closed iron core on which are wound a primary and a secondary coil.

$$\frac{\textbf{Voltage across secondary}}{\textbf{Voltage across primary}} = \frac{\textbf{Number of turns in secondary}}{\textbf{Number of turns in primary}}.$$

Power may be transmitted with far less loss in the form of heat in the cables at a high voltage and small current than at a low voltage and a high current.

$$\text{Power loss in cable} = i^2R \text{ watts,}$$

where i = current, R = resistance of cable.

The current in the Grid is three-phase A.C.

QUESTIONS

1. What is the voltage drop in a cable carrying a current of 100 amperes if the resistance of the cable is 0·2 ohm?

2. Power is supplied to a factory by two cables two miles long, and at the power station the potential difference between the ends of the cable is maintained at 220 volts. The potential difference between the two ends at the factory must not fall below 200 volts and a maximum current of 40 amperes is needed. What is the greatest permissible resistance per mile of the cable? (O. and C.)

3. A power station maintains a difference of potential between trolley-wire and rail of 550 volts at the station end. If a car requires 35 amperes at 500 volts to work it, how far from the station can it go, assuming it to be the only car on the track? (Resistance of trolley-wire = 0·52 ohm per mile, and of rail = 0·04 ohm per mile.) (O. and C.)

4. A power station A is connected with a station B, 15 kilometres away, by two copper cables each of resistance 0·5 ohm per kilometre. If the power input at A is 4000 kilowatts at 50,000 volts, what is (*a*) the current in the cables, (*b*) the power lost in the cables owing to the generation of heat, (*c*) the difference of potential between the ends of the cable at B? (C.)

5. Power amounting to 100 kilowatts is delivered by direct current through a cable having a total resistance of 0·001 ohm. Calculate the power in kilowatts lost in the cable if the voltage at the point of

delivery is (*a*) 50, (*b*) 250 volts. If the cost of the power is ½*d*. a unit, calculate the saving effected in 8 hours, by using the higher voltage. (C.)

6. A motor taking a current of 100 amperes at a potential difference of 2000 volts is supplied with current by cable from a dynamo several miles away. If the loss of power in the cable is 10 per cent, what is (*a*) the P.D. at the dynamo, (*b*) the resistance of the cable? (O. and C.)

7. Compare the design and operation of the induction coil and the A.C. transformer.

8. An X-ray tube takes 5 milliamperes from the secondary coil of a transformer. If the voltage on the primary is 100 volts and the current through it is 8 amperes, what is the approximate voltage across the X-ray tube? (Durham.)

9. The current in the L.T. (or primary) of a transformer which is stepping up the voltage from 11,000 to 132,000 volts is 2000 amperes. If the efficiency of the transformer is 90 per cent, what is the current in the H.T. (or secondary)?

10. Write a short essay on the Grid.

Chapter XIV

ELECTRIC TELEGRAPHY AND TELEPHONY

The use of electricity as a means of communication is a factor of such importance in world history that we shall devote this chapter to a simple consideration of the sending of messages along wires.

In Napoleonic days signalling was done by semaphore: along the Dover and Portsmouth roads, situated on hill tops, were tall masts with arms, for the purpose of transmitting news from these ports to London. The process broke down completely in fog and was not applicable across the sea. Indeed the news of the battle of Waterloo took several days to reach London.

We may contrast this with the fact that we can now send a message from London to America, South Africa or Australia, by submarine cable, in a few seconds.

Telegraphy

The first electric telegraphs in England.

Commercial telegraphy only became practicable after the invention in 1836 of a steady and moderately strong source of electric current, the Daniell cell.

The current, switched on at the transmitting end of the line, was detected at the receiving end by means of its magnetic effect. It was made to pass round a coil of a number of turns, at the centre of which was pivoted a magnetic needle. The needle, under the action of the magnetic field set up by the current in the coil, tended to set itself at right angles to the plane of the coil, along the lines of force. The receiver was thus similar to a simple moving-magnet galvanometer (see p. 87).

In 1820 Ritchie had suggested a telegraph comprising twenty-six coils (and needles), each corresponding to a letter of the alphabet—a letter being signalled by the deflection of the needle within the appropriate coil. Since this system entailed twenty-six pairs of telegraph wires and twenty-six switches or keys at the transmitting end it never became a commercial proposition.

In 1837 Wheatstone and Cooke invented a telegraph in which only five coils and needles (and five single telegraph wires, plus a common 'return' wire) were needed. They could signal, with this instrument, any one of the twenty commonest letters of the alphabet, by causing two of the needles (see Fig. 195) to point to it. This was done by depressing two appropriate keys at the transmitting end. Suppose, for example, the letter *B* was signalled, the left-hand needle in Fig. 195 would be deflected to the right, and the fourth needle from the left would be deflected to the left. (In order to deflect a needle to right or left reverse currents were needed.) The coils and magnetic needles are behind the board and hence invisible in Fig. 195: the needles shown are merely indicators.

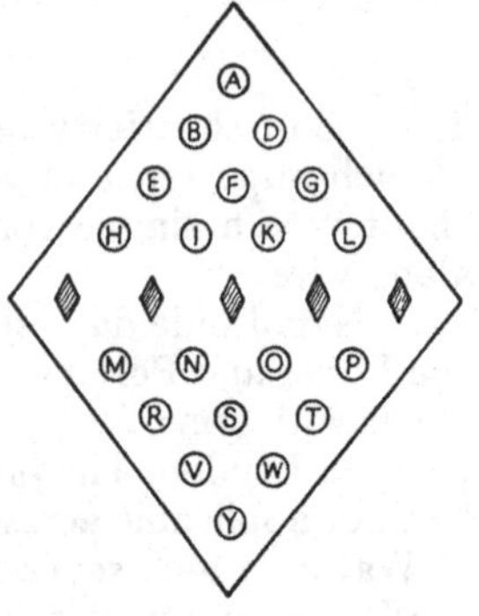

Fig. 195

Wheatstone and Cooke obtained permission to set up their telegraph between Euston and Camden Town, a distance of 1½ miles, on the London and Birmingham Railway, which had just been opened. Wheatstone wrote of the first message sent through the wires on 20 July 1837: "Never did I feel such a tumultuous sensation before, as when, alone in the still room, I heard the needles click."

Despite its success the directors refused to buy the telegraph and the line was removed. However, the Great Western Railway had a line installed between Paddington and Slough, and it was the dramatic use of this telegraph to capture a murderer that first brought electric telegraphy before the public eye. The murderer, dressed as a Quaker, was seen to board a train at Slough. While the train was making its way a description of him was telegraphed to Paddington, with the result that he was arrested on arrival there.

Needle instruments are still used on the railways, but are obsolete elsewhere. They contain only a single needle and the Morse code is employed, a deflection to the left corresponding to a dot, and to the right, a dash.

Samuel Morse.

One of the chief pioneers of electric telegraphy was Samuel F. B. Morse, originally an artist by profession. In October 1832 he was crossing the Atlantic in the packet ship *Sully*, on his way to New York, when he met a certain Dr Jackson who had an electromagnet in his possession. As a result of his conversations with Dr Jackson Morse became keenly interested in the possibilities of signalling by electricity, and during the next three years he perfected his first receiver, an electromagnet which attracted an iron armature when a current passed through its coils. The armature carried a pencil which made a permanent record on a paper tape moving by means of clockwork, at right angles to the direction of movement of the pencil: hence the word telegraph (*tele*, far; *grapho*, I write).

During the early years of his career as an inventor Morse endured all the hardships of poverty: he was almost at the end of his resources when he obtained the authorisation of the U.S. Senate to lay a telegraph line from Washington to Baltimore. The line was laid underground at first, but its insulation was defective, and almost in despair, Morse hastily constructed an aerial line. As happened in England, a dramatic use of the telegraph—the transmission of some election results from Baltimore to Washington—captured the public imagination.

The Morse code.

Morse's most famous achievement was the invention of the Morse code, the representation of the letters of the alphabet by dots and dashes. His collaborator and financial backer Vail is said to have examined a printer's type case to discover the relative frequency with which the different letters are used. The most common letter of all, *e*, is represented in the Morse code by a single dot, the next most common letter, *t*, by a single dash, *a* by a dot and a dash, and so on. (See table on p. 254.)

Letters

A · —	J · — — —	S · · ·
B — · · ·	K — · —	T —
C — · —	L · — · ·	U · · —
D — · ·	M — —	V · · · —
E ·	N — ·	W · — —
F · · — ·	O — — —	X — · · —
G — — ·	P · — — ·	Y — · — —
H · · · ·	Q — — · —	Z — — · ·
I · ·	R · — ·	

Numerals

1 · — — — —	6 — · · · ·
2 · · — — —	7 — — · · ·
3 · · · — —	8 — — — · ·
4 · · · ·	9 — — — — ·
5 · · · · ·	0 — — — — —

The Morse sounder.

Vail hit upon the idea that messages could be read by the ear as well as the eye, and Morse invented a receiver called a *sounder*, which made two clicks at a very short interval to represent a dot and two clicks at a longer interval to represent a dash.

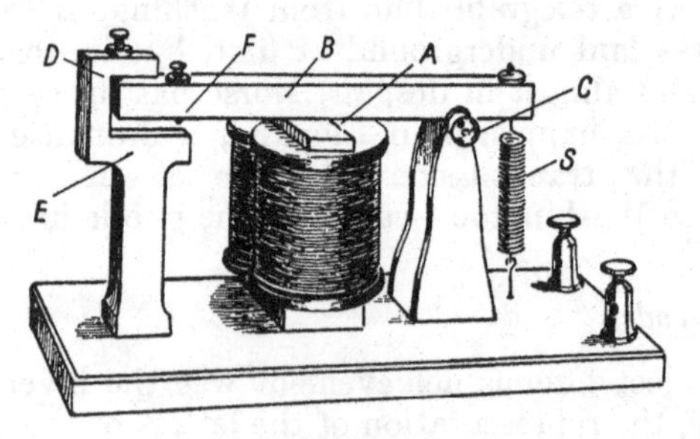

Fig. 196. The Morse sounder.

The instrument (see Fig. 196), consists of an electromagnet comprising two coils (through which the current passes), wound round two iron cores. The electromagnet attracts a soft-iron armature *A* which is carried by, and at right angles to, a bar *B* pivoted at *C* and free to move between the two coils. A spring *S* keeps the bar up against the stop *D* when no current is flowing. When the current does flow the stop *F* hits the "elbow" *E* before the armature reaches the core. This prevents the armature from sticking to the core.

The Morse sounder is still used in certain parts of the world. It was, until only a few years ago, employed by the British Post Office for sending telegrams, but has now been superseded by automatic receiving apparatus. The buzzer, an instrument similar to a trembler bell, is still used by army signallers.

The Morse key.

The transmitter for sending messages by hand in the Morse code is a simple tapping key (see Fig. 197). It must be well balanced and move easily. Normally the lever is held against the "back stop" *B* by a spring. When the knob is depressed contact is made at the "front stop" *A* and a current passes along the telegraph wire or line.

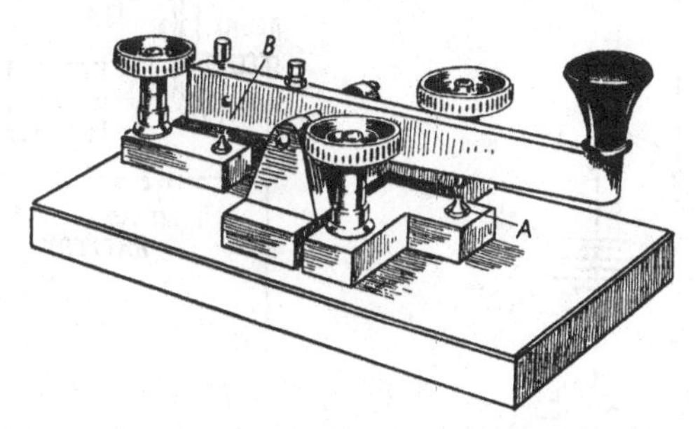

Fig. 197. Morse tapping key.

Simple telegraph circuit.

Fig. 198 represents a simple telegraph circuit to work in either direction. Note that the line consists of a single wire only: the current completes the circuit through the earth. Steinheil, of Germany, discovered by accident that one wire only is necessary. He was using two wires when one of them broke, and he found that the signals continued to come through.

If the key at either station in Fig. 198 is depressed a current flows through the line and through the sounder at the other station.

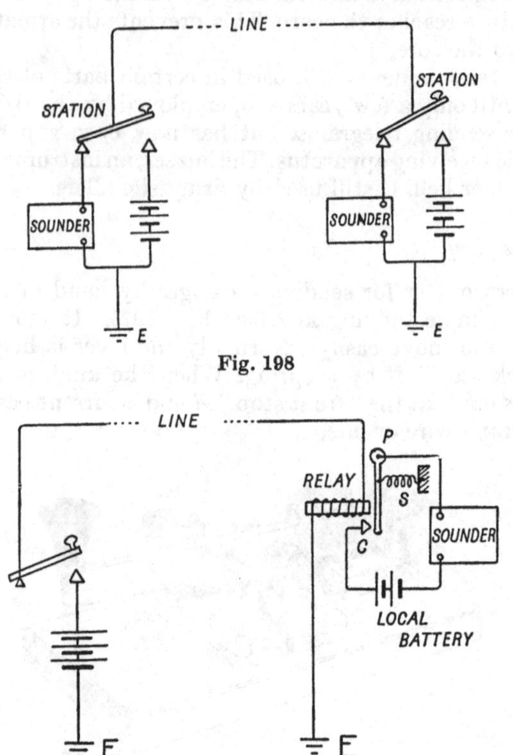

Fig. 198

Fig. 199

The use of the relay.

When a signal has to be sent through the resistance of hundreds of miles of wire, the current may be too weak to work the sounder or other apparatus at the receiving end. In such cases a relay is used, whereby the weak incoming current is made merely to complete a local circuit in which a comparatively large current, supplied by a local battery, then flows. A call, say from New York to San Francisco, or from London to Cape Town, requires a series of relays situated at intervals across the continent or on islands in mid-ocean, and the current is thereby continually renewed.

Fig. 199 shows the use of a relay. The relay consists of an electromagnet, through the coil of which the line current flows, causing the armature, pivoted at P and normally held back by the spring S, to be attracted and make contact with C. The local battery can then send a comparatively large current through the sounder.

Duplex and multiplex telegraphy.

The most costly part of a telegraphic system is the line conductor, just as the greatest capital outlay of a railway is the track. In order to run a railway as economically as possible, several trains must be run on the same track simultaneously, and the speed of the trains must be high. In a similar way the efficiency of a telegraph system can be increased

(1) by sending several messages along each line simultaneously,
(2) by increasing the speed of transmission.

Between the years 1850 and 1870 several methods of Duplex telegraphy, enabling two messages, one in each direction, to be sent along a single line simultaneously, were perfected.

Fig. 200 is a simplified diagram of the differential duplex transmission. The essential feature of this system is a relay with two exactly similar windings, one over the other. If two equal currents are passed through the coils in opposite directions, so that one current circulates round the core in a clockwise direction and the other in an anti-clockwise direction, it is clear that the currents will nullify each other's magnetising effects and the relay will not operate. The relays are shown in Fig. 200 in end-on positions to show clearly the directions of the currents in the windings. R_1 and R_2 are resistances equal in value to the resistance of the line and one coil of the relay.

When the tapping key at station I is depressed a current flows up to A and divides into two equal parts, since the resistances of the two possible routes to earth are equal. One-half of the current flows through the outer coil of relay I in a clockwise direction, through the line and through the outer coil of relay II to earth. The outer half of the current flows through the inner coil of relay I in an anti-clockwise direction and through R_1 (which has a resistance equal to that of the line and the outer coil of relay II) to earth. Thus relay II is energised and attracts its armature (not shown), whereas relay I is not.

It is clear that while this is happening a current could be sent from station II which would operate, independently, relay I, and hence two messages can be sent along the line in opposite

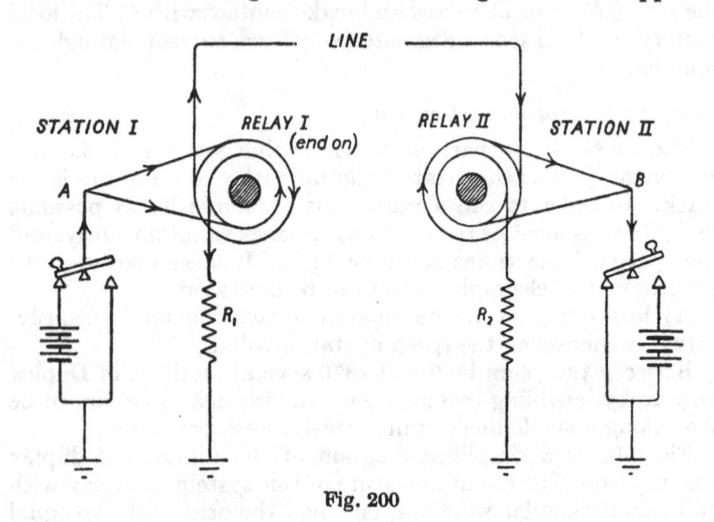

Fig. 200

directions simultaneously. At any particular instant the two opposite currents in the line may cancel each other, so that no current actually flows, but nevertheless the relays will be operated by the half currents flowing round their inner coils.

More elaborate multiplex systems have also been invented which enable as many as ten messages to be sent simultaneously through a single line.

Automatic telegraphic apparatus.

We have stated that the second method of increasing the efficiency of a telegraphic system is to increase the speed at which messages are sent along the lines.

So far we have described two distinct methods, which are still in use, by means of which two operators can send and receive messages along a telegraph line. Both use the Morse code, and the first may be termed directional and the other durational:

(1) Needle left and right by reversals of the current (used in railway signal cabins and also in submarine telegraphy).

(2) Short and long attractions of the armature of an electro-magnet—the sounder and the buzzer.

Now if an operator is transmitting at one end and another operator is receiving by ear or eye at the other, the maximum rate of transmission is about thirty words per minute.

The rate of transmission can be increased to as many as several hundred words per minute by the use of machines instead of operators for transmitting and receiving. The first automatic transmitter and receiver was invented by Sir Charles Wheatstone in 1867 and is still used very much in its original form.

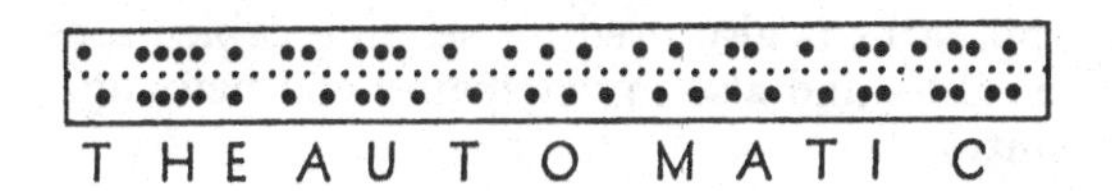

Fig. 201

The Creed high-speed automatic printing telegraph.

When a telegram is handed in to a post office the message is no longer sent by an operator using a tapping key but by means of a Creed automatic transmitter. The operator taps out the message on a machine with a keyboard like a typewriter which perforates a paper tape similar to that in Fig. 201. The paper tape is then run through a sending machine capable of sending 200 words per minute, with the result that several typewriter operators serve one line. As the tape is passed through the transmitting machine two contacts touch through each perforation and currents are sent through the line. The currents operate a receiver at the other end which reproduces a perforated tape identical with that at the transmitter. This tape then passes through a machine called a teleprinter which prints the message in Roman characters on another paper tape. The operator tears off the printed tape and gums it on to a telegram form.

Sometimes telegrams are received from a country post office without gummed strips. This is because the message has been telephoned from the nearest town telegraph office and then written down on a form at the country office.

The perforations on the paper ribbon are really a form of Morse code, as can readily be seen by examining Fig. 201. The continuous row of small holes in the centre are the feed holes for feeding through the machine. Two big holes immediately above and below a feed hole represent a dot: a big hole above one feed hole and another below the next feed hole represents a dash. When there is no large hole above or below a feed hole this represents a blank. A single blank is left between letters and a double blank between words.

Creed apparatus is also used for sending cablegrams through submarine cables and also in newspaper and news-agency work in this country.

The laying of the first Atlantic cable.

The story of the laying of the first submarine (i.e. underwater) cable across the bed of the Atlantic, to link up telegraphically the continents of Europe and America, is one of dogged persistence in the face of repeated disaster and widespread ridicule. At the outset many people thought that it would be impossible to signal through a wire over 2000 miles long. Others maintained that it would be impossible to lay a cable at the great depth of the Atlantic.

In 1851, a cable had been laid between England and France and in 1856 the Atlantic Telegraph Company was formed. A cable was made weighing about 1 ton per nautical mile, and since there was no ship afloat capable of carrying the complete cable, the British and American Governments were induced to lend two strong warships, the *Agamemnon* and the *Niagara*, to lay it.

The *Agamemnon* took aboard half the cable at London, and the *Niagara* the other half at Liverpool. They both sailed to

Valentia in the south-west of Ireland, from which the laying was to begin. The idea was that the *Niagara* should lay the first half accompanied by the *Agamemnon*, that a splice should then be made in mid-Atlantic, and that the *Agamemnon* should complete the laying. After considerable festive ceremonies at Valentia, the laying began, but when five miles had been paid out the cable broke. A fresh start was made and 235 miles were laid, but the cable broke again. The flags of the ships were put at half mast and they returned to Ireland.

Reproduced from Post Office Green Paper No. 7 by permission of the Controller of H.M. Stationery Office

Fig. 202. The frigate *Agamemnon* laying one of the first cables.

In the summer of the next year, after considerable work had been done to improve the paying-out gear, the two ships set out for the middle of the Atlantic. This time a splice was to be effected first, and then the two ships were to lay the cable in opposite directions. But there arose a terrible storm in which the *Agamemnon* nearly capsized. The cable in her holds began to shift and it was feared that her sides might be knocked out. Forty-five of the crew were injured. However, when the storm subsided the *Agamemnon* was still afloat, and the tangle in her holds was straightened out. A splice was effected and the two ships started laying. The cable broke three times, and since on the third occasion the ships were 300 miles apart, they both returned to harbour.

The directors of the company met in London. The chairman recommended that the enterprise should be abandoned and that the company should cut its losses. It was decided, however, to make one further attempt.

Reproduced from Post Office Green Paper No. 7 by permission of the Controller of H.M. Stationery Office

Fig. 203. The *Great Eastern* in a gale in the Atlantic, September 1861. (From the original water-colour by one of the ship's company.)

The *Agamemnon* and the *Niagara* started again from mid-Atlantic and at last, in August 1858, a complete cable lay between Ireland and Newfoundland.

Great were the rejoicings on both sides of the Atlantic. Queen Victoria and the President of the United States exchanged the first messages on August 16. But by September 1, the day chosen for general celebration, the cable stopped working. Too great a potential difference, about 2000 volts, had been used to drive the current, and the cable had burnt out. (Nowadays a potential difference of about 60 volts is usual.)

By this time the English and American peoples were fully determined that there should be telegraphic communication between them. In 1866, at its second attempt, the *Great Eastern*, the largest and most famous ship of her day, successfully laid another cable.

By courtesy of Messrs Siemens Bros. and Co. Ltd.

Fig. 204. The Australian-Tasmanian cable being conveyed over pulleys from a tank at the works where it was made into a similar tank in the cable ship *Faraday* in September 1935. The tank holds 40 miles of cable.

Modern submarine cables.

There are now some twenty cables on the bed of the North Atlantic, and the total length of the cables in the oceans of the world is 400,000 miles (see map inside cover).

It will be seen from the map how useful to telegraph companies are the islands in mid-ocean. A message, for example, from London to Capetown passes first by land line from London to Porthcurnow in Cornwall, thence to the Azores, Cape Verde Islands, Ascension Island and Capetown. At each of the intermediate stations the current operates a relay and a fresh current is supplied by local batteries for the next span.

The cable between India and England was laid in 1870. The first Pacific cable, between Vancouver and Australia and New Zealand, via the Fiji Islands and Norfolk Island, was completed in 1902. The great depths, sometimes as much as 6 miles, and the great distances, made laying exceedingly difficult.

The British post office employs two special cable ships, the *Monarch* and the *Alert*, for grappling and splicing broken cables, which is often a long, difficult, and perhaps dangerous task. In Fig. 205 a broken cable has been raised by the grapnel near the ship's side (top right), and a man has been lowered in "the bosun's chair" to bind and secure it with chains. This is a particularly perilous operation, especially in rough weather, since the cable may break.

Cable ships are kept fully employed: in 1929, for instance, there was an eruption on that comparatively flat part of the bed of the Atlantic known as Telegraph Plateau, where all the cables are laid; all the cables were broken. Cable vessels may lay two or three new cables a year and, in that time, make 100 repairs.

From P. O. Green Paper No. 7

Fig. 205.
The "bosun's chair".

Fig. 206 shows the construction of a submarine cable. The copper conductor is insulated with gutta-percha (a gum similar to india-rubber), round which are wound several layers of jute tape and compound (a mixture of pitch, tar and powdered silica), a layer of steel armouring and one of hempen rope. These are necessary to give it strength, to keep out the water, and to protect it from the depredations of creatures in the sea. The largest submarine cable in the world was laid between this country and Belgium in 1932. It weighs 33 tons per nautical mile and contains 120 conductors. Such a cable costs £2000 per nautical

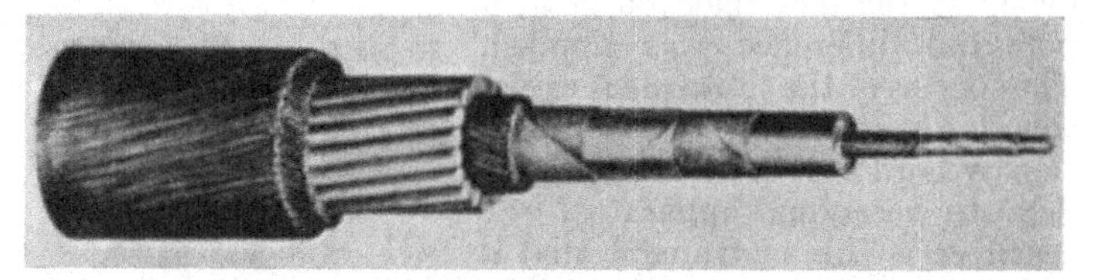

By courtesy of Messrs Siemens Bros. and Co. Ltd.

Fig. 206. A submarine telegraph cable. The conductor is of copper which is "loaded" with a thin wrapping of a nickel-iron alloy, thereby enabling messages to be sent more rapidly through the cable. Surrounding this are various layers of insulation, mainly gutta-percha, steel armouring, and finally an outer layer of hempen rope.

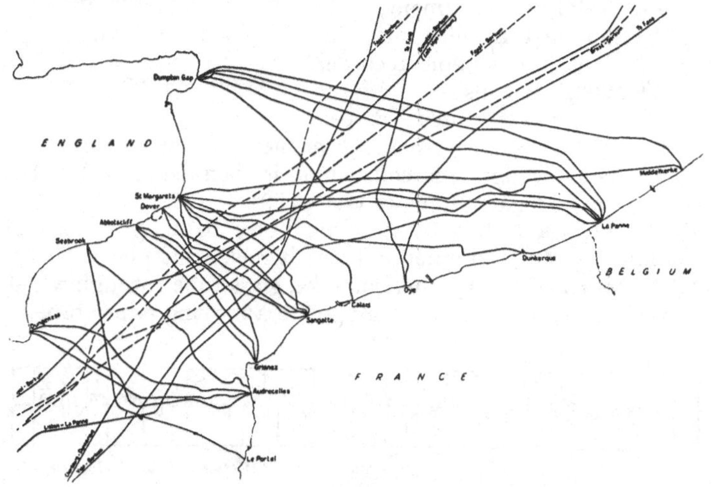

Reproduced from Post Office Green Paper No. 7 by permission of the Controller of H.M. Stationery Office

Fig. 207. Cable chart of the Straits of Dover. Grappling for a broken cable here is a difficult operation and a wrong cable is often raised.

mile, and hence it is desirable that it should be laid on the shortest practicable route: a smooth sea bed with no sharp rocks is essential.

The siphon recorder.

We have mentioned that the first Atlantic cable was burnt out owing to the fact that too great a potential difference was applied. The success of the subsequent cables was due very largely to the invention by Lord Kelvin of an extremely delicate receiving apparatus. So sensitive is this instrument that it will detect at the further end of the cable, 2000 miles away, a current generated by a small rod of zinc dipping into dilute sulphuric acid contained in a silver thimble.

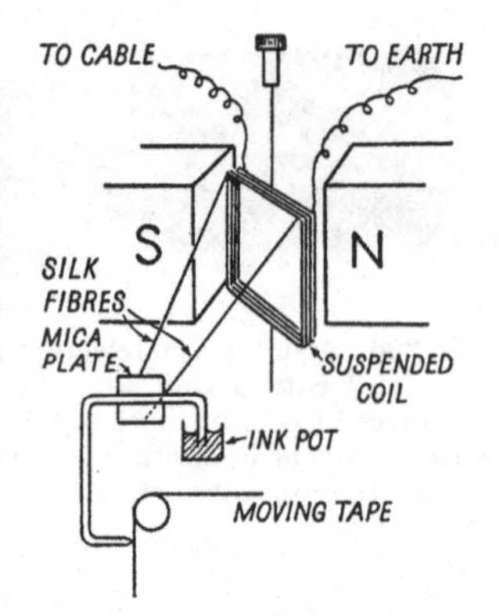

Fig. 208. Siphon Recorder

Lord Kelvin's instrument, only slightly modified, is still in use and is known as the siphon recorder (see Fig. 208). It consists of a light coil of fine wire suspended between the powerful poles of a permanent magnet. The current from the cable passes through this coil, which is then deflected like the moving coil of a galvanometer (see p. 89).

The coil is attached by fine silk fibres to a light mica or aluminium plate, and when it turns it rocks this plate. To the plate is attached a very fine, hair-like, glass tube through which ink is siphoned over from an inkpot to a moving paper tape.

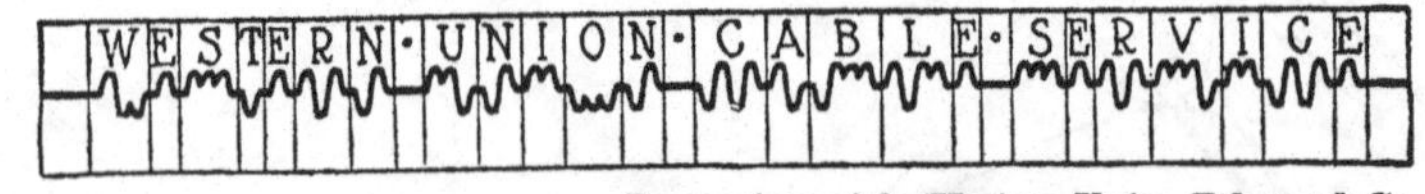

By courtesy of the Western Union Telegraph Co.

Fig. 209. A message, as recorded by a siphon recorder, after being transmitted through a submarine telegraph cable. The Morse code is employed: upward kinks represent dots, and downward kinks dashes.

Dots are represented by a deflection of the coil in one direction and dashes by a deflection in the other. The trace made on the paper tape is similar to that shown in Fig. 209.

By courtesy of the Western Union Telegraph Co.

Fig. 210. An operator punching a paper tape, on an instrument with a keyboard similar to that of a typewriter, for transmitting messages automatically through a submarine telegraph cable.

By courtesy of the Western Union Telegraph Co.

Fig. 211. An operator reading a siphon recorder tape and typing the message as the tape moves across his line of sight.

The speed of signalling through the first Atlantic cable was only 1·8 words per minute and the messages of greeting between Queen Victoria and President Buchanan took thirty hours to transmit.

The difficulty is that the currents charge up the cable and it takes some time to discharge before the current can be sent in the reverse direction.* Signals, if sent too rapidly, tend to become jumbled together.

However, by "loading" the cable with an iron-nickel alloy, called permalloy (see Fig. 206), it is now possible to transmit at the rate of 1500 words per minute.

Telephony

Telephony means the transmission of speech and sounds (Greek *tele*, far; *phone*, sound). We owe the invention of the telephone to Alexander Graham Bell.

Bell's father and grandfather had been professors of elocution. Bell himself opened a school for deaf mutes in Boston, and took a scientific interest in speech. He conceived the idea of transmitting sounds by electricity with the use of a vibrating steel reed, acting as the armature of an electromagnet, rather on the lines of an electric bell (see Fig. 212). One day in June 1875, Bell's assistant, Watson, plucked a reed of an instrument in one room which was connected to a similar instrument in another room. Bell noticed that the sound of the plucking was reproduced in the second instrument. He rushed in to see what Watson had been doing, and "for a long time that day little was done but plucking reeds and observing the effect". The effect, as we shall see, was due to electromagnetic induction. The principle of the telephone had been discovered.

Bell's telephone.

When a person speaks he causes sound waves to be propagated in the air, and it is the varying pressure of these waves falling on the ear-drum which gives the sensation of sound. Bell's idea was to cause these sound waves to make a reed or "diaphragm", as we now call it, vibrate, convert the vibrations into varying electric currents (see Fig. 213), pass the currents along wires and

* The cable with its insulated core surrounded by water acts as a condenser (see p. 347).

By courtesy of the American Telephone and Telegraph Co.

Fig. 212. Bell's vibrating reed, the precursor of the telephone. When the steel reed was plucked it caused variations in the magnetic field due to the electromagnet, and hence gave rise to induced currents in the coil. These currents were passed through the coil of a similar instrument, and caused its reed to vibrate like the first.

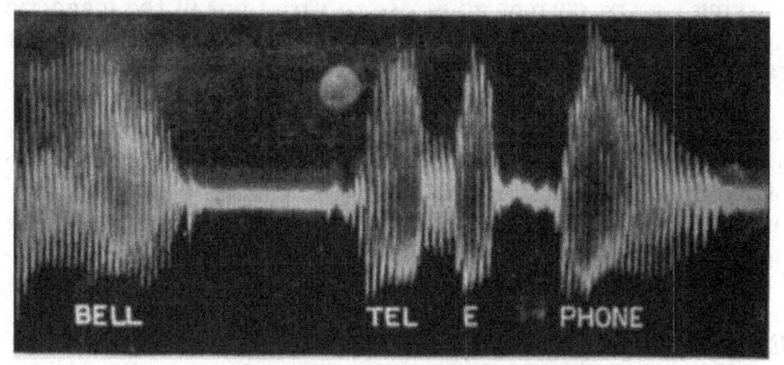

By courtesy of the American Telephone and Telegraph Co.

Fig. 213. Oscillograph record showing the fluctuations in current when the words "Bell Telephone" are spoken into a telephone transmitter. The record is made on a film moving at about 1 foot per sec.

cause them to make another diaphragm vibrate at the receiving end and hence to reproduce the sound waves.

The principle of his transmitter is shown in Fig. 214. A thin steel plate *D*, the diaphragm, is made to vibrate by the sound waves in the air. The diaphragm is held very close to two pieces of soft iron, *AA*, round which pass coils of wire and which are fixed to the poles of a permanent horse-shoe magnet *M*. When the diaphragm vibrates it causes slight changes in the magnetic field in the neighbourhood of the coil, and hence gives rise to varying induced currents.

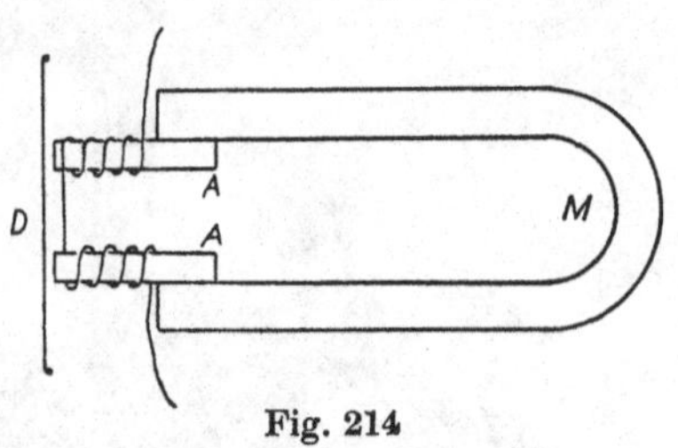

Fig. 214

The changes in the magnetic field can be illustrated (in a much magnified form) by moving a soft-iron keeper in the neighbourhood of the poles of a horse-shoe magnet under a card on which iron filings are sprinkled. The changes in the position of the lines of force will easily be visible. The pole pieces of Bell's telephone, round which the coil is wrapped, are made of soft iron because the magnetisation of soft iron changes much more easily than does that of steel. The magnet must be made of steel in order to retain its magnetism.

Bell's receiver is identical with his transmitter. But here the varying electric current gives rise to variations in the magnetic field: the corresponding changes of attraction of the diaphragm cause it to vibrate, and hence generate sound waves.

It will be noticed that no battery is required. Bell's arrange ment may be set up with two pairs of wireless headphones which are simply Bell receivers. Connect up the two wires of one pair of headphones with the two wires of the other and persuade a friend to speak into one headphone. On putting on the other headphones you will be able to hear what he is saying.

In 1876 Bell's apparatus was publicly shown in an exhibition in Philadelphia, but attracted no attention until Don Pedro, the Emperor of Brazil, visited the exhibition. Don Pedro was a friend of Bell's, and asked to be shown the instrument. On picking up the receiver and listening to Bell's assistant Watson speaking in another room, the Emperor exclaimed excitedly:

"It speaks, it speaks." Lord Kelvin, the great English physicist, was present at the time and remarked that it was the most wonderful thing he had seen in America.

Afterwards Bell and Watson went on a lecture tour and Watson used to shout into the transmitter from another room

By courtesy of the American Telephone and Telegraph Co.

Fig. 215. A model of Bell's first telephone—compare with Fig. 214. The steel reed or armature is attached to a stretched diaphragm. When used as a transmitter the diaphragm vibrates under the action of sound waves: when used as a receiver its vibrations give rise to sound waves.

to the excited audience, "How do you do? Good evening. What do you think of the telephone?" So (to indulge in a platitude) do the marvels of yesterday become the commonplaces of to-day.

Several companies, including the famous Bell Telephone Company, were floated and Bell lived to see the widespread adoption of his invention. To-day there are over 30,000,000 telephones in use throughout the world.

The microphone.

Bell's transmitter is really a very weak dynamo, which converts the energy of sound waves into electricity, and his receiver is a form of electric motor which converts the electricity back again to mechanical motion.

The electric current generated by the transmitter is, however, minute and it is incapable of transmitting speech for any considerable distance. All telephone transmitters to-day are made on a different principle, that of the microphone, invented by Hughes. Modern receivers, however, are similar to Bell's original instrument.

The microphone, as its name implies, was originally invented for making audible very minute sounds.

It makes use of a peculiar property of carbon. If two pieces of carbon are pressed more tightly together a much better contact is made and the resistance decreases considerably. A simple experiment to show this may be performed with two sticks of arc carbon placed in series with an accumulator and ammeter. The reading of the ammeter increases as the carbon sticks are pressed gradually together and decreases again when the pressure is removed.

The modern telephone transmitter.

The modern telephone transmitter, as we have said, is a form of microphone. Fig. 216 shows the principle of its action. A thin flexible diaphragm, *D*, is in contact with small granules of carbon, *C*, which are packed between it and a carbon block, *B*. When sound waves fall on the diaphragm it vibrates and compresses the carbon granules, thus altering their resistance. A current flowing between *B* and *D* is thus made to fluctuate and reproduce the sounds in a receiver.

It will be seen that this transmitter differs from Bell's instrument in that it does not generate its own current. The current is supplied by a battery and is therefore much stronger.

Figs. 217 and 218 show the exterior and interior of a modern combined receiver and transmitter.

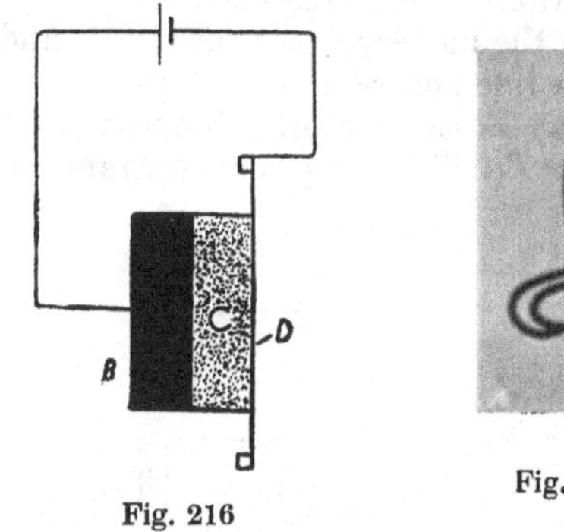

Fig. 216

By courtesy of the G.P.O.

Fig. 217. A hand micro-telephone (see also Fig. 218).

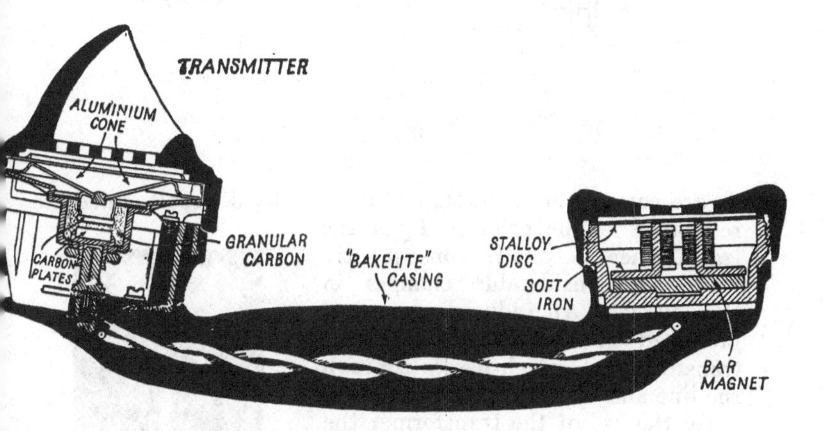

Fig. 218. A hand micro-telephone—a combined transmitter and receiver (see also Fig. 217.)

The Edison telephone circuit.

The simple circuit consisting of transmitter, battery, line and receiver has a serious defect if the resistance of the line is at all considerable; for the change in the resistance of the microphone transmitter due to the compression of the carbon granules will only produce an appreciable change in the current if the resistance of the rest of the circuit is small. Edison eliminated this defect by using a transformer (or induction coil as it is called in telephony) with its primary in series with the battery and transmitter and its secondary in series with the line and receiver.

Fig. 219 represents a two-way Edison circuit. Suppose some one speaks into the transmitter T_1. The changes of resistance of

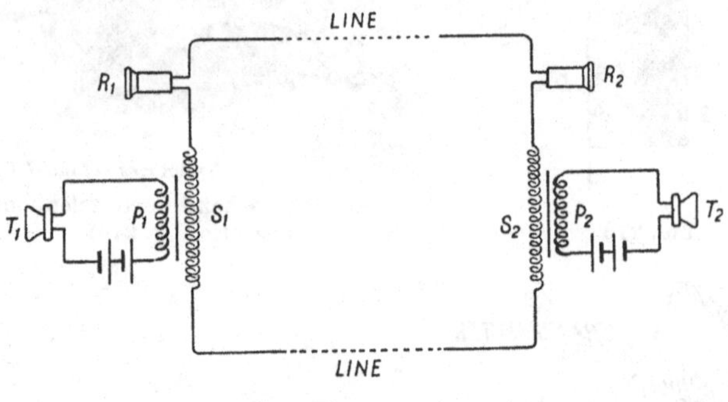

Fig. 219

T_1 are comparable with that of its own local circuit, since the resistance of the primary P_1 of the transformer is small: consequently there are considerable changes in the current. An induced current is produced in the secondary of the transformer S_1 which passes through the line and operates the receiver R_2.

By the use of the transformer the resistance of the line does not affect the current in T_1; furthermore, the induced current in the secondary is produced only by the variations of current in the primary and is unaffected by the steady flow.

American T. and T. Co.

Fig. 220. A telephone cable containing 1200 pairs of wires fanned out at the top: the correct orderly arrangement is seen below.

It will be noticed that two wires are used for the lines and no use is made of the earth, as in telegraphy. It was found, when earth connections were tried, that owing to induction effects, people were able to overhear the conversation on other lines and the receiver emitted objectionable and untraceable noises.

The exchange.

For one subscriber to "get through" to another he must first call the exchange. When he lifts his telephone receiver from the hook the hook rises and completes a circuit causing a lamp to light or a tiny door, called a drop, to open at the exchange.

Reproduced from P.O. Green Paper No. 3

Fig. 221. Operator at work on a continental circuit. Note the jacks plugged into the sockets in the panel.

This attracts the attention of the operator, who inserts one of a pair of plugs into a hole (or jack) connected to the subscriber's line. Having ascertained the required number the operator then pushes the second plug into the jack corresponding to the required number and thereby connects the two lines. When the subscribers ring off the operator takes out the two plugs.

It is common nowadays for a single large battery at the

exchange to replace all local batteries. When the receiver is taken off its hook a current from this central battery begins to flow in the transmitter and the primary of the transformer.

Reproduced from Post Office Green Paper No. 3 by permission of the Controller of H.M. Stationery Office

Fig. 222. Operators at work in the London International Switchroom. When a London subscriber wishes to make a continental call he asks for Trunks: particulars of his call are taken and sent to one of these operators. The subscriber is advised of the probable delay, if any. As soon as a continental circuit is available, the operator rings the subscriber and connects his telephone to the continental exchange by plugging in two jacks at the end of a short wire (see also Fig. 221).

The automatic system.

Automatic have now superseded hand-operated exchanges in most districts. The following is a much simplified description of their mode of operation. The reader should refer constantly to Fig. 223.

Let us suppose that a certain telephone subscriber wishes to call No. 2368. He lifts his receiver: the receiver hook completes a circuit and connects him to his line switch at the exchange. The line switch consists of a row of twenty-five pairs of contacts in a semicircle over which a rotary double-ended "wiper" can move. The wiper rotates (under the action of an electromagnet and ratchet) until it reaches a pair of contacts connected to a disengaged selector, when it stops. The selectors, as we shall see, are the instruments by means of which the subscriber is connected to the number he requires. Since it would be very expensive to provide each subscriber with a set of selectors, a sufficient number are shared by all the subscribers, and each subscriber has a line switch which finds a disengaged selector. Should the selectors all be engaged the subscriber receives the "busy tone" signal.

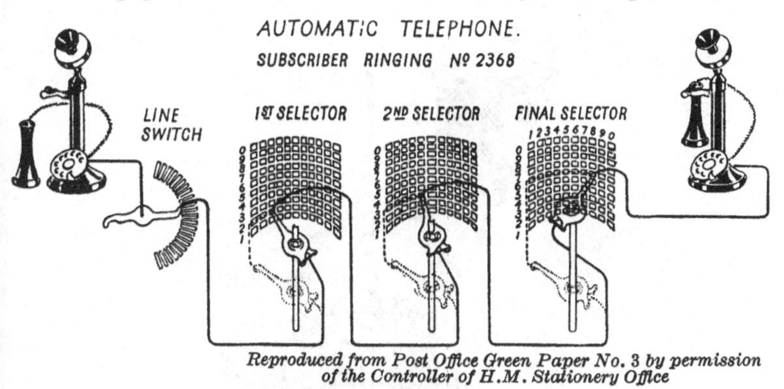

Reproduced from Post Office Green Paper No. 3 by permission of the Controller of H.M. Stationery Office

Fig. 223

When his line switch has found a disengaged selector, the subscriber receives "dialling tone". He proceeds to dial 2. As the disc of the calling dial moves back it breaks the circuit twice and two impulses are sent which raise the wiper of the 1st selector (see Fig. 223) to level two of the ten horizontal rows of contacts. The wiper automatically rotates until it finds a disengaged link to a 2nd selector on that level. The subscriber then dials 3. The wiper of the 2nd selector is raised to the third level of contacts, and automatically rotates until it finds a disengaged link to a final selector on that level.

The subscriber now dials 6 and the final selector wiper is raised to level six. The subscriber finally dials 8 and the wiper moves round to the number 8 contact of level 6 which is connected to the required subscriber. The latter, if his line is disengaged, receives a ringing current and the former a "ringing tone" signal.

Fig. 224 shows how the wiper of a selector is raised to the requisite level of contacts. At each impulse, set up by the interruptions of the current at the subscriber's calling dial, the armature A is attracted by the electromagnet EM and the pawl B raises the vertical shaft carrying the wiper through one notch. The arm C holds the shaft in place until the subscriber rings off, when it is attracted by an electromagnet (not shown) and the shaft falls. Note that this diagram does not show how the wiper is rotated.

In the final selector there are vertical notches round a drum carried by the vertical shaft and a pawl operated by the armature of a suitably placed electromagnet rotates the shaft and wiper horizontally to the necessary position.

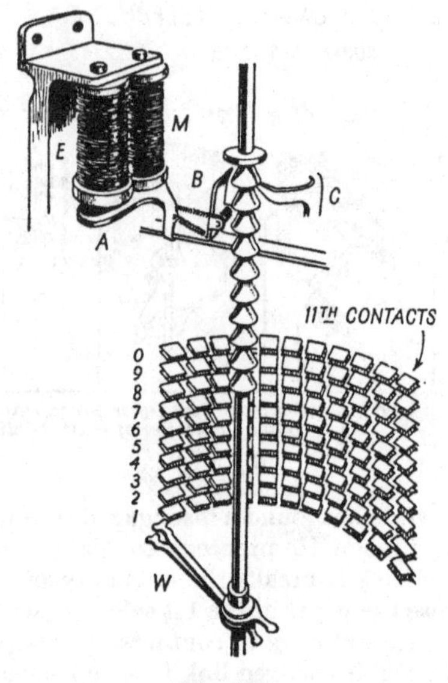

Fig. 224

Summary

Manually operated electric *telegraphs* utilise the Morse code: they may be (1) durational (sounder or buzzer), or (2) directional (single needle telegraph or siphon recorder). When the signals are sent for long distances the current is renewed at intervals by relays.

Several signals may be sent simultaneously along a single line —a system known as multiplex telegraphy.

Most telegraph transmitters and receivers are now automatic: in this way messages are sent very much more rapidly than by hand. A tape perforated with dots representing the Morse code is used.

The modern *telephone* transmitter consists of a carbon microphone which utilises the marked decrease in resistance of carbon granules when compressed: the receiver works on the principle of electromagnetic induction. The varying current through the microphone transmitter passes through the primary of a transformer and induces a current in the secondary: the induced current passes along the line and operates the receiver.

QUESTIONS

1. Describe the construction and operation of a telegraph sounder, and draw a diagram of connections for a simple telegraph circuit in which it is used. (O. and C.)

2. Describe, with the aid of diagrams, the action of (*a*) the carbon microphone telephone transmitter, (*b*) the telephone receiver.

3. Describe, with a clear circuit diagram, the use of a relay in electric telegraphy.

4. Draw a diagram of a telephone circuit including an induction coil. Explain the advantages obtained by the use of the induction coil.

Chapter XV

THE MEASUREMENT OF MAGNETISM

In this chapter we shall consider how magnetism is measured.

The simplest method of estimating the strength of a magnet and that used by Faraday and the early experimenters, is to see how large a weight of iron it can lift.

Now the iron is lifted because of the attraction of the induced pole in the iron by the pole of the magnet. And any method of finding the strength of a magnet entails the measurement of the force between two poles. Hence a preliminary investigation as to how the force between two magnetic poles depends upon their distance apart is necessary.

The inverse square law.

The man who carried out the first successful experiments on this problem was Charles Augustin Coulomb (1736–1806), an engineer in the French army. He suspended a long ball-ended magnet by a fine wire and brought near to one of its poles the pole of another long ball-ended magnet. In Fig. 225 the repulsion between the two N. poles causes the wire to twist. Coulomb noted the twist, which is a measure of the force of repulsion, and also determined the distance between the two poles, *d*. The poles were assumed to be at the centre of the balls. We have no space to discuss the details of Coulomb's apparatus and procedure, but it should be said that he showed great experimental skill.

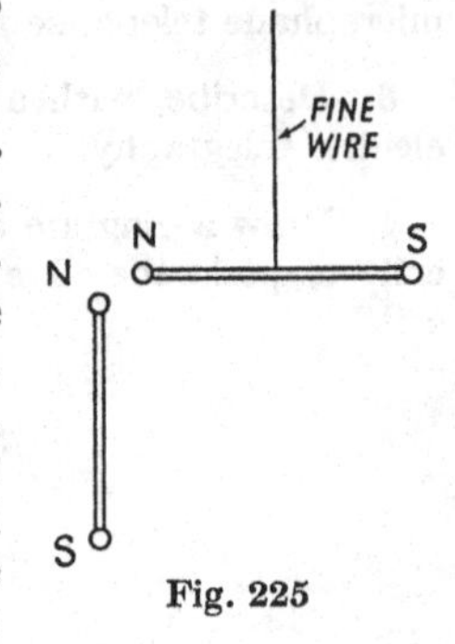

Fig. 225

He found that on doubling the distance, d, the force, F, was reduced, not to $\frac{1}{2}F$ but to $\frac{1}{4}F$: on trebling the distance d the force was reduced to $\frac{1}{9}F$. He thus discovered what is known as the **Inverse Square law. The force between two magnetic**

poles is inversely proportional to the square of the distance between them:

$$\mathbf{F} \propto \frac{1}{\mathbf{d}^2}.$$

It is of interest to note that Newton believed an inverse cube law to be probable, i.e. $F \propto \frac{1}{d^3}$.

The centimetre-gramme-second system of units.

The next step towards a method of making magnetic measurements is to devise a unit of pole strength.

In 1832 Carl Friedrich Gauss, of Göttingen University, proposed a system of units for use throughout the whole of physics which has been universally adopted. This is the c.g.s. system, based on the three fundamental units, the centimetre, gramme and second. On this system, the unit of force is the dyne. 1 *dyne is the force required to give a mass of* 1 *gramme an acceleration of* 1 *cm. per sec. per sec.* It is equal to $\frac{1}{981}$ gram wt.

Using the dyne, we can define a unit of pole strength:

A pole of unit strength placed 1 cm. from a similar pole in air, repels it with a force of 1 dyne.

A pole of strength 2 c.g.s. units behaves like two unit poles and would repel a pole of strength 1 c.g.s. unit, 1 cm. away, with a force of 2×1 dynes: it would repel a pole of strength 3 c.g.s. units 1 cm. away, with a force $2 \times 3 = 6$ dynes. The force is clearly proportional to the product of the pole strength.

We can combine this fact with the inverse square law in the following fundamental equation,

$$\mathbf{F} = \frac{\mathbf{m}_1 \mathbf{m}_2}{\mathbf{d}^2},$$

where F dynes is the force between two poles of strengths m_1 and m_2 c.g.s. units situated d cm. apart in air (see Fig. 226).

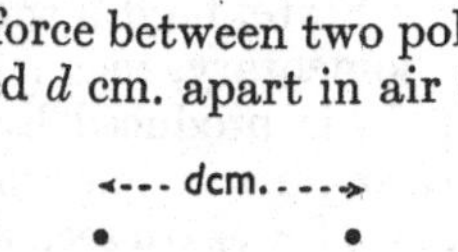

Fig. 226

Example. What force will a pole of strength 50 c.g.s. units exert on a pole of strength 20 c.g.s. units 5 cm. distant in air?

$$F=\frac{50\times 20}{5^2}=40 \text{ dynes.}$$

The strength of a magnetic field.

How can we measure the strength of a magnetic field?

A magnetic field exerts a force on a magnet. If we could produce an isolated N. pole (actually an impossibility), and suspend it from a little balloon, it would tend to set off towards the magnetic N. pole of the earth. This suggests a method of measuring the strength of a magnetic field. **The strength** (or **intensity**) **of a magnetic field is the force in dynes which it exerts on a unit pole placed in it.** The unit of field strength is the oersted, formerly called the gauss. **A field which exerts a force of 1 dyne on a unit pole is said to have a strength of 1 oersted.**

Example. Find the force exerted on a pole of strength 24 units in a field of 0·2 oersted.

A field of 1 oersted on pole of 1 c.g.s. unit exerts force of 1 dyne,

∴ A field of 0·2 oersted on pole of 1 c.g.s. unit exerts force of 0·2 dyne,

∴ A field of 0·2 oersted on pole of 24 c.g.s. units exerts force of $24\times 0{\cdot}2$ dyne $=4{\cdot}8$ dynes.

Intense magnetic fields.

The horizontal component of the earth's magnetic field in England is 0·18 oersted. The field between the poles of the most powerful electromagnet is of the order of 50,000 oersted.

Dr Kapitza, who, a few years ago, worked at Cambridge University but is now detained compulsorily in Russia because of his value to the Soviet State, built a special dynamo capable of producing intense, momentary, magnetic fields of strength 320,000 oersted. The field is produced by short-circuiting the dynamo momentarily through a magnetising coil and obtaining a current of 72,000 amperes for $\frac{1}{100}$th sec., during which the wire in the coil has no time to melt. The magnetic forces are so tremendous that the coil tends to burst, and it must be designed to withstand a force of 100 tons.

When the generator is short-circuited the electric forces jar it and shake the building: "the energy released in a single experiment approaches that of a field gun."

Now the magnetic field inside an atom due to the rotations of the electrons is of the order of several million oersted. Kapitza's idea is to produce a magnetic field strong enough to disrupt, or at least affect, the atom.

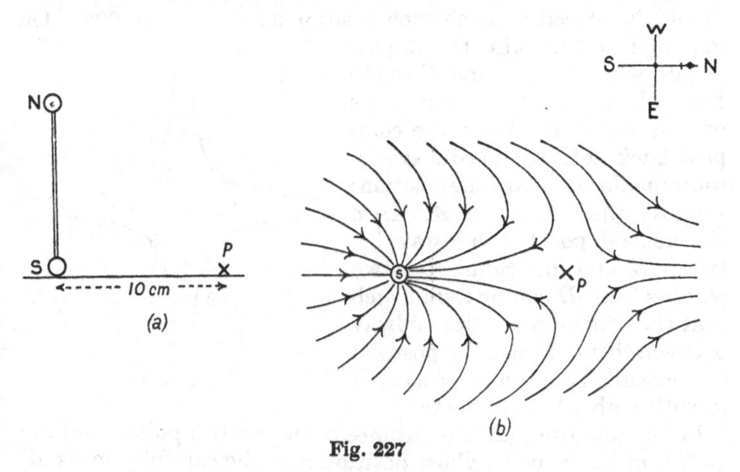

Fig. 227

Determination of pole strength by plotting the lines of force due to a single pole in the earth's field, given that the earth's horizontal component is 0·18 *oersted.*

Set up a long ball-ended magnet vertically with one end, say the S. pole, resting on a sheet of plain white paper on a table (see Fig. 227 (*a*)).

Plot the lines of force due to the combined fields of the earth and the magnet's pole on the table, using a small compass (see p. 10). We can assume, if the magnet is a long one, that the upper N. pole has no effect: it is far away and its field is inclined to the horizontal so that only the resolved part is effective. The field will look like that in Fig. 227 (*b*).

At the point *P*, marked with a ×, the field due to the magnet's S. pole and the earth's field are equal and opposite: they annul each other. The point is called a *neutral point*.

It is necessary in this experiment to find the position of the neutral point as accurately as possible. To do this the lines of force in its neighbourhood should be plotted as symmetrically as possible. Near the neutral point the resultant magnetic field is very weak and it is difficult to plot lines nearer than about 1 cm. to it.

To fix the exact position of the neutral point the compass should be placed in some such position as *A* (see Fig. 228). On moving it to the right the compass needle will swing round through 180°. Mark the position where this occurs, say at *B*. Move the compass back until the needle swings round again and mark the position: suppose this occurs at *A*. Then the neutral point is midway between *A* and *B*. Similarly, two points *C* and *D* can be found such that the neutral point lies midway between them. Hence the position of the neutral point may be located to within about a millimetre.

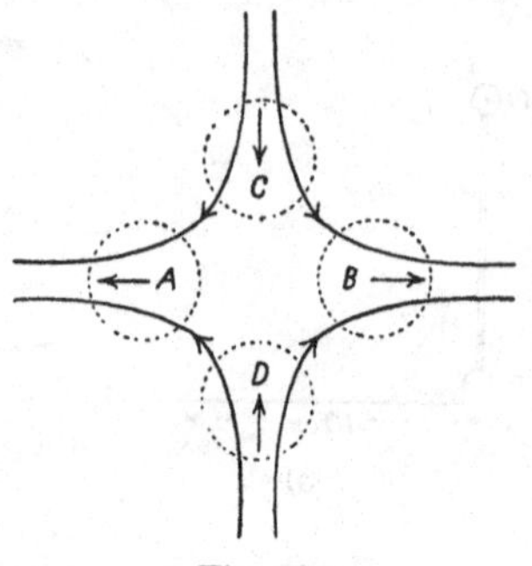

Fig. 228

Let us suppose that the distance of the neutral point from the middle of the S. pole (whose position must be carefully marked) is 10 cm. (see Fig. 227 (*a*)).

Let m c.g.s. units = pole strength of the S. pole.

Strength of field at *P* due to S. pole

$$=\text{Force exerted on a unit pole imagined at } P$$

$$=\frac{m\times 1}{10^2}=\frac{m}{100} \text{ oersted (or dynes per unit pole).}$$

But strength of field at *P* due to the earth $=0{\cdot}18$ oersted (given).

$$\therefore \frac{m}{100}=0{\cdot}18,$$

$$m=18 \text{ c.g.s. units.}$$

Determination of the pole strength of a bar magnet by finding the position of the neutral points when it lies horizontally in the earth's field with its axis N. and S. and its N. pole pointing S.

Lay the bar magnet N. and S., with its N. pole pointing S., on a sheet of paper. Plot the lines of force and locate the positions of the neutral points, which will be in the positions shown in Fig. 229, by the method described on p. 284.

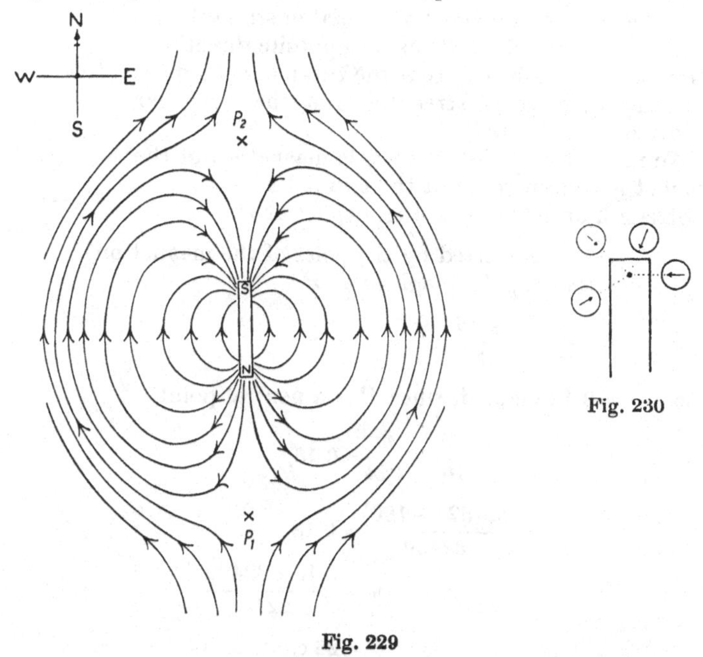

Fig. 230

Fig. 229

Find the positions of the poles of the magnet by placing a compass needle in several positions very near to the magnet (so that the other pole of the magnet and the earth's field will have a negligible influence upon the needle), as in Fig. 230: the point where the directions of the needle intersect may be taken as the position of the pole. This point will, as a rule, lie well within the magnet.

Measure the distances of the neutral points from the nearest poles, P_1N and P_2S, and take the average. Let us suppose that this average distance is 10 cm. (see Fig. 231), and also that the magnetic length of the magnet (the distance between the poles) is 8 cm.

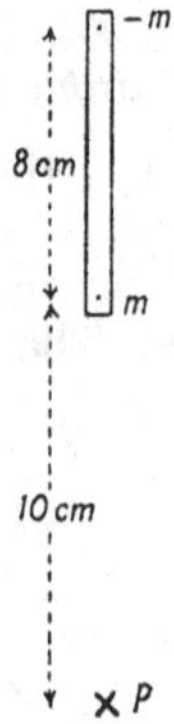

Let m c.g.s. units = pole strength of the N. pole.

Then $-m$ c.g.s. units = pole strength of the S. pole,

since the poles of a magnet are equal in strength and exert forces, or set up fields, in opposite directions. (How would you justify that the two poles of a magnet must be of equal strength, using the molecular theory of magnetism?)

$-m$

8 cm

m

10 cm

× P

Fig. 231

We may assume that the whole magnetism of the magnet is concentrated at the two poles.

Strength of field due to the magnet at P

= Force exerted by the poles of the magnet on a unit pole imagined at P

$$= \frac{m \times 1}{10^2} - \frac{m \times 1}{18^2};$$

also = 0·18 oersted, since P is a neutral point.

$$\therefore \frac{m}{100} - \frac{m}{324} = 0{\cdot}18,$$

$$\frac{m\,(324 - 100)}{32400} = 0{\cdot}18,$$

$$m = \frac{0{\cdot}18 \times 32400}{224}$$

$$= 26 \text{ c.g.s. units.}$$

The moment of a magnet.

The moment of a magnet is the product of its magnetic length and pole strength:

$$\mathbf{M = 2lm,}$$

where M = moment of magnet, m = pole strength of magnet, $2l$ = magnetic length of magnet.

Thus in the case of the magnet on p. 286, whose pole strength is 26 c.g.s. units, and magnetic length 8 cm.,

$$\text{Moment of the magnet} = 8 \times 26$$
$$= 208 \text{ c.g.s. units.}$$

The principle of the deflection magnetometer.

The deflection magnetometer is an instrument for measuring the strength of the magnetic field at any point due to a bar magnet, from which the pole strength and magnetic moment of the magnet may be calculated.

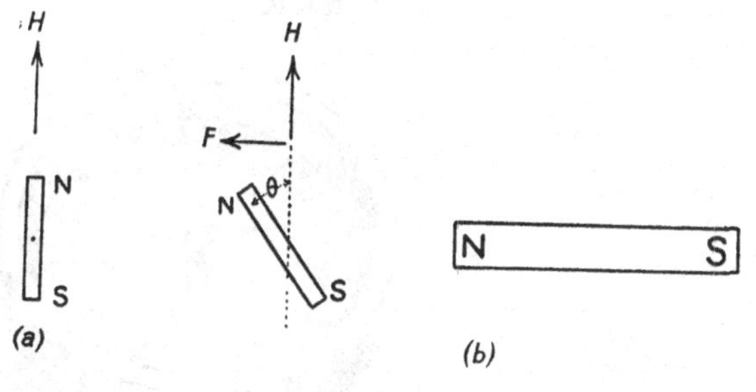

Fig. 232

The principle of the instrument is as follows.

A small pivoted magnet or compass needle sets N. and S. in the earth's field alone (see Fig. 232 (*a*)). If now a bar magnet is brought up to the E. of it, as in Fig. 232 (*b*), it will be deflected, say, through an angle θ. The earth's field, H, is trying to make it set N. and S., while the magnet's field, F, is trying to make it set at right angles, E. and W. The angle θ depends on the relative strengths of H and F. We can show that, so long as the two fields are at right angles,

$$\mathbf{F = H \tan \theta.}$$

Hence knowing H and measuring θ we can find F.

Proof of $F = H \tan \theta$.

Since the strength of a field is measured by the force it exerts on a unit pole, we can combine two fields by the principle of the parallelogram of forces. In the parallelogram *ABCD* (see Fig. 233), if *DA* and *DC* are proportional to *H* and *F*, respectively, then *DB* represents the resultant field in magnitude and direction. The small pivoted magnet will set itself along the resultant field and hence $A\hat{D}B = \theta$.

In the right-angled triangle *ADB*,

$$\tan \theta = \frac{AB}{AD} = \frac{F}{H}.$$

$$\therefore F = H \tan \theta.$$

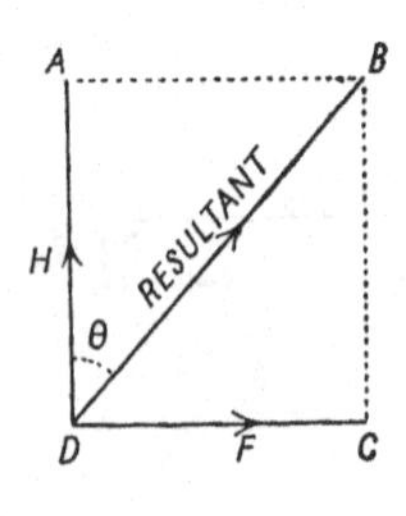

Fig. 233

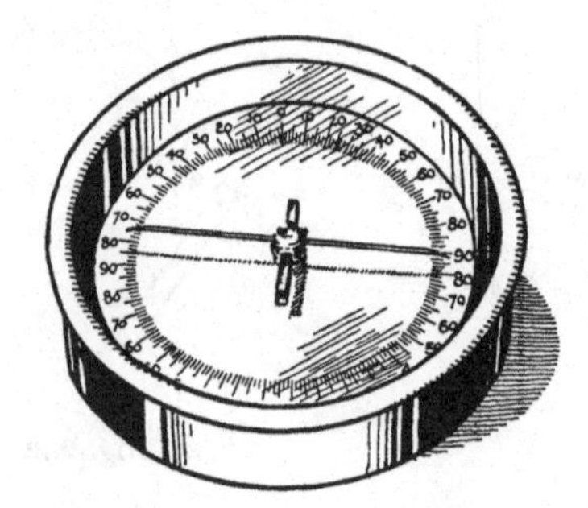

Fig. 234. Deflection magnetometer.

Details of the instrument.

The deflection magnetometer consists of a short pivoted magnet to which is attached a long aluminium pointer moving over a scale marked in degrees (see Fig. 234). Under the aluminium pointer is a mirror and, when taking a reading, the eye should be placed in such a position that the pointer exactly covers its image in the mirror, in order to avoid the error due to parallax.

On either side of the magnetometer are two arms, not shown in the figure, consisting of rulers marked in centimetres to enable one to determine the distance of the poles of a magnet from the centre of the magnetometer.

Use of the deflection magnetometer

(*a*) *to find the strength of the field at a point x cm. from the nearer end of a bar magnet on the axis of the magnet produced;*

(*b*) *hence to calculate the pole strength and magnetic moment of the magnet.*

The magnetometer must first be set with its arms pointing E. and W., since the field of the bar magnet, when laid on one of the arms, must be at right angles to the earth's field. The bar magnet is then laid at the correct distance from the centre of the magnetometer and the deflections of both ends of the pointer are read (in case the pointer is not pivoted at the centre of the circular scale).

The bar magnet is now turned through 180° so that the pole which was nearer to the centre of the magnetometer is now the further away, and two deflections again read, making a total of four.

The magnet is transferred to the other arm of the magnetometer and four more readings again taken. The average of all eight readings is calculated, and in this way errors such as that due to imperfect setting of the magnetometer are eliminated.

It should be added that it is desirable for the deflection to be about 45° (say between 30° and 60°), since, in the neighbourhood of 45°, a small error in the angle causes a minimum error in the tangent ($\tan\theta$).

Let us suppose that the average deflection is 50° and that the strength of the earth's field is 0·18 oersted.

Strength of field at centre of magnetometer due to bar magnet,

$$F = H \tan\theta = 0{\cdot}18 \tan 50° = 0{\cdot}214 \text{ oersted.}$$

To determine the pole strength of the bar magnet, a method exactly similar to that on p. 286 is used. But in this case the strength of the field due to the magnet is 0·214 oersted and not 0·18 oersted.

The poles of the magnet must be located with a compass.

Let us suppose that the distance of the nearer pole of the bar magnet from the centre of the magnetometer is 10 cm., and that the magnetic length of the magnet is 5 cm. (see Fig. 235).

If m = pole strength of the magnet,

Strength of field due to the magnet at the centre of the magnetometer

$$=\frac{m}{10^2}-\frac{m}{15^2}=0{\cdot}214.$$

$$\therefore\ m=38{\cdot}5 \text{ c.g.s. units,}$$

$$M=2lm=5\times 38{\cdot}5$$

$$=193 \text{ c.g.s. units.}$$

To compare the moments of two magnets, the moment of each may be obtained by the method described above. But in this case it is not really essential to know the value of the earth's field. It could be denoted by the letter H, which will cancel out in the final calculation.

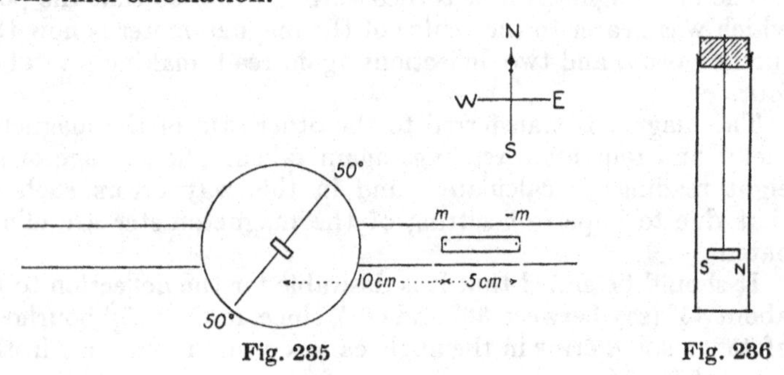

Fig. 235 Fig. 236

The vibration magnetometer.

The vibration magnetometer is used to compare the strengths of magnetic fields. It consists of a small magnet, usually suspended by a thread of unspun silk, in a glass vessel to protect it from draughts (see Fig. 236).

Now a suspended magnet when given a slight displacement or twist will begin to vibrate in a magnetic field, just as a pendulum, when displaced, begins to swing. The pendulum tends to return to a vertical position but overshoots the mark, swings back and again overshoots the mark, and so on. Similarly, a magnet tends to return to a N. and S. position (in the earth's field) when displaced and let go, but overshoots the mark and so continues to vibrate.

A pendulum when taken to the top of a very high mountain swings more slowly than at sea level, since the earth's gravitational field is smaller there. Similarly, the rate of vibration of a magnet depends on the strength of the magnetic field. It is an experimental fact that

$$n^2 \propto F,$$

where F oersted = strength of field, n = number of vibrations per minute, so long as the vibrations of the magnet are small.

Thus if a magnet makes n_1 and n_2 swings per minute in fields of strength F_1 and F_2 oersted, respectively,

$$\frac{F_1}{F_2} = \frac{n_1^2}{n_2^2}.$$

It must be the *same* magnet, since the strength of magnet also affects the frequency.

Example. It is required to find the strength of the horizontal component of the magnetic field at a certain place in a laboratory where there is much iron. It is known that the horizontal component of the earth's field in the open is 0·18 gauss. A vibration magnetometer makes 20 swings per min. in the laboratory and 25 swings per min. in the open.

Let Strength of field in laboratory = F oersted.

Then

$$\frac{F}{0{\cdot}18} = \frac{20^2}{25^2} = \frac{16}{25}.$$

$$\therefore F = 0{\cdot}12 \text{ oersted.}$$

Other factors affecting the rate of vibration of a magnet.

The rate of vibration of a magnet depends not only upon the strength of the magnetic field in which it is vibrating but also upon its magnetic moment, and its mass and dimensions or moment of inertia. A powerfully magnetised magnet vibrates more quickly than a weak one: a large, ponderous magnet vibrates more slowly than a small, light one.

When comparing the strengths of two magnetic fields the same magnet vibrates in the two fields, and hence we can ignore the magnetic moment and moment of inertia of the magnet.

The moments of two magnets of the same moment of inertia may be compared by timing their rate of vibration in the same magnetic field:

$$\frac{M_1}{M_2}=\frac{n_1^2}{n_2^2},$$

where n_1 = number of swings per min. of magnet moment M_1, n_2 = number of swings per min. of magnet moment M_2.

The general expression for the strength of the magnetic field at a point on the axis produced of a bar magnet, known as the end-on position.

We will now obtain, by the method used on pp. 286 and 290, the general expression for the strength of the magnetic field due to a bar magnet, of magnetic length $2l$, and pole strength m, at a point distant d from its centre along its axis produced (see Fig. 237).

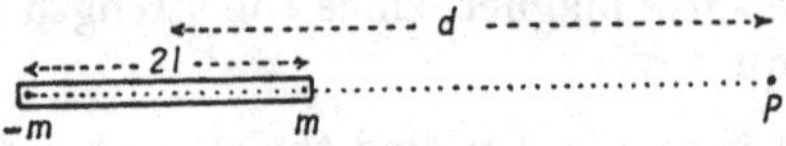

Fig. 237

Strength of field at P due to the magnet

$$= \text{Force exerted by the magnet on a unit pole imagined at } P$$

$$=\frac{m\times 1}{(d-l)^2}-\frac{m\times 1}{(d+l)^2}.$$

$$=\frac{m\,(d^2+2dl+l^2-d^2+2dl-l^2)}{(d-l)^2\,(d+l)^2}$$

$$=\frac{4mdl}{(d^2-l^2)^2}=\frac{2\mathrm{Md}}{(\mathrm{d}^2-\mathrm{l}^2)^2}$$

where M = moment of magnet $=2lm$.

If l is small compared with d we can ignore l^2 compared with d^2 and write d^2 instead of d^2-l^2.

Thus if

$$d=20 \text{ cm.},$$

$$l=2 \text{ cm.},$$

$$d^2-l^2=400-4=396,$$

$$d^2=400.$$

Hence for a short bar magnet at a considerable distance along its axis

$$\text{Strength of magnetic field}=\frac{2Md}{(d^2)^2}=\frac{2\mathrm{M}}{\mathrm{d}^3}.$$

The strength of the magnetic field at a point on the perpendicular bisector of a bar magnet, known as the broadside position.

We have just obtained an expression for the strength of the field due to a bar magnet in the simplest position—the end-on position. We shall now consider the strength of the field in the next simplest case—the broadside position. In Fig. 238, let the point P be distant d from the centre of the magnet.

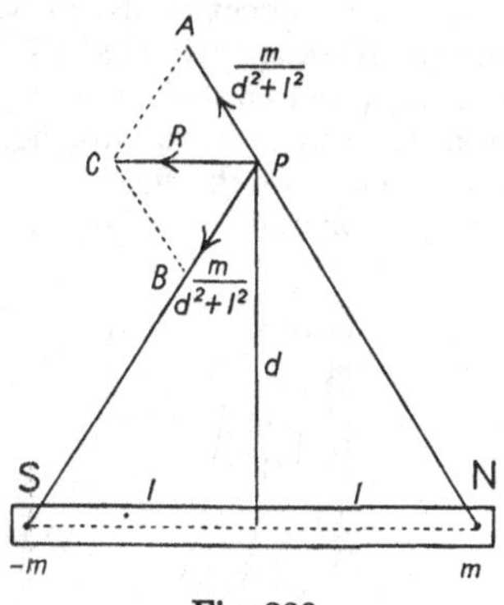

Fig. 238

To find the strength of the field due to the magnet at P we must find the resultant force exerted by the two poles of the magnet on a unit pole imagined at P. These two forces act along NP and PS, respectively (the lines joining the unit pole to the two poles), and they are equal to

$$\frac{m\times 1}{PN^2} \text{ and } \frac{m\times 1}{PS^2}.$$

By the theorem of Pythagoras:

$$PN^2=PS^2=d^2+l^2.$$

Hence the forces are $$\frac{m}{d^2+l^2}.$$

Since the two forces are inclined we must find their resultant by applying the principle of the parallelogram of forces. In the parallelogram $PACB$, if PA and PB are proportional to the forces $\frac{m}{d^2+l^2}$, then PC represents their resultant, R, in magnitude and direction. The triangles ACP and PSN are equiangular (why?), and therefore similar.

$$\therefore \frac{CP}{AP}=\frac{SN}{PN},$$

$$\text{i.e. } \frac{R}{\frac{m}{d^2+l^2}}=\frac{2l}{\sqrt{d^2+l^2}}.$$

$$\therefore R=\frac{2lm}{(d^2+l^2)^{\frac{3}{2}}}=\frac{M}{(d^2+l^2)^{\frac{3}{2}}}.$$

$$\therefore \text{Strength of field at } P=\frac{\mathbf{M}}{(\mathbf{d}^2+\mathbf{l}^2)^{\frac{3}{2}}},$$

where M = moment of the magnet $=2lm$.

If l is small compared with d,

$$\text{Strength of field due to a short bar magnet} = \frac{M}{(d^2)^{\frac{3}{2}}} = \frac{\mathbf{M}}{\mathbf{d}^3}.$$

If a bar magnet is placed N. and S. with its N. pole pointing N. in the earth's magnetic field the neutral points will be in the broadside position (see Fig. 239). If d represents the distance of the neutral points from the centre of the magnet, by equating the above expression for the strength of the field due to the magnet to 0·18 oersted the pole strength or moment of the magnet may be determined.

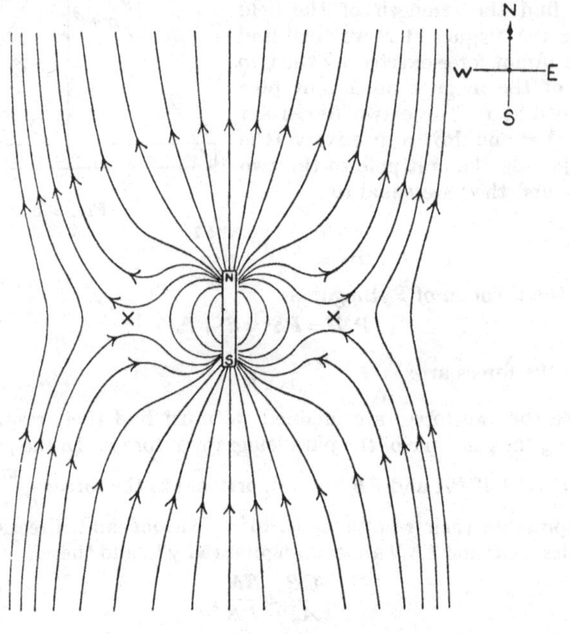

Fig. 239

The strength of the field at any point due to a bar magnet.

In order to find the strength of the field due to a bar magnet at *any* point P, the parallelogram construction is again necessary.

We will take a definite case. In Fig. 240, suppose $PN = 5$ cm., $PS = 8$ cm., $NS = 10$ cm., and the pole strength of the magnet is

30 c.g.s. units. To find the strength of the field due to the magnet at P we must find the resultant force exerted by the poles of the magnet on a unit pole imagined at P:

$$\text{Force along } NP = \frac{30 \times 1}{5^2} = \frac{6}{5} \text{ dynes,}$$

$$\text{,,} \qquad PS = \frac{30 \times 1}{8^2} = \frac{15}{32} \text{ dynes.}$$

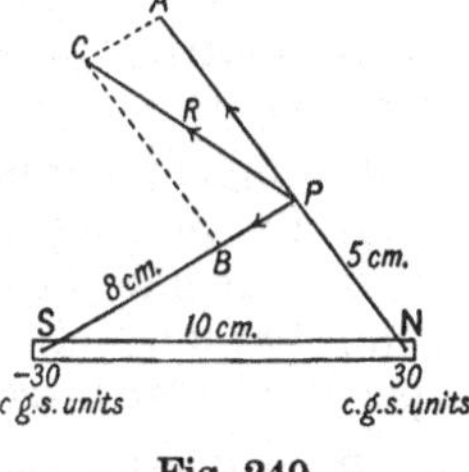

Fig. 240

Thus the parallelogram $PACB$ must be constructed so that PA and PB are proportional to $\frac{6}{5}$ and $\frac{15}{32}$ dynes respectively. Then the length PC will be proportional to the resultant force, and hence the resultant field due to the magnet at P.

The reader should obtain the answer by drawing a figure to scale. (Answer 1·35 oersted.)

SUMMARY

Inverse square law. The force between two magnetic poles is inversely proportional to the square of the distance between them:

$$F \propto \frac{1}{d^2}.$$

Unit of pole strength. A pole of unit strength placed 1 cm. from a similar pole in air repels it with a force of 1 dyne:

$$F = \frac{m_1 m_2}{d^2}.$$

The strength (or intensity) of a magnetic field is the force in dynes which it exerts on a unit pole placed in it. A field which exerts a force of 1 dyne on a unit pole is said to have a strength of 1 oersted.

The moment of a magnet is the product of its magnetic length and pole strength:

$$M = 2lm.$$

Experimental determination of pole strength and magnetic moment (given the value of H).

1. By plotting lines of force and finding the position of the neutral points.

(*a*) Single pole; one neutral point.

(*b*) Bar magnet with S. pole pointing N.: two neutral points on axis of magnet produced (end-on position).

(*c*) Bar magnet with N. pole pointing N.: two neutral points on perpendicular bisector of magnet (broadside position).

2. By deflection magnetometer.

Find strength of field at point along axis (produced) of magnet: $F = H \tan \theta$.

Then calculate as in 1 (*b*) above.

Comparison of magnetic fields.

1. Deflection magnetometer: fields must be at right angles:

$$F = H \tan \theta.$$

2. Vibration magnetometer:

$$\frac{F_1}{F_2} = \frac{n_1^2}{n_2^2}$$ (see p. 291).

Strength of field due to a short bar magnet:

(*a*) End-on position, $F = \frac{2M}{d^3}$.

(*b*) Broadside position, $F = \frac{M}{d^3}$.

QUESTIONS

1. Two poles of strengths 100 and 120 c.g.s. units are placed 2 cm. apart in air. Find the force of repulsion between them (*a*) in dynes, (*b*) in gm. wt. (1 gm. wt. = 980 dynes.)

2. A pole of strength 80 c.g.s. units attracts another pole 4 cm. from it with a force of 250 dynes. Find the strength of the other pole.

3. Two poles of strengths 50 and 64 c.g.s. units repel each other with a force of 8 dynes. What is their distance apart?

4. Two similar poles placed 3 cm. apart in air repel one another with a force of 400 dynes. Find the strengths of the poles.

5. Find the strength of the field due to a pole of strength 40 c.g.s. units at a distance of 10 cm. from the pole.

6. At what distance from a pole of strength 25 c.g.s. units has the field a strength of 0·16 oersted?

7. What is the strength of the magnetic field midway between two N. poles of strengths 40 and 60 c.g.s. units, when they are placed 6 cm. apart?

8. A long ball-end magnet is placed vertically with one end resting on a table. Draw the lines of force on the table, showing carefully any neutral points, and indicating the direction of the earth's magnetic field.

If a neutral point is situated 12 cm. from the pole, find the pole strength, given $H=0{\cdot}16$ oersted.

9. What is meant by a line of magnetic force?

A bar magnet of length 20 cm. is placed in the magnetic meridian with its S. pole pointing N. Draw a diagram to show the distribution of the lines of force. If a neutral point is found 10 cm. from the end of the magnet, deduce the pole strength of the magnet. ($H=0{\cdot}18$ oersted.) (C.)

10. Find the position of the neutral point between two N. poles of strengths 20 and 30 c.g.s. units respectively, 10 cm. apart. Calculate the position of the neutral point if the weaker pole had been a S. pole.

11. How would you magnetise a needle so that there is a N. pole at each end? If placed in the earth's field, where would you expect the neutral points to be? Such a magnet, 10 cm. long, has a neutral point 10 cm. from one end along the axis. Find the strengths of the poles, assuming those at the ends to be equal. ($H=0{\cdot}18$ oersted.)

12. Two bar magnets, each of unit pole strength, one 6 cm. and the other 8 cm. long, are placed in a line with their N.-seeking poles facing one another, their centres being 12 cm. apart.

Calculate the force they exert on a unit magnetic pole placed on a line between the magnets at the point 3 cm. from the N.-seeking pole of the shorter magnet. (O.)

13. The magnetic moment of a bar magnet is 500 units: the strength of the horizontal component of the earth's magnetic field is 0·18 oersted. Explain these statements. If the magnet is 50 cm. long and is placed vertically with its N.-seeking pole just passing through a hole in the table, in what direction will a compass needle point when it is placed on the table 10 cm. E. of the hole? (N.)

14. If a magnet has a pole strength of 25 c.g.s. units and a length between the poles of 16 cm., find the deflection produced in a magnetometer, in the "end-on" position, at a distance of 16 cm. from the centre of the magnet. ($H = 0·18$ oersted.) (O. and C.)

15. When finding the moment of a magnet by means of a deflection magnetometer it is found that, on reversing the magnet, a larger deflection of the magnetometer needle is obtained. Will it be right to conclude that one pole of the magnet is stronger than the other? Give reasons and suggest any other possible causes. (L.)

16. Describe the magnetometer.

A bar magnet has a magnetic moment of 200 units and its magnetic length is 10 cm. It is placed with its centre 15 cm. from the centre of the magnetometer in such a position that the force due to the magnet is at right angles to that due to the earth. Calculate the intensity of the horizontal component of the earth's magnetic force if the angle of deflection is 45°.

Show the arrangement in a diagram. (L.)

17. A suspended magnetic needle makes one oscillation in 4 sec. at a place where the intensity of magnetic field is 0·20 oersted. Find the time of oscillation at a place where the intensity of field is 0·30 oersted. (C.)

18. A vibration magnetometer makes 20 oscillations in 90 sec. in a magnetic field of strength 0·18 oersted, and 30 oscillations in 80 sec. in another field. Find the strength of the second field. (O. and C.)

19. How would you determine the values of H, the horizontal component of the earth's magnetic field, at various points in the laboratory, if its value at some given point were known?

A laboratory bench is supplied with gas by a vertical iron pipe coming up from the floor. State and explain how this will affect the indications of a compass needle placed on the bench near the gas supply. (O. and C.)

20. Describe in detail how you would make a comparison of the magnetic moments of two short bar magnets.

Two short bar magnets, *A* and *B*, produce equal deflections of a magnetometer needle when their distances from it are respectively 15 and 20 cm. What is the ratio of their magnetic moments? (C.)

21. A magnet of length 16 cm. is placed with its N. pole pointing N. Two neutral points are found each at a distance of 12 cm. from the centre of the magnet. Find the pole strength of the magnet. ($H = 0{\cdot}18$ oersted.) (O. and C.)

22. Explain the terms pole strength, intensity of a magnetic field.

A short magnet of moment 200 units is placed in the magnetic meridian (*a*) with its S. pole pointing N., (*b*) with its N. pole pointing N. Find the distances of the neutral points from the centre of the magnet in the two cases, assuming that H is 0·2 oersted. Indicate on a diagram the positions of the neutral points in each case. (C.)

23. Define moment of a magnet.

At a place where the vertical component of the earth's field is 0·42 (dyne per unit pole) it is found that a mass of 20 milligrams placed 8 cm. from the axle of a dip needle will keep it horizontal. Find the magnetic moment of the dip needle. (Take the weight of a milligram as approximately 1 dyne.) (L.)

24. Define pole strength and magnetic moment of a magnet.

Two magnets are taken, each of pole strength 2. One, of length $2\sqrt{3}$ cm., is placed in a N.S. position, and the other, of length 2 cm., is placed in an E.W. position with its nearer pole 1 cm. from the mid-point of the first magnet. Calculate the resultant forces exerted by the poles of the first magnet on each of the poles of the second magnet, and indicate their directions. (O.)

25. Define unit magnetic pole and magnetic field strength.

Two magnets, *AB*, *BC*, form two sides of a square *ABCD*, and have their S. poles at *B*. The length of the magnet is 10 cm. and the magnetic moment is 200 in each case. What is the field strength at the corner *D*? Draw a diagram, marking the component forces and the resultant force. (L.)

26. The poles of a bar magnet are 10 cm. apart and each has a strength of 72 units. Determine the magnitude and direction of the resultant magnetic force due to the magnet alone at a point which is

6 cm. from one pole, 8 cm. from the other, and lies in the same horizontal plane as the magnetic axis of the magnet. Show the result on a carefully drawn scale diagram.

Explain how you would test experimentally the accuracy of the direction obtained. (L.)

27. The poles of a magnet N. and S. are 15 cm. apart. Give a construction for finding theoretically the direction of the force due to these poles at a point P in the same horizontal plane when $PN = 9$ cm., $PS = 12$ cm. If this direction happens to be at right angles to the earth's horizontal field ($H = 0{\cdot}18$ oersted), show how by the help of a small compass needle, on a graduated card, you might find the magnitude of the force. (L.)

28. Explain what is meant by the following statements:

(*a*) A certain magnetic pole is of unit strength.

(*b*) The intensity of a magnetic field at a certain point is 0·2 oersted.

A magnet of pole strength 50 c.g.s. units and magnetic length 10 cm., suspended so that it is able to vibrate in a horizontal field of intensity 0·2 oersted, is held so that its axis is at an angle of 30° to the field. Show on a diagram the forces acting on the poles of the magnet, and state what would happen if the hold on the magnet were released.

What experimental evidence have we that the poles of a magnet are equal in strength? (N.)

29. A bar magnet of pole strength 30 units and length 10 cm. is placed horizontally in the earth's field. Find the couple tending to turn it in a horizontal plane if the angle between its axis and the magnetic meridian is 30°. ($H = 0{\cdot}18$ oersted.) (O. and C.)

30. A small magnetic needle oscillating in the earth's field makes 50 oscillations in 200 sec. It is placed at a point due S. of a bar magnet which is lying in the magnetic meridian with its N. pole pointing S., and it is then found that the needle oscillates very slowly indeed. Explain how this could occur. If the bar magnet is taken up, and replaced in the same position but with its two poles reversed, what will now be the approximate time of one vibration of the needle? (N.)

31. A small magnet vibrating horizontally in the earth's magnetic field makes 25 vibrations per minute. When a bar magnet is placed near so that its horizontal magnetic field is in the same direction as

that due to the earth, the number of vibrations per minute is 30. Calculate the intensity of the horizontal magnetic force due to the magnet, given that $H = 0{\cdot}18$ oersted. How many vibrations per minute would the small magnet make if the bar magnet were turned end to end so that its field opposed that of the earth? (L.)

32. A short bar magnet is suspended by a loop of thread so as to vibrate horizontally in the earth's field. What are the factors which determine its time of oscillation?

If an exactly similar magnet is placed on top of the first with the corresponding poles in contact, how will the time of oscillation be affected? (O. and C.)

33. The moment of a magnet is sometimes defined as the product of the pole strength and the distance between the poles, sometimes as the moment of the couple required to keep the magnet at right angles to a field of unit intensity. Show that the two definitions are equivalent.

A bar magnet, suspended by a thread so as to be in a horizontal plane, is observed to make 12 small oscillations per minute. If the magnet is replaced by another of the same linear dimensions and mass, the number of small oscillations per minute is found to be 15. Compare the magnetic moments of the two magnets. (O. and C.)

Chapter XVI

THE EARTH'S MAGNETISM

Declination.

A magnetic compass does not, in most places, point true geographic N. In England, at the present time, it points about 12° W. of N. (see Fig. 241). This angle—between geographic N. and magnetic N.—is called the *angle of declination*. The angle of declination may be most neatly defined as *the angle between the magnetic and geographic meridians* (see p. 15). It must be allowed for when finding one's way with a compass or when

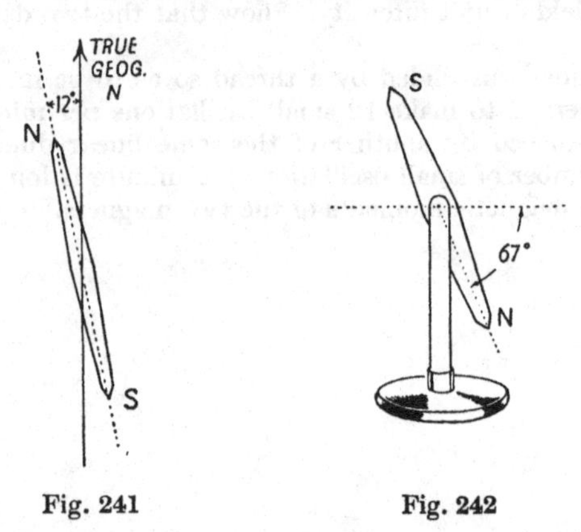

Fig. 241 Fig. 242

setting a map. Sailors call it by another name, the variation of the compass, and its importance in navigation is obvious.

The Chinese knew of it as early as the second century A.D.: "The soothsayers rub a needle with the magnet stone, so that it may mark the south; however it declines a little to the east."

Dip.

An ordinary compass needle can swing only in a horizontal plane. If a magnet is pivoted so that it can swing in a vertical plane (see Fig. 242), which runs N. and S., i.e. the magnetic meridian, the magnet does not remain horizontal. In England at the present time the N. pole dips at an angle of about 67° with the horizontal. This angle is known as the *angle of dip*.

To see a magnetic needle dipping like this makes one instantly suspicious that it is not properly balanced about its axis, and that one end is weighed down because it is heavier than the other. But even perfectly balanced magnetic needles dip. We can prove that the dip is not due to lack of balance by magnetising the needle in the opposite direction. What was the S. pole becomes the N. pole and this end now dips.

The earth's field.

We saw at the end of Chapter I that Gilbert explained the setting of a magnetic compass needle by assuming that the earth is a gigantic magnet.

Since a magnetic needle, when suspended freely so that it can turn in both a horizontal and a vertical plane, sets itself (in England) in a vertical plane 12° W. of N., with its N. pole dipping 67° below the horizontal, it follows that the lines of magnetic force due to the earth must run in this direction.

Now the magnetic needles in most of our instruments are pivoted so that they can turn only in a horizontal plane. Hence we are usually most interested in the effective horizontal field at a particular place, called the *horizontal component* of the earth's magnetic field.

A field is measured in terms of a force per unit pole (see p. 282), and a force can be resolved, by the principle of the parallelogram of forces, into two components. Similarly, the complete earth's field, called *the total intensity*, may be resolved into *a horizontal component* and *a vertical component* (see Fig. 243).

The horizontal component in England at the present time is about 0·18 oersted. Knowing the angle of dip we can determine the vertical component and total intensity either by drawing a right-angled triangle to scale like Fig. 243 or by trigonometry.

$$\text{Vertical component} = 0{\cdot}18 \tan 67^\circ$$

$$= 0{\cdot}42 \text{ oersted.}$$

$$\text{Total intensity} = 0{\cdot}18 \sec 67^\circ$$

$$= 0{\cdot}46 \text{ oersted.}$$

The angles of declination and dip, the total intensity, and the horizontal and vertical components, at a particular place, are called the *magnetic elements*. The angle of declination and any two of the others suffice to denote the magnitude and direction of the earth's magnetic field at that place.

Determination of declination.

The determination of the angle of declination at a place involves finding accurately two directions, geographic N. and magnetic N.

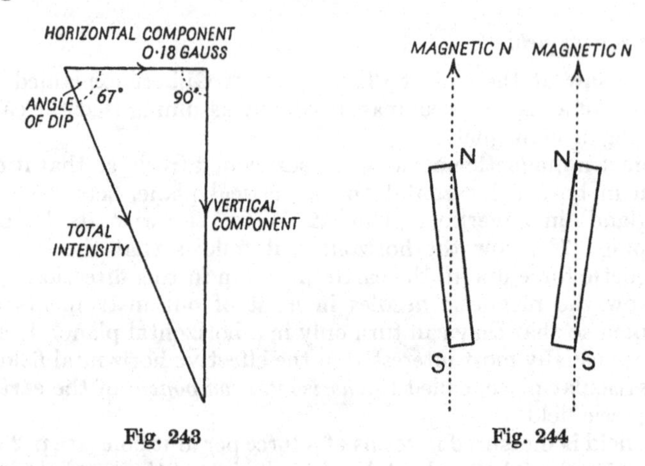

Fig. 243 Fig. 244

The former can be found accurately only by an astronomical method—observation of the sun or stars. It can be found roughly from the fact that the shadow of a vertical stick cast by the sun at midday is due N.

Magnetic N. is found by suspending a bar magnet or needle freely on a vertical axis. There is a serious source of possible error here, however, since the magnetic axis of the magnet (i.e. the line joining its poles) may not coincide with the geometric axis. It is the direction of the geometric axis which we determine, whereas it is the direction of the magnetic axis that we require.

The error may be eliminated by removing the magnet from its suspension, turning it over and replacing it with what was the bottom face now uppermost.

Fig. 244 shows the positions in which a bar magnet, whose magnetic axis is a diagonal (instead of a line parallel to the sides), sets itself before and after being turned over. It is clear that magnetic N. may be obtained by bisecting the angle between the sides (or geometric axes) of the magnet in the two positions.

Then the angle of declination is the angle between the directions of magnetic N. and true N.

Determination of dip.

The angle of dip is measured by means of an instrument called a dip circle (see Fig. 245). This consists of a magnetic needle freely pivoted on a horizontal axis at the centre of a vertical, circular

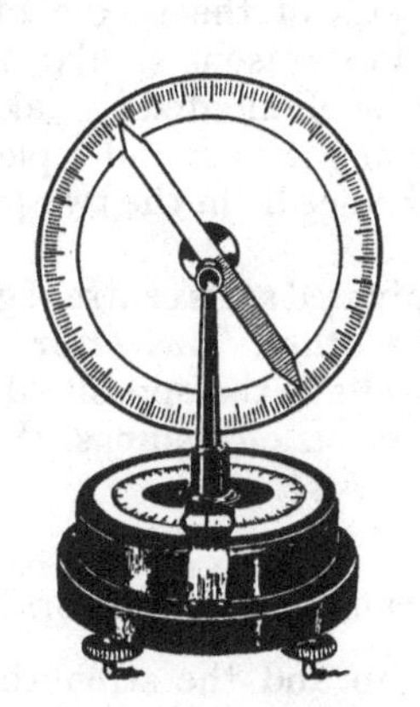

Fig. 245. The dip circle.

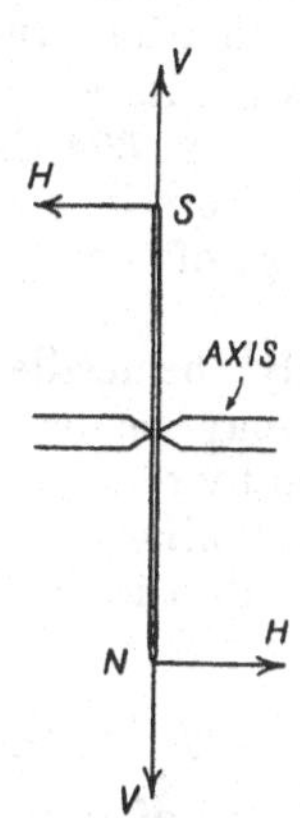

Fig. 246

scale marked in degrees. The whole instrument can be revolved about a vertical axis and the angle turned through measured on a horizontal scale.

The instrument is first levelled with the aid of a spirit level, usually incorporated in the instrument. This is of vital importance, since it is just as necessary to obtain a true horizontal as the true direction of the earth's lines of force.

The needle has now to be set so that it swings in the magnetic meridian, i.e. the plane of the vertical dip circle must be magnetic N. and S. The dip circle is turned about its vertical axis until the needle is absolutely vertical. When this occurs, the plane of the dip circle must be at right angles to the magnetic meridian. For the vertical component of the earth's field V pulls the needle into a vertical position, but the horizontal component H is trying to pull the poles of the needle N. and S., whereas they are free to move only E. and W. (see Fig. 246); hence the horizontal component has no effect and the needle stays vertical. The whole dip circle is now turned through exactly 90° (measured on the horizontal circular scale), and it is then in the magnetic meridian.

The angle the needle makes with the horizontal gives the angle of dip.

There are several sources of error: we shall mention two and explain how to eliminate them. Firstly, as in the determination of the declination, the magnetic axis of the needle may not coincide with its geometric axis. For reasons similar to those described on p. 304 the error may be eliminated by taking out the needle from its suspension, turning it over and replacing it. The average of the readings with the needle in the two positions is taken.

Secondly, the needle may not be pivoted at its centre of gravity: one end may be heavier than the other. This error may be eliminated by remagnetising the needle in the opposite direction and again taking the average of two sets of readings. Why does this eliminate the error?

The charting of the magnetic elements throughout the world.

The angles of declination and dip and the strength of the earth's magnetic field vary widely over the surface of the world. Columbus is said to have discovered the gradual change in declination as he sailed across the Atlantic, and we are told that Magellan's sailors were greatly alarmed when their magnetic compass began to point E. instead of W. of N. as they passed from the Atlantic to the Pacific.

In 1698 Edmund Halley was sent on a government expedition across the Atlantic, where he made important observations of the variation of declination. But it was mainly between the years 1840 and 1850 that a great international effort was made to collect magnetic data over a wide area. Stations were set up all over Europe. So impressed was Dr Whewell, a historian of science of the last century, that he wrote: "The manner in which the business of magnetic observation has been taken up by the governments of our time makes this by far the greatest scientific undertaking which the world has ever seen." To-day we are more used to the universality and international spirit of science and we should not speak in such superlatives.

In 1904 the Carnegie Institute of Washington founded a Department of Terrestrial Magnetism. A special wooden, non-magnetic ship, the *Carnegie*, was built for the special purpose of making magnetic observations. This ship, unfortunately, was destroyed by fire at Apia, Samoa, in 1929. The British Admiralty is now constructing a new non-magnetic ship. Fresh magnetic surveys are necessary every 15–20 years.

Variation of dip over the earth's surface.

The angle of dip is 0° approximately at the equator, and as we go northward or southward it increases steadily until it becomes 90° at the poles. In the northern hemisphere the N. pole dips and in the southern hemisphere, the S. pole. Fig. 248 is a map showing isoclinic lines, which are lines joining places at which the angle of dip is the same.

Gilbert constructed a model spherical magnet of lodestone, which he called a *terrella* or "little earth", with poles as shown in Fig. 247. By means of this model he was able to show that the angle of dip should vary over the surface of the earth in the way we have described.

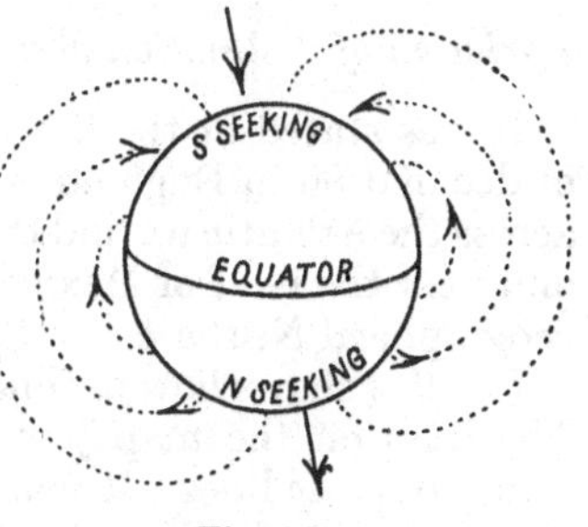

Fig. 247

The dotted lines in the figure represent the lines of magnetic force, and the arrows represent compass needles, the arrowheads being the N. poles. Remembering that "the horizontal" at any point on this model earth is parallel to the surface, it will be seen that the compass is horizontal at the equator and dips more and more as the poles are approached.

It will be seen from Fig. 247 that Gilbert predicted the existence of a pole of S.-seeking polarity near the geographical N. pole, and one of N.-seeking polarity near the geographical S. pole.

These poles were subsequently discovered in the Arctic and Antarctic. In 1831 Sir James Ross found the N. magnetic pole in the far north of Canada. He was able to fix its exact position, because at that spot his dip needle pointed vertically downwards to the earth. It will be found marked in any atlas. We are told that in his enthusiasm Ross built a large cairn and planted the British flag to mark the position of the pole. We believe, however, that the earth's magnetic poles are slowly moving and the magnetic N. pole is now probably some hundreds of miles from Ross's cairn.

The S. magnetic pole was discovered in 1909 by Sir Ernest Shackleton. Here a compass needle dips vertically with its S. pole pointing downwards. You should look up the position of this pole, also, in an atlas. It, too, will have moved slightly since 1909.

Near the magnetic poles the horizontal component is very small, since the earth's field is nearly vertical. Shackleton observed this, and in his book, *With Shackleton in the Antarctic*, wrote: "The compass by this time was very sluggish and would scarcely work at all....On 16th January the compass was useless."

Variation of declination over the earth's surface.

It was stated at the beginning of this chapter that the angle of declination in England is 12° W. of N. On sailing westwards across the Atlantic we find that the angle of declination decreases, and near the Gulf of Mexico it is zero, i.e. a magnet points true geographical N. there.

Fig. 249 shows how declination varies over the earth's surface. The lines on the map joining places of equal declination are called *isogonic* lines. It will be seen that the two specially dark lines are lines of zero declination. Both start at the N. magnetic pole of the earth and end up at the S. magnetic pole, but while one runs fairly straight, across North and South America, the other describes a large loop over Siberia. At all places between these two lines in Fig. 249 the declination is W. of N., while at places on either side of them the declination is E. of N. There is

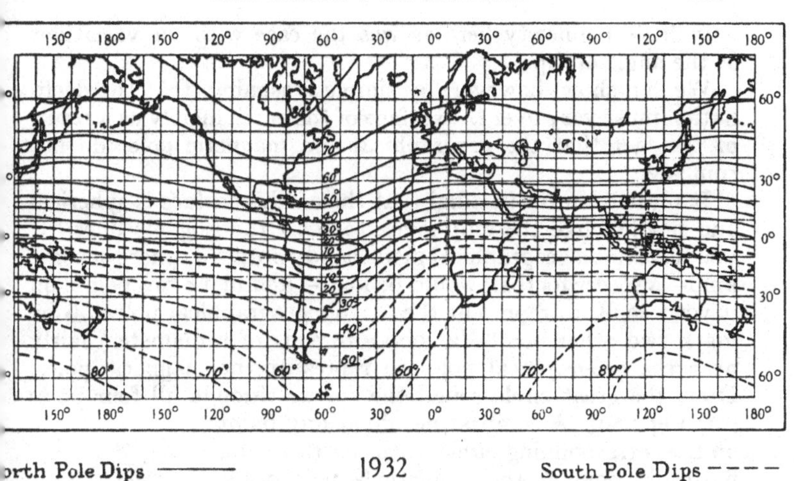

Fig. 248. Isoclinic lines, joining places of equal dip

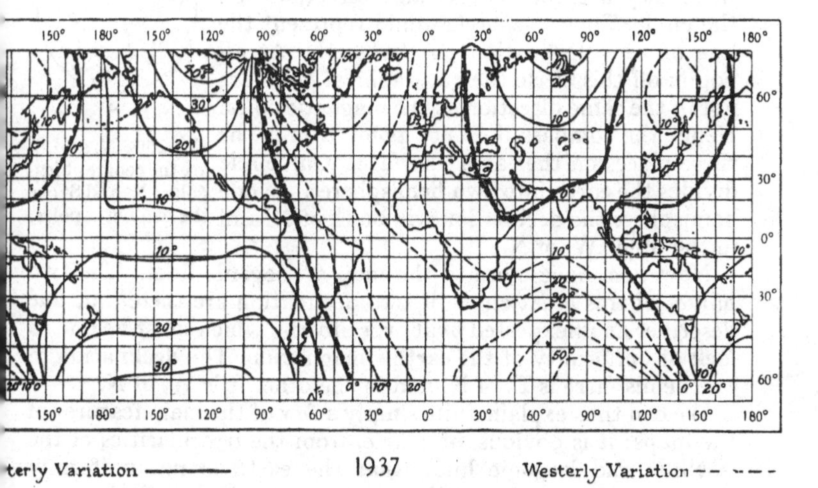

Fig. 249. Isogonic lines, joining places of equal declination.

[*Based upon British Admiralty Charts, with the permission of the Controller of H.M. Stationery Office and of the Hydrographer of the Navy.*]

very little regularity here as was the case with the variations of the angle of dip.

We can show, however, by simple reasoning, that the declination must vary over the surface of the earth merely as a result of the fact that the geographical and magnetic poles do not coincide.

It must be borne in mind that the earth's surface is spherical and not flat like the map given to show the isogonic lines. If you have not a geographical globe handy obtain a sphere of some kind—a soccer football would be ideal—and mark two points on it with chalk to represent the earth's geographical poles. Now mark two other points (at opposite ends of a diameter) fairly near the geographical poles to represent the magnetic poles. Draw the great circle round the sphere through all four poles (see Fig. 250). A compass needle at any point on the corresponding circle on the earth would, naturally, point to the magnetic N. pole, but it would also point to the geographical N. pole, since they are both in the same direction—on the circle. Thus this circle would represent the line of zero declination on the earth. At all points off this circle the geographical poles are not in the same direction as the magnetic poles. Hence the compass does not point true N. at these places on the earth's surface. Our circle divides the earth into two hemispheres. In one hemisphere the declination will be E. of N. and in the other W. of N.

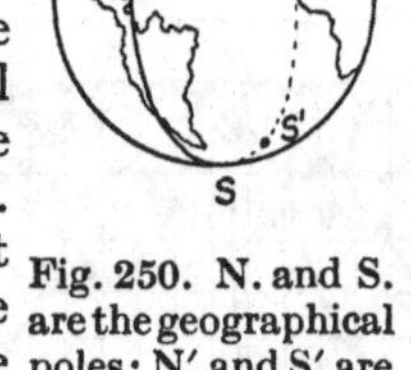

Fig. 250. N. and S. are the geographical poles: N′ and S′ are the magnetic poles.

Now turn once more to the map of isogonic lines. The two halves of this great circle of zero declination are there, but one describes an unexpected oval over Siberia, which must be due to lack of uniformity of the earth's magnetism. The declination in one hemisphere is E. of N. and in the other it is W. of N.

We can thus explain quite simply a few of the main features of the maps: it is obvious, of course, from the irregularities of the isoclinic and isogonic lines, that the earth is not uniformly magnetised. Indeed, a delicately balanced dip needle is sometimes used by the geologist to detect large local deposits of the (magnetic) ores of iron and nickel.

Secular variations.

Besides varying in space over the surface of the world, the magnetic elements are slowly varying in time. Measurements of the angle of declination in London have been made since the year 1580. The following table shows the variation from that date to 1934:

1580	**11° E.**	**1816**	**24° W.**
1622	**6° E.**	**1900**	**17° W.**
1658	**0°**	**1934**	**11° 41′ W.**

The angle of declination is now decreasing, and it seems likely that it will, in due course, become E. of N. again. It has been estimated that a complete cycle should be completed in about 480 years, that is, the value will be the same as in 1580 in the year 2060.

The angle of dip is also varying, but to a smaller extent than the declination. In 1580 its value in London was 72°, in 1658 74°, in 1816 71°, and in 1934 66° 40′.

Besides these slow secular changes, as they are called, there occur at intervals what are known as *magnetic storms*, when the needles of all magnetic recording instruments are affected in a comparatively violent manner. We shall have more to say about these later.

Theories of the origin of the earth's magnetism.

We have seen that the assumption that the earth is a uniformly magnetised sphere (with some irregularities) enables us to account for the main features of the earth's magnetic field.

We must now ask:

(1) Is the earth capable of being magnetised sufficiently strongly to produce the field?

(2) How did it come to be magnetised?

(3) Is it a coincidence that the magnetic axis nearly coincides with the axis of rotation?

(4) What is the cause of the secular variation of declination which suggests that the magnetic poles are slowly moving round the geographical poles?

(5) What is the cause of magnetic storms?

It is believed that the central core of the earth is largely composed of iron, and if it were magnetised only to $\frac{1}{1000}$th of its full intensity it would produce a field as strong as that of the earth. But this iron is at a very high temperature, and iron at a high temperature ceases to be magnetic. Indeed, at depths greater than only 20 kilometres the earth is too hot to be magnetic. Hence the magnetism must reside in a shell 20 kilometres thick. To account for the strength of the earth's field the magnetisation of this shell needs to be only a half the possible maximum.

So far, then, our theory is satisfactory. But no one has yet put forward a satisfactory explanation of the origin of the earth's magnetism, of the approximate coincidence of the earth's magnetic axis and axis of rotation, and of the slow revolution of the magnetic poles round the geographic poles.

That the earth's magnetism has some intimate connection with its rotation seems very likely in view of the discovery, from the analysis of its light, that the sun has a magnetic field. The sun is very much too hot to contain any magnetic ore. Moreover, the sun's magnetic axis is inclined at only 5° to the axis of rotation, and it is revolving round the axis of rotation in a period of 8 months (as compared with 480 years in the case of the earth). The sun's magnetic field varies between 50 and 3000 oersted (in sun-spots), while the earth's total intensity varies between 0·25 and 0·65 oersted.

The sun's magnetic field is probably due to the circulation of electric currents in it. Hence it has been suggested that the earth's magnetic field may be due to electric currents flowing inside the earth. A large current flowing round the earth from E. to W. under the equator would give rise to lines of force similar to those of the earth's field (see Fig. 251). However, owing to the resistance of the earth the current would be reduced to one-third of its value in three days, and no one has been able to suggest a source to maintain it.

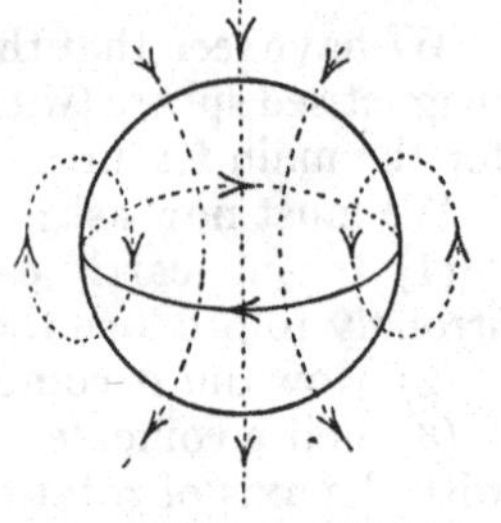

Fig. 251

We are now confident that electric currents in the upper atmosphere are responsible for the magnetic storms. Electrons are shot out of the sun from vast spouts in its surface, called *sun-spots* (which can sometimes be seen through coloured glass with the naked eye), and on reaching the upper atmosphere are deflected by the earth's lines of magnetic force, from the equator towards the poles. The electron stream gets lower and lower in the atmosphere towards the poles, and when it reaches a certain pressure of air there is a most beautiful and spectacular glow in

Dr Carl Størmer

Fig. 252. The *Aurora Borealis* or Northern Lights.

the sky called the *Aurora Borealis* or Northern Lights (see Fig. 252). It is significant that the appearances of the *Aurora Borealis* coincide with magnetic storms. Furthermore, magnetic storms show an 11-year cycle of activity which corresponds with the 11-year sun-spot cycle of the sun: they also tend to recur every 27·3 days, the period of rotation of the sun.

A stream of electrons is an electric current and an electric current sets up a magnetic field. Magnetic storms are caused by the magnetic field set up when an intense stream of electrons reaches the earth from the sun.

There are always some electrons in the upper atmosphere which constitute what is called the Heaviside layer, a reflector of wireless waves making possible radio communication between places at opposite ends of the earth.

We must add, to save the reader from valueless speculation, that we can prove mathematically that electrons in the upper atmosphere can be responsible for no more than 1 per cent of the earth's normal, steady field.

The mariner's compass.

For many hundreds of years the mariner's compass has consisted of a pivoted magnet fastened to a circular disc which is marked with the thirty-two points of the compass. Since the disc moves with the magnet, all the directions marked on it are automatically correct if the N.-S. line coincides with the axis of the magnet. In its modern form, several short magnets are attached to the card.

The chief difficulty with which the designers of ships' compasses have had to cope is unsteadiness in a rough sea and especially during heavy naval gunfire.

Anyone who possesses a small pocket compass made in the form of a moving circular magnetic disc will readily understand this. Before a bearing can be taken it must be held horizontally and perfectly still until it has settled down. On board ship in a rough sea it would never settle down.

A ship's compass is not fixed rigidly to the vessel but is mounted so that it can swing about two axes at right angles. This is known as gimbal mounting and a bird's eye view of it is shown in Fig. 253. By this means the compass remains horizontal even in the stormiest weather. If the ship rolls from side to side the ring *ABCD* swings about the axis *AB*: if the ship pitches so that its keel is executing a see-saw motion the compass swings about the axis *CD*, its weight in each case keeping it comparatively horizontal.

The compass itself consists of a light card made of aluminium or mica beneath which are fixed symmetrically a number of light magnets. It is pivoted on a hard jewelled bearing and contained in a bowl full of water with which alcohol has been mixed to prevent it from freezing. The liquid damps the movement of the card, i.e. prevents it from swinging about too much, and also by its buoyancy takes some of the weight of the card from its pivot, thereby reducing friction. In the old days the compass used to stick so badly that sailors were in the habit of kicking the binnacle, the wooden pillar on which it was mounted, to make it move.

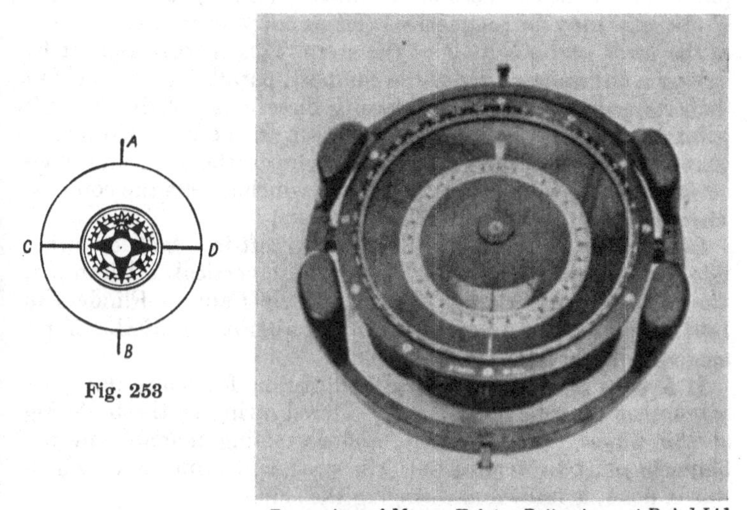

Fig. 253

By courtesy of Messrs Kelvin, Bottomley and Baird Ltd.

Fig. 254. A ship's compass, showing gimbal mounting.

Deviations of the compass.

As long as ships were made of wood and were propelled by sails the compass was very little affected by the small amount of iron, such as the anchor, on board. However, when at the end of the eighteenth century steam-boats began to be built and then, later, ships were themselves made of iron, it became necessary to devise some means of correcting for the magnetism of the ship. For an iron ship is bound to be magnetised. A soft-iron rod held N. and S. (preferably inclined to the horizontal at the angle of dip), when tapped gently with a hammer for some time, is found to be slightly magnetised. This is due to the influence of the earth's field; the molecules of the iron are assisted into alignment by the tapping. Now a ship during construction receives a considerable amount of hammering, especially while rivets are being driven in, and consequently becomes magnetised.

The magnetism of a ship is counteracted by means of magnets placed under the compass in the binnacle (see Fig. 255). The keel of the ship may be magnetised: let us suppose it has a N. pole at the prow and a S. pole at the stern. This is corrected for by having a bar magnet under the compass, parallel to the keel but with its poles pointing in opposite directions, i.e. its N. pole pointing towards the stern. Magnetism of the cross-thwarts is counteracted by magnets at right angles to the keel, and magnetism of funnels and vertical girders, which affects the compass when the ship heels, by vertical magnets.

Outside the binnacle may be seen two soft-iron balls on either side of the compass, and also at the back a vertical soft-iron bar, known as a Flinder's bar (introduced by Captain Flinders in 1801). These are to counteract slight alterations of the ship's magnetism due to the earth's field.

If a ship sets in a particular direction for some days, its magnetism may be considerably altered owing to the buffeting of the waves. Consequently, the correcting magnets in the binnacle must be readjusted. The process, by means of which this is done, is known as swinging the ship.

A landmark is sighted perhaps some miles away and its bearing observed on the compass while the ship is swung completely round in a circle.

For simplicity let us suppose that the landmark is due N. of the ship and that the magnetism of the keel is not sufficiently corrected for (see Fig. 256). When the ship is pointing N. and S. the compass bearing will be correct, due N. However, when the ship is pointing E. and W. the N. pole of the compass will be slightly repelled by the N. pole of the keel and the landmark will appear to be a little to the W. or E. of N.

The correcting magnet in the binnacle must therefore be raised slightly until the compass bearing of the landmark is always the same whatever the direction of the ship.

By courtesy of Messrs Kelvin, Bottomley and Baird Ltd

Fig. 255. A binnacle. The compass is at the top and below it, inside the black cylindrical body, are the various correcting magnets to counteract the ship's permanent magnetism. The two soft-iron spheres, on either side of the compass, counteract the varying, induced magnetism of the ship. As the ship's magnetism varies, so does that of the spheres.

Swinging the ship is done periodically, the magnets in the binnacle being supported on chains for adjustment.

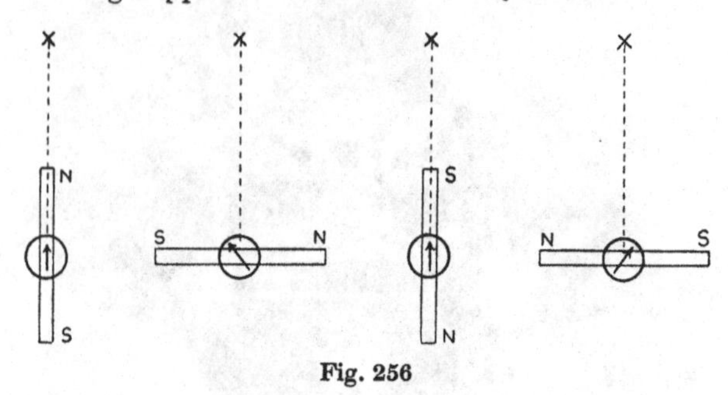

Fig. 256

Summary

The *magnetic meridian* at a place is the vertical plane through the magnetic axis of a freely suspended compass needle (i.e. magnetic N. and S.). The *geographic meridian* at a place is the vertical plane which passes through the geographic pole (i.e. geographic N. and S.).

The angle of declination is the angle between the magnetic and geographic meridians.

The angle of dip is the angle between the horizontal and the magnetic axis of a magnet free to swing about a horizontal axis in the magnetic meridian. It is the angle between the true direction of the earth's field and the horizontal.

The *horizontal component*, *vertical component* and *total intensity* of the earth's magnetic field are represented in Fig. 243. These, together with declination and dip, are called the *magnetic elements* at a place.

The *dip circle* is an instrument for measuring dip.

The magnetic elements vary in space over the surface of the earth and also in time.

The earth's magnetic field may roughly be accounted for by assuming that the earth is an irregularly magnetised sphere with poles of S.-seeking and N.-seeking polarity near the N. and S. geographic poles, respectively. The origin of the earth's magnetism has not yet been explained satisfactorily.

The mariner's compass is the most important practical application of terrestrial magnetism.

QUESTIONS

1. A nautical almanack for the year 1910 gave the following magnetic data for London (Greenwich): Declination 15° 41′ W., Dip 66° 57′ N., Horizontal Force (*H*) 0·1851 oersted. Explain fully what you understand by these figures.

A rod of iron is hammered (*a*) when in a vertical position, (*b*) when in a horizontal position, in London. In which case would you expect it to become the stronger magnet? Give reasons for your answer.

(O. and C.)

2. What is meant by the statement that the variation (or declination) is 12° W.?

(*a*) The true bearing of a place, obtained from a map, is 80° E. of N. On what compass bearing must one walk to reach the place?

(*b*) The magnetic bearing of a distant tower, obtained with a prismatic compass, is 164° (i.e. 74° S. of E.). What is the true bearing?

3. Write a brief account of the earth's magnetism.

At a place where the horizontal strength of the earth's field is 0·20 dyne per unit pole, and the angle of dip is 60°, what is the total intensity and the vertical intensity? (N.)

4. Assuming the earth to be a uniformly magnetised sphere, and that its poles lie at some distance from the geographical poles, how would you expect the angle of declination to vary as you sail round the equator?

5. Examine the map of isogonic lines on p. 309.

(*a*) Indicate the regions where the compass points (i) true N., (ii) N.W., (iii) N.E.

(*b*) What is the approximate value of the declination at (i) Cape Town, (ii) New York, (iii) San Francisco, (iv) Buenos Aires, (v) Bombay, (vi) Sydney?

6. Examine the map of isoclinic lines on p. 309. Find the approximate angle of dip at the places mentioned in Question 5 (*b*).

7. Define declination, dip. The magnetic N. pole is in Northern Canada. What would you expect the declination to be a few miles (*a*) North, (*b*) West, (*c*) South, (*d*) East of it? Why would it be difficult to measure the declination in such a region? (L.)

8. What explanations can be given of the change from place to place of (*a*) the angle between the magnetic and geographical meridians, (*b*) the angle of dip, at different places on the earth's surface? (N.)

9. Explain the meanings of the terms declination (or variation) and dip; and show that these phenomena would occur if a comparatively short bar magnet occupied a central position in the earth. (N.)

10. Define (*a*) the magnetic axis of a magnet, (*b*) the magnetic meridian at a point on the earth's surface. Being provided with a circular disc of steel known to be magnetised along a diameter and means of supporting it horizontally, describe how you could determine (i) its magnetic axis, (ii) the direction of the earth's magnetic meridian. (L.)

11. Given a bar magnet (only), how would you determine (*a*) the magnetic meridian, (*b*) the magnetic axis of the magnet? (C.)

12. Define dip.
Describe and explain the behaviour of a dip needle as the bar supporting it is gradually turned through 360° in a horizontal plane. (L.)

13. Write a short account of the theories which have been put forward to explain the nature, origin and fluctuations of the earth's magnetic field.

14. An iron ship is built with its bow to the north. How would you expect its permanent magnetism to affect a compass needle on deck amidships: (*a*) when sailing East; (*b*) when sailing South?

15. A long bar of soft iron is placed N. and S. horizontally and tapped for some time. It is then placed with the S. end near to and E. of a magnetometer and there is a small deflection. Explain.

If it is tapped with its axis vertical and its lower end placed in the same position with respect to the magnetometer needle there is a larger deflection than before in the opposite direction. Explain.

If the deflections were 8° and 20°, respectively, calculate the angle of dip.

16. A small magnet, suspended so as to vibrate in a horizontal plane, makes 10 vibrations per minute at a place where the angle of dip is 60°, and 12 vibrations per minute at a place where the angle of dip is 65°. If the earth's horizontal field at the first place is 0·20 oersted, find the horizontal field at the second place and also the vertical fields at the two places. (O. and C.)

Chapter XVII

THE MEASUREMENT OF ELECTRICITY

We described in Chapter xv how the c.g.s. system of units (based on the centimetre, gramme and second), first proposed by the great German mathematician Gauss, was adopted for the measurement of magnetism. We shall now explain how the system was extended by the colleague of Gauss at Göttingen, Wilhelm Weber, to include the measurement of electricity.

The electromagnetic unit of current.

So far in this book, while we have had much to say of amperes, we have not considered how the ampere is derived. Someone had to think out the unit: that man was Weber. His suggestions, with slight modifications, were adopted in 1881 at an international congress of physicists in Paris: at this conference the practical unit of current was given its name in honour of André Marie Ampère.

Weber's idea was to utilise the magnetic effect of an electric current and to measure this effect in terms of the magnetic units proposed by Gauss.

Now it is an experimental fact that the strength of the magnetic field, F, at the centre of a circle, due to an electric current flowing in an arc of the circle, is proportional to (1) the strength of the current, i, (2) the length of the wire, l, (3) the inverse square of the radius of the circle, r:

i.e. $F \propto i$ when r and l are constant, $F \propto l$ when i and r are constant,

$F \propto \frac{1}{r^2}$ when i and l are constant,

$$\text{i.e. } F \propto \frac{il}{r^2} \quad \text{or} \quad F = k\,\frac{il}{r^2},$$

where k is a constant.

The reason why the wire must be an arc of a circle with the point at which the magnetic field is being considered as centre is that every part of the wire must be equidistant from the point under consideration.

Weber obtained his unit of current by making, in the formula $F = k\frac{il}{r^2}$, $i = 1$ c.g.s. unit when $F = 1$ oersted, $l = 1$ cm. and $r = 1$ cm.: then, of course, $k = 1$, and $\mathbf{F} = \frac{\mathbf{li}}{\mathbf{r^2}}$.

Definition of the electromagnetic unit of current.

The electromagnetic unit of current is that current which, flowing in an arc, 1 cm. long, of a circle of 1 cm. radius, in air, creates a field of 1 oersted at the centre.

This unit of current is called the electromagnetic unit (e.m.u.), since it is based, ultimately, on the definition of the unit magnetic pole.

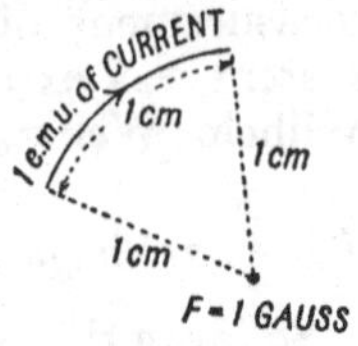

1 e.m.u. of CURRENT
1cm
1cm
1cm
F = 1 GAUSS

Fig. 257

The electromagnetic unit of current was found to be rather large for ordinary practical purposes. Hence, at the conference in 1881, it was decided to define **1 ampere as $\frac{1}{10}$th of an electromagnetic unit.**

The tangent galvanometer.

The tangent galvanometer is an instrument for measuring an electric current from first principles. An ammeter must, of course, be calibrated by comparing it with a standard instrument. But anyone, who has reasonable manual skill and obtains a grasp of the following theory, could construct a tangent galvanometer, starting only with a short magnet and a length of wire, and measure an electric current in amperes.

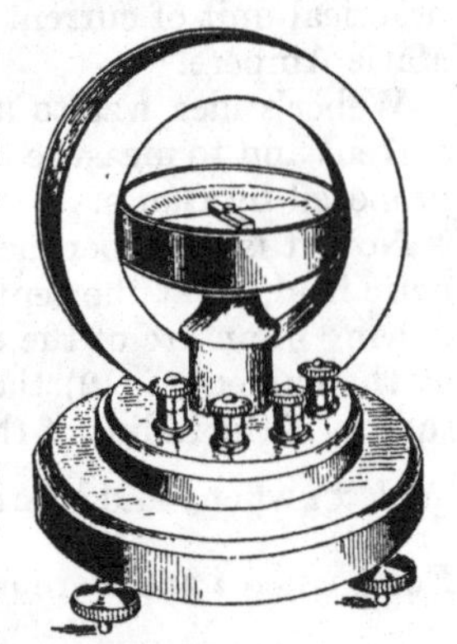

Fig. 258. The tangent galvanometer.

The tangent galvanometer consists of a circular coil through which the current is passed and a deflection magnetometer for measuring the strength of the magnetic field due to the current at the centre of the coil (see Fig. 258). The coil must be set with its plane N. and S., so that its magnetic field lies E. and W., at right angles to the earth's field.

Then if F is the strength of the magnetic field due to the current in the coil, H is the horizontal component of the earth's field, and θ is the deflection,

$$F = H \tan \theta \qquad \text{(see p. 288).}$$

But

$$F = \frac{li}{r^2} \qquad \text{(see p. 322).}$$

If the coil of the galvanometer consists of n turns of radius r cm.,

$$l = 2\pi rn,$$

$$F = \frac{2\pi rni}{r^2} = \frac{2\pi ni}{r}.$$

$$\therefore \frac{2\pi ni}{r} = H \tan \theta,$$

$$i = \frac{Hr}{2\pi n} \cdot \tan \theta \text{ e.m.u.}$$

$$= \frac{10\,Hr}{2\pi n} \tan \theta \text{ amperes}$$

since 1 e.m.u. = 10 amperes.

The term $\frac{10Hr}{2\pi n}$ is called the **reduction factor** of the galvanometer and is denoted by K.

Thus

$$\mathbf{i = K \tan \theta}.$$

This makes for convenience. The term K is found once and for all (at a particular place), and then is used as a multiplying factor to determine currents from the tangents of the deflections.

To check the reading of an ammeter using a tangent galvanometer.

In conjunction with a tangent galvanometer a commutator (see Fig. 259) is always used: the commutator reverses the current in the galvanometer without affecting the current in the rest of the circuit. By averaging the readings of both ends of the pointer before and after the current is reversed, errors due to a slightly incorrect setting of the galvanometer coil (with its plane N. and S.) may be eliminated.

To check the readings of an ammeter by means of a tangent galvanometer a circuit such as that in Fig. 260 is set up. A suitable current is sent through the circuit by means of a battery and variable resistance, and the reading of the ammeter and average of the readings of the galvanometer are determined. Several sets of readings may be obtained using different currents.

The reduction factor of the galvanometer is then calculated, $K=\frac{10Hr}{2\pi n}$, and hence the current $i=K\tan\theta$ can be found.

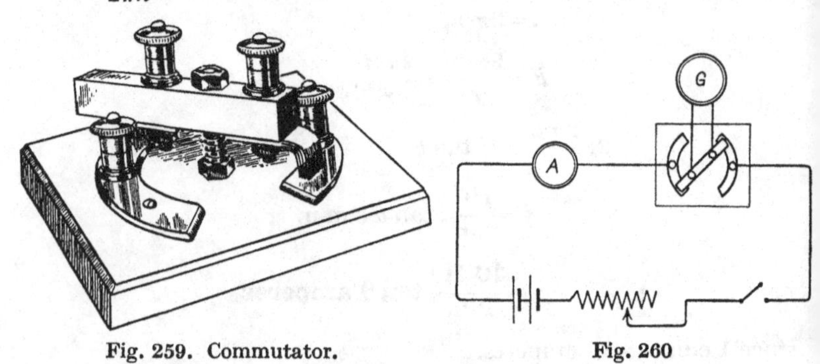

Fig. 259. Commutator. Fig. 260

The following is a set of readings:

Readings of galvanometer = 43°, 44·5°, 47·5°, 46°.

∴ Mean value of θ = 45·25°.

Number of turns of coil, $n=2$.

Radius of coil, r = 7·2 cm.

$$H=0{\cdot}18 \text{ oersted}.$$

$$\therefore K=\frac{10Hr}{2\pi n}=\frac{10\times 0{\cdot}18\times 7{\cdot}2}{2\pi\times 2}$$

$$=1{\cdot}03,$$

$$\therefore i=K\tan\theta$$

$$=1{\cdot}03\tan 45^\circ\,15'$$

$$=1{\cdot}04 \text{ amperes}.$$

Actual reading of ammeter = 1·02 amperes.

Thus the error of the ammeter was about 2 per cent.

Note that to do this check, H must be known. If on the other hand the ammeter were calibrated in another way, e.g. by electrolysis, the method could be used to find H.

Theoretical definitions of the volt and the ohm.

The volt and the ohm are based on the ampere. Before we can define the volt, however, we must define the units of work and power on the c.g.s. system.

Units of work:

1 *erg* of work is done when a force of 1 dyne moves its point of application 1 cm.

1 *joule* $=10^7$ ergs.

Unit of power:

1 *watt* = 1 joule per second.

The theoretical definitions of the volt and the ohm are given in the table below.

Practical standards.

At an international conference held in London in 1908 it was decided, for the purpose of convenience, to define practical standards and constantly to adjust them, should more accurate experimental evidence become available, to agree with the theoretical standards. These international standards are given in the table below.

	Ampere	Volt	Ohm
Theoretical Units	1 ampere $=\frac{1}{10}$ e.m.u. 1 e.m.u. of current when flowing through an arc 1 cm. long of a circle of radius 1 cm. sets up a field of 1 oersted at the centre	Two points are at a P.D. of 1 volt if work is done at the rate of 1 watt when a current of 1 ampere flows between them	A conductor has a resistance of 1 ohm if a current of 1 ampere flows through it when a P.D. of 1 volt is applied across it
Practical Standards	The international ampere is that current which, flowing through a solution of silver nitrate in water, deposits silver at the rate of 0·001118 gm. per sec.	The international volt is obtained from the Weston Cadmium cell, which has an E.M.F. of 1·0183 volts at 20° C.	The international ohm is the resistance at 0° C. of a column of mercury of mass 14·4521 gm. and length 106·300 cm.

SUMMARY

The electromagnetic unit of current is that current which, flowing in an arc, 1 cm. long, of a circle of 1 cm. radius, creates a field of 1 oersted at the centre.

$$\mathbf{F = \frac{li}{r^2} \text{ oersted,}}$$

where i = current in e.m.u.

1 ampere = $\frac{1}{10}$th e.m.u.

The tangent galvanometer is an instrument for measuring the strength of current from first principles:

$$\mathbf{i = \frac{10Hr}{2\pi n} \tan \theta \text{ amperes}}$$

$$\mathbf{= K \tan \theta \text{ amperes.}}$$

$\mathbf{K \left(= \frac{10Hr}{2\pi n}\right)}$ **is called the reduction factor of the galvanometer.**

QUESTIONS

1. Describe the construction of a tangent galvanometer and how you would set it up for use, and explain what you understand by its reduction factor.

A tangent galvanometer with 10 turns of wire of mean radius 15 cm. gives a deflection of 60° with a certain current in a place where $H = 0{\cdot}18$ oersted. Calculate the strength of the current and the reduction factor of the galvanometer. (O. and C.)

2. When a current of 2 amperes passes through the coil of a tangent galvanometer the deflection is 45°. What is the reduction factor of the galvanometer? What current will cause a deflection of 60°?

3. Explain clearly:

(*a*) In a tangent galvanometer the magnet must be short, and when in use the coil must be set with its plane in the magnetic meridian.

(*b*) The sensitivity of a tangent galvanometer may be increased by placing a bar magnet over the coil with its N. pole pointing N.

4. Explain how a tangent galvanometer can be placed so that even a large current through its coil produces no deflection of the needle. What would happen if the current were reversed?

5. If one tangent galvanometer has 40 turns of 5 cm. radius and another 30 turns of 3 cm. radius, which will be the more sensitive? Give reasons for your answer. (L.)

6. Describe the tangent galvanometer.

When a battery of 10 ohms' resistance is connected in series with a galvanometer of 100 turns and of 40 ohms' resistance, the deflection is 30°. What would be the deflection if only 50 turns of the galvanometer were connected in series with the battery? (C.)

7. A tangent galvanometer which gives a deflection of 20° for a current of 0·01 ampere in England gives a deflection of only 12° for the same current when in Ceylon. How do you account for this? (O. and C.)

8. A tangent galvanometer has 9 coils of 12 cm. radius, and is used at a place where the horizontal component of the earth's magnetic field is 0·2.

If the plane of the coils is east and west and a current of 0·84 ampere is sent round the coils in a clockwise direction viewed from the N., what is the strength of the field at the centre of the coil? In what direction would the needle point, if the plane of the coils were in the magnetic meridian and the same current as before were used in a clockwise direction viewed from the east? (O.)

9. Why is it desirable to obtain a deflection in a tangent galvanometer as nearly as possible in the neighbourhood of 45°?

The reduction factor of a tangent galvanometer is 1·00. Find the current to produce a deflection of (*a*) 45°, (*b*) 46°. Hence find the increase in current required to increase the deflection from 45° to 46°. Similarly, find the increase in current to increase the deflection (*c*) from 10° to 11°, (*d*) from 80° to 81°.

Point out the significance of your answers.

10. A tangent galvanometer and a copper voltameter are joined in series and a current which causes a deflection of 60° in the galvanometer is passed through for 30 minutes. If 0·45 gm. of copper is deposited in the voltameter, calculate the reduction factor of the galvanometer. The electrochemical equivalent of copper = 0·000329 gm. per coulomb. (O. and C.)

11. A circular coil of wire is placed with its axis in the magnetic meridian. A compass needle suspended at its centre by a silk fibre makes 5 vibrations per minute in the earth's field and 10 vibrations per minute when a current passes through the coil. If the horizontal component of the earth's magnetic field is 0·2 oersted, find the magnetic field due to the current in the coil. (C.)

Chapter XVIII

ELECTROSTATICS

We shall now continue our account, begun in Chapter II, of the behaviour of electricity at rest, a branch of the subject known as electrostatics.

Much of our knowledge of electrostatics was acquired before the discovery of current electricity, but since its interest is largely theoretical we have postponed the discussion of it until now.

Charging by rubbing.

In Chapter II we explained that bodies can be charged with electricity by rubbing and that there are two kinds of electricity.

Ebonite rubbed with fur acquires a negative charge and the fur acquires an equal positive charge. We believe that electrons, tiny negative charges of electricity, pass from the fur into the ebonite, leaving the former with a deficiency and the latter with an excess of negative electricity. Only the negative electricity moves.

Similarly, glass rubbed with silk acquires a positive charge and the silk a negative charge.

Any substance can, in fact, be charged by rubbing with another substance, but a good conductor like a metal must be held with an insulating handle: otherwise the charge passes swiftly through the hand and body to the earth.

Airships acquire so large a charge by friction with the air that they have to be discharged by lowering a wire to the ground before anyone alights. Otherwise the electricity might flow through the first person leaving the airship and give him a considerable shock.

Passengers occasionally complain of shocks when boarding or alighting from an omnibus on a dry day. The bus acquires a charge by friction with the air, and since its tyres are made of rubber it is insulated from the earth. Two cases of fires, when aeroplanes began to refuel, were caused by sparks passing between the earthed petrol pipe and the tank of the aeroplane. A large percentage of carbon is now incorporated in the rubber of which aeroplane tyres are made, to render the tyres conducting and prevent the repetition of such accidents.

Distribution of charge on conductors.

(*a*) *The charge on a conductor resides on the outside surface.*

This can be demonstrated by the following three experiments.

1. Insulate a deep metal can by standing it on a slab of paraffin wax. Give it a positive charge by induction by holding near to it a rubbed ebonite rod and touching it momentarily to earth. Now test whether there is a charge on the inside of the can by touching the inside with a proof plane (a metal disc on an insulating handle) (see Fig. 261), and presenting the proof plane to an electroscope.

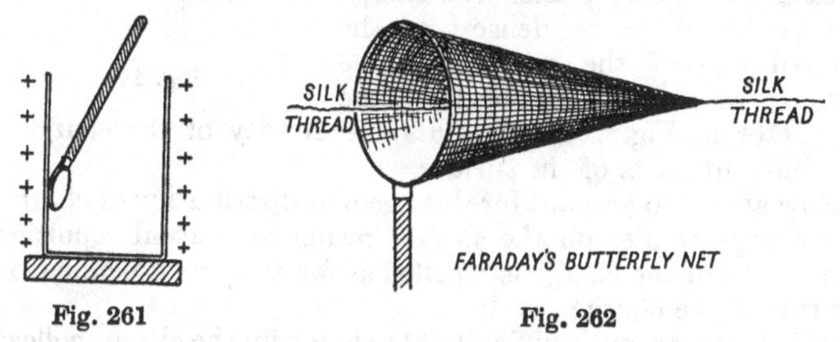

Fig. 261 Fig. 262

In this way a sample, as it were, of the charge on the inside is carried away and tested. It will be found that there is no charge on the proof plane, showing that no charge resides on the interior of the can.

Repeat the experiment, taking away a sample of the charge from the outside of the can. It will be found that there is a charge on the outside of the can.

2. Faraday made a large cubical conductor of 12-foot side, insulated on glass, which he himself entered. The cube was then so highly charged that sparks could be drawn from the outside. However, inside, Faraday could obtain no evidence at all of a charge: the whole charge was on the outside.

3. Another ingenious method of demonstrating the same phenomenon is Faraday's butterfly-net experiment. A butterfly net, made of linen, on an insulating handle (see Fig. 262) is

charged. By means of a proof plane a charge is shown to reside on the outside and none on the inside. The butterfly net is now pulled inside out by means of an insulating silk thread. The charge is again found to reside on the outside, despite the fact that what is now the outside was formerly the inside.

(*b*) *The charge accumulates on the more curved parts of the surface of a conductor.*

A convenient piece of apparatus for demonstrating this phenomenon is a pear-shaped conductor, supported on an insulating handle. Charge the conductor by induction with an ebonite rod, and test with an electroscope samples of charge taken from the surface on a proof plane. The charge will be found to be densest at the pointed end of the conductor. The distance of the dotted line from the conductor in Fig. 263 represents the density of the charge at the different parts of the surface.

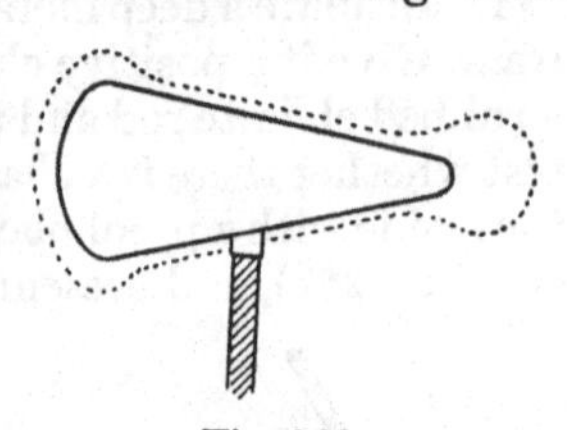

Fig. 263

How are we to account for this peculiar distribution of charge? The charge resides on the surface owing to mutual repulsion. Every part of the charge is repelled as far away as possible from the rest of the charge.

It is, however, more difficult to explain why the charge collects at the more curved parts of the surface, although this again is due to the mutual repulsion of each part of the charge. When the surface is flat the repulsion has a greater component along the surface than when the surface is curved. There is therefore a tendency for a thinning out of the charge on the flat parts of the surface and a bunching up of the charge on the curved parts.

The action of points.

If a charged conductor is pointed, there is a great curvature at the point and therefore a great accumulation of charge. The air near the point is subjected to an intense electric force and becomes charged:* it is then repelled by the rest of the charge and streams away as an "electric wind".

This electric wind can be demonstrated by attaching a point

* Actually the air is ionised, see p. 367.

B. F. Brown

Fig. 264. A Wimshurst machine, for generating electric charges by induction. The two ebonite discs, carrying thin metal sectors, revolve in opposite directions. Opposite charges are induced on diametrically opposite sectors when the latter are connected (momentarily) by the (bright) rod having copper brushes at its ends, by a charged sector on the other disc. These charges are separated as the disc revolves and collect on the knobs and Leyden jars in the foreground. There is a similar rod with brushes on the other side.

to a knob of a Wimshurst machine which generates a continuous charge (see Fig. 264). The wind will blow out a candle flame (see Fig. 265).

A pointed conductor soon loses its charge. Indeed a smooth conductor, on the surface of which are particles of dust, will slowly discharge in this manner.

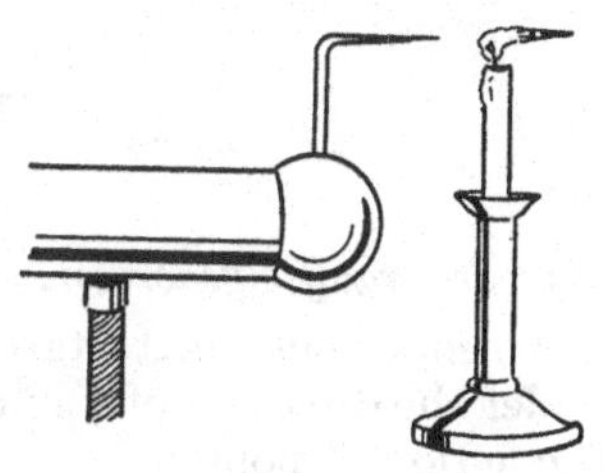

Fig. 265. An electric wind.

Collecting charges by points.

The charges from both conductors and insulators can be collected by means of points. In the case of insulators this is sometimes of definite use, since an insulator will not readily part with its charge by contact.

Suppose a pointed conductor is held with the point near to a positively charged glass rod. A negative charge is induced on the point and a positive charge at the other end (see Fig. 266). A negatively charged electric wind streams from the point and is attracted to the glass rod, where it tends to neutralise the positive charge there. Thus the glass rod loses its positive charge and the pointed conductor gains a positive charge.

We shall see that this method of collecting charges is employed in the Van de Graaff generator, and also in lightning conductors for discharging clouds.

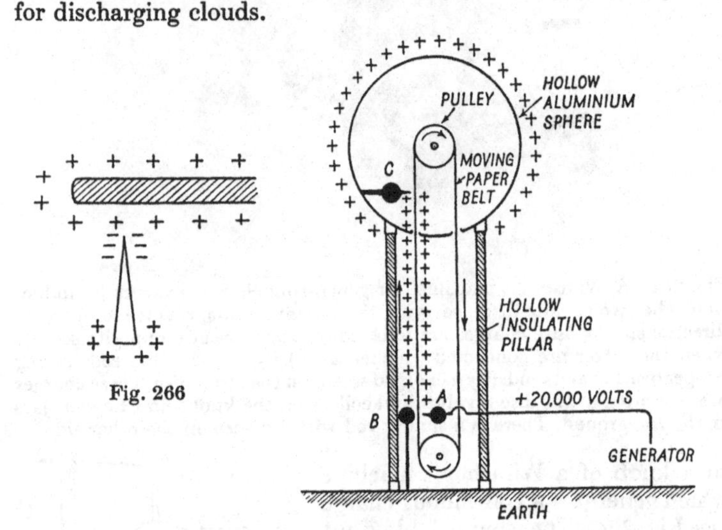

Fig. 266

Fig. 267. The Van de Graaff generator.

Van de Graaff's generator.

An enormous electrostatic generator recently built in America, under the direction of Van de Graaff, depends for its working on the action of points.

A pointed conductor *A* (see Fig. 267) is attached to the positive terminal of a generator producing D.C. at 20,000 volts. A positively charged electric wind flows from the point on to a moving paper belt which collects the charge and carries it up into an aluminium sphere, 15 feet in diameter, supported on hollow insulating columns 25 feet high and 6 feet in diameter. (The earthed conductor *B* becomes negatively charged by induction and helps to attract the positively charged wind on to the paper belt.)

The positive charge on the paper belt is carried up inside the sphere and is there collected by the pointed conductor *C* in the manner described on p. 332.

By using two spheres and charging one positively and the other negatively, it is possible to obtain a potential difference of about 10,000,000 volts. This tremendous P.D. is used for producing a beam of electrons of very high speed: the apparatus is contained in an insulating tube connecting the two spheres. The experimenters take up a position inside one of the spheres in which there is a room. Here they are perfectly safe since, although the sphere is charged to a potential of several million volts, all its charge resides on the outside.

The work done in charging the spheres is performed by the motor driving the belt; for there is a strong repulsion between the positive charge on the belt and the positive charge on the sphere.

The whole apparatus is housed in an airship hangar: it can be moved outside the hangar on trucks which run on rails.

Thunder and lightning.

Lightning and thunder consist of a flash and a bang and it was only natural that they should be ascribed, at one time, to an explosion in the upper atmosphere. However, as soon as machines were invented for generating electric sparks on a fairly large scale the similarity between lightning and the electric spark became apparent.

If lightning is a large electric spark something in the sky must be charged—the clouds, and these must spark to the earth or between themselves.

The fact that the clouds are charged during a thunderstorm was proved in an experiment performed by Benjamin Franklin,

By courtesy of Dr R. J. Van de Graaff

Fig. 268. The Van de Graaff generator showing the trucks and rails which enable it to be run out of its hangar. The aluminium spheres are hollow and 15 feet in diameter: the insulating columns supporting them are 25 feet high and 6 feet in diameter.

By courtesy of Dr R. J. Van de Graaff

Fig. 269. Looking up through one of the insulating columns of the Van de Graaff generator, showing the endless paper belt. There is a room at the top inside the aluminium sphere.

By courtesy of Dr R. J. Van de Graaff

Fig. 270. Streamers of sparks from the Van de Graaff generator to the airship hangar in which it is housed.

a great American, famous alike as scientist, statesman and philosopher. His idea was to draw off some of the charge from a cloud by means of a pointed conductor. At first he thought of erecting a sentry box at the top of a tall spire in Philadelphia. He then realised that a better method of getting his pointed conductor near to a cloud was by sending it up attached to a kite.

Accordingly he made a strong kite of cedar wood and a silk handkerchief which would not be torn by the wind and the rain. At the upper end of the string he attached a pointed conductor and at the lower end a key upon which to collect the charge. To prevent himself receiving a shock he held this key with a strip of silk.

It was on a day in June 1752 that Franklin flew his kite in a thunderstorm and performed his famous experiment. At first, on presenting his knuckle to the key he could draw from it no spark. Then when the string became wet and therefore a better conductor, he noticed that the tiny bristles on it stood up on end, and he was able to draw a spark from the key.

The experiment is a dangerous one, and while repeating it a Russian scientist was killed.

The potential of the clouds is of the order of a thousand million (10^9) volts. When a spark has passed its path remains conducting for some time, and usually several flashes, each lasting about $\frac{1}{10000}$th sec., follow each other in rapid succession. This explains the flickering of a lightning discharge. The average quantity of electricity in a discharge is 20 coulombs. It has been estimated that there are 16,000,000 thunderstorms in the world per annum, that is 1800 per hour, and that, on an average, there are 6000 flashes every minute.

As it proceeds, a lightning flash, which may be a mile long, branches. Hence we can see whether the discharge is from a cloud to the earth or vice versa. The latter is the more unusual and the more dangerous, since the main trunk of the discharge is at the earth. Most lightning is of this forked variety. So-called sheet lightning is the reflection by clouds of forked lightning not directly visible. However, there is a rare and completely mysterious phenomenon known as globular lightning. A luminous ball, as big as a man's fist or head, is observed to travel slowly through the air, making a buzzing sound according to some observers, and then to burst with violence.

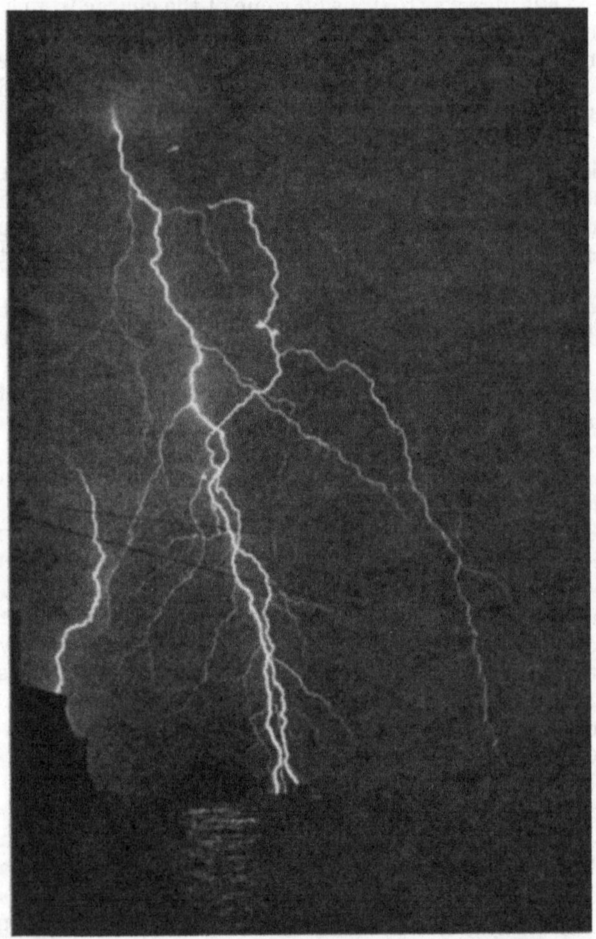

By courtesy of the Royal Meteorological Society

Fig. 271. A lightning discharge from a cloud to the earth.

Thunder is caused by the violent expansion of heated air along the path of the discharge—like the bursting of a balloon. The roll of thunder is due to echoes, the reflections of the sound from clouds and hills.

Lightning conductors.

Benjamin Franklin, being a practical man of affairs, soon began to apply his knowledge of the action of points to protect buildings from being struck by lightning. It is interesting to recall that certain theologians denounced this as an impious attempt to thwart divine wrath.

A lightning flash, like a river, takes the line of least resistance. That is why its path is not straight. It takes the easiest path to earth and is very likely to strike a prominent building. The Boston stump, the high church tower standing in the flat fens of Lincolnshire, was many times struck before it was equipped with an efficient lightning conductor.

A lightning conductor consists of a metal rod attached to the outside of a building, efficiently earthed, and with its upper end as high as possible and pointed. Its action is twofold, preventive and protective. A highly charged cloud induces a strong charge on the point which flows away in the form of an electric wind and tends to neutralise the cloud in the vicinity of the building. If this action fails to avert a flash the discharge will pass down the conductor, the easiest route to earth, instead of the building.

How the clouds become charged.

The way in which the clouds become charged is still the subject of controversy. There are two theories. Dr C. G. Simpson, director of the Meteorological Office, London, believes that the charges are produced by the breaking up of drops carried upwards by the vertical air currents. Professor C. T. R. Wilson, of Cambridge University, maintains that there is a separation of charges inside a cloud due to falling drops, a mechanism similar to that of the moving belt in Van de Graaff's generator. We can say no more about these theories, but mention them as presenting one of the interesting problems in electrostatics to-day.

Electric potential.

When a negatively charged electroscope is touched with the finger the leaf collapses. We have explained this by saying that, in its charged state, the electroscope has more than its complement of free electrons and that the excess electrons flow to the earth. Again, a positively charged electroscope has less than its full complement of free electrons and when it is touched with the finger electrons flow up into it from the earth.

This flow of electrons is rather like the flow of water from rivers into the sea or from the sea into wells or holes below sea level. The earth is a vast reservoir of free electrons, just as the sea is a vast reservoir of water.

Now the direction of the flow of water to or from the sea depends upon whether the place with which the sea communicates is *above* or *below sea level.* In a similar way the direction of the flow of electricity when a body is connected to the earth depends upon whether the body is at a *higher* or a *lower potential than the earth. A body is said to be at a positive potential when electrons flow up into it from the earth, and at a negative potential when electrons flow from it to the earth.* The potential of the earth is said to be zero.

The more charge you give a body the higher (or lower, according to the sign of the charge) its potential becomes, just as the more you heat a body the higher its temperature becomes. This is somewhat obvious, since the greater the excess or deficiency of electrons in a body, the greater the tendency for electrons to flow out of or into the body when it is connected to the earth.

But the problem of potential is not quite as simple as this. For *the potential of an uncharged conductor can be raised above that of the earth merely by bringing a positively charged body near to it.*

Let us consider a simple case of electrostatic induction. In Fig. 24, p. 26, a positively charged body is brought near to an uncharged conductor. The induced positive charge in the conductor disappears when the conductor is earthed: electrons flow up from the earth and neutralise it. The conductor must therefore have been at a positive potential: electrons flowed into it from the earth until it had acquired sufficient negative charge to counteract the influence of the positively charged body near to it and its potential had become zero, the same as that of the earth.

The gold-leaf electroscope measures potential rather than charge.

The gold-leaf electroscope is a form of voltmeter: it measures potential rather than charge. This can be shown by the following experiment.

Take a large can and a small can, insulate them on slabs of paraffin wax and give them *equal* charges. This may be done by charging (twice) a small electrophorus cover (see p. 28), and allowing it to make contact right inside each can: all its charge will then, on each occasion, go to the outside of the can. Now connect the cans to two similar electroscopes. The electroscope connected to the smaller can will show a greater divergence than the other (see Fig. 272).

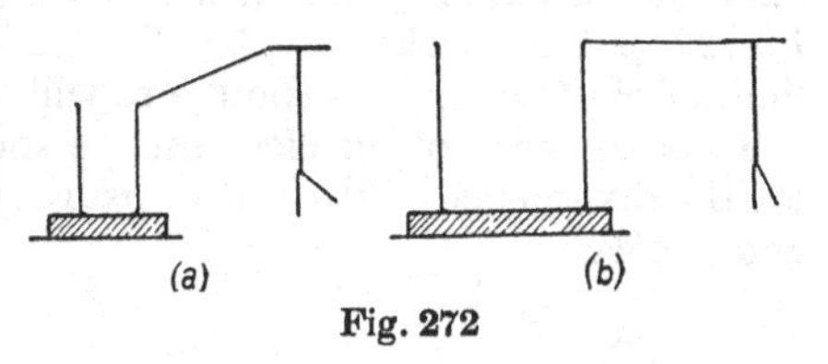

Fig. 272

But the cans have equal charges. The electroscopes cannot therefore be registering the charges. It seems probable that the electroscopes are registering the potentials of the cans. For when equal charges are given to large and small cans we should expect the latter to have the higher potential, just as, when equal quantities of water are poured into cans of different cross-section, the level of the water in the smaller can rises higher than that in the larger (see Fig. 273(*a*)).

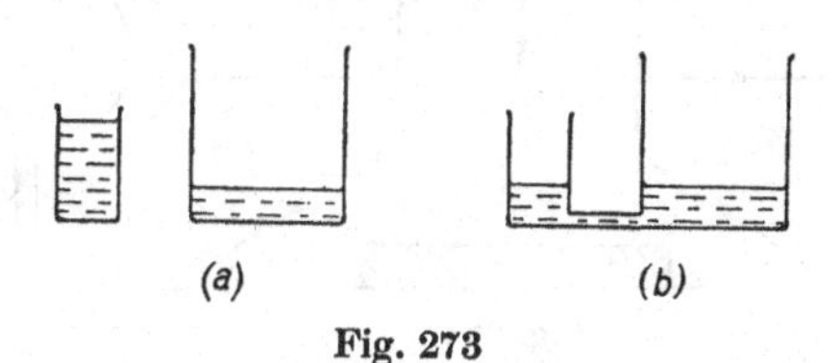

Fig. 273

We can settle the matter by connecting the two cans with a wire. Electricity will then flow from the can at the higher potential to the one at the lower until the potentials of the two cans are equal, just as, in our water analogy, water will flow from one can to the other until the levels are the same, when they are connected with a pipe as shown in Fig. 273(*b*). Note that the cans now have equal potentials but unequal charges. If our supposition that the electroscopes register potential rather than charge is correct we ought to find that the divergences of the leaves of the two electroscopes are now equal. And this is precisely what we do find.

The effect of earthing the case of an electroscope.

Connect the cap of an electroscope to the case by a wire and stand the instrument on an insulating slab of paraffin wax (see Fig. 274). Charge the cap. No divergence of the leaf results. The lines of electric force in the figure will be understood later.

Thus the case of an electroscope should always be earthed: for the divergence of the leaf measures the P.D. between the leaf and the case.

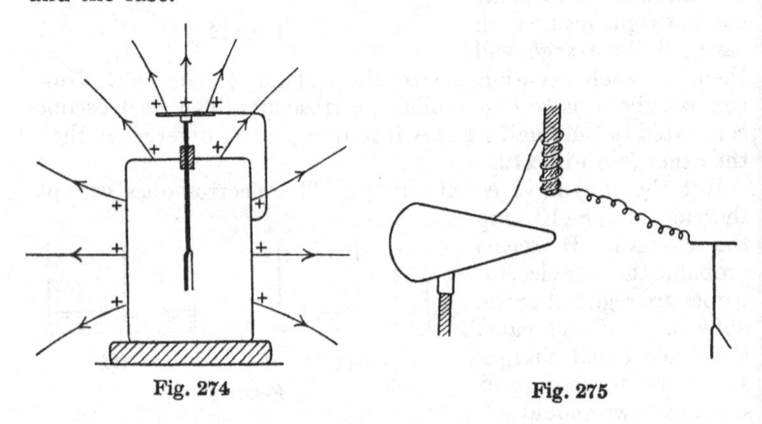

Fig. 274 Fig. 275

The potential throughout a charged conductor is uniform.

By connecting an electroscope permanently with a wire to different parts of the pear-shaped conductor (see Fig. 275), we can demonstrate, by the constant divergence of the leaf, that the potential throughout the pear-shaped conductor is uniform. Of course, if this were not the case, electricity would be bound to flow in the pear-shaped conductor until the potential was uniform: electricity always flows along a conductor if there is a difference of potential.

On p. 330 we explained how, by taking sample charges on a proof plane from various parts of a pear-shaped conductor, we could demonstrate with the self-same instrument—the gold-leaf electroscope—that the charge on the pear-shaped conductor tends to collect at the pointed end.

The vital difference between these two experiments is this. When testing the potential of the pear-shaped conductor a wire connects it to the electroscope permanently: when testing the distribution of the charge on it there is no permanent connection. The proof plane, on being taken away from the pear-shaped conductor, acquires a potential proportional to the sample of charge which it carries.

The electric field.

A charged body brought near to a gold-leaf electroscope affects the divergence of the leaf: it exerts an electric influence around it.

A magnet also exerts an influence in the space around it, since it can affect a compass needle at a distance. The space in which a magnetic influence can be detected is called a magnetic field and may be represented by lines of force.

Similarly, the space in which an electrical influence can be detected is called an *electric field*, and an electric field can also be represented by lines of force.

Lines of magnetic force may be plotted with a compass needle or iron filings, but there is no simple method of plotting lines of electric force. However, if a small paper needle, pivoted on a pin and held by an insulating handle of sealing wax, is brought near to a strongly charged body, such as the knob of a Wimshurst machine, it will set itself along the lines of force. At one end of the paper needle there will be an induced positive charge and at the other end an induced negative charge. The needle will behave as a kind of "electric compass".

We may define a *line of electric force as the path along which a free positive charge would tend to move.* The analogy with magnetism is obvious.

Now a free positive charge tends to move from a point at a higher potential to one at a lower, just as a ball tends to roll downhill. Thus down a line of electric force, i.e. in an electric field, there is a continuous drop of potential. We may regard the P.D. between its ends as the "pulling power" of a line of force. For lines of electric force may be regarded as being in tension in the same way as lines of magnetic force.

The unit of potential is the volt, and two points are said to be at a P.D. of 1 volt if 1 joule of work is done when a charge of 1 coulomb passes from one point to the other (see p. 162). The potential of the earth, as we have already mentioned, is taken as zero.

Fig. 276 shows various sets of lines of electric force.

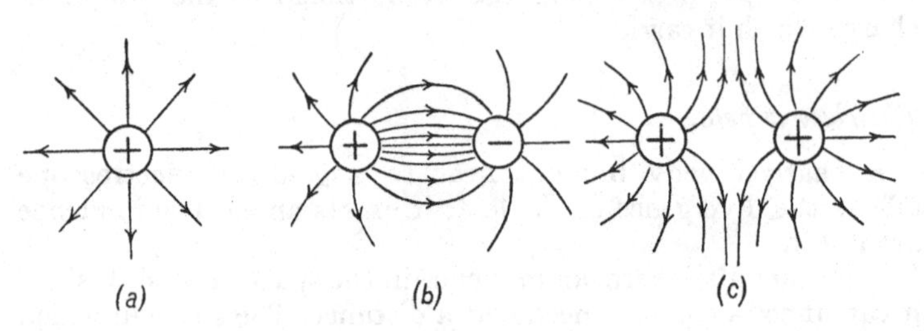

Fig. 276

Properties of lines of electric force.

1. A line of electric force starts at a positive charge and ends at an equal negative charge.

2. There is a drop of potential down lines of electric force, and hence charges tend to move along them. There is no tendency for charges to move at right angles to them. Thus lines of force are always at right angles to the surface of a conductor, since charges do not tend to move along the surface of the conductor (which is at a uniform potential).

3. Lines of electric force can pass through insulators but not conductors.

4. The number of lines of electric force per unit area (perpendicular to the direction of the lines) is proportional to the strength of the electric field at any point.

5. Lines of force may be regarded as being in tension and tending to repel each other sideways.

Use of lines of force.

When a positively charged rod is brought near to the cap of an uncharged electroscope, lines of force pass between it and the cap (see Fig. 277 (*b*)). Lines of force also appear between the leaf and the earthed case, owing to the induced charge on the leaf, and we may regard the divergence of the leaf as being due to the tension of these lines of force.

Suppose now a negatively charged body is brought up (see Fig. 277 (*c*)). Fewer lines of force pass from the positively charged

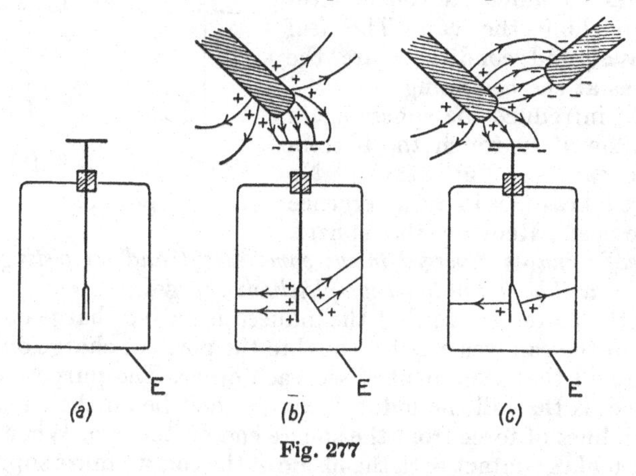

Fig. 277

body to the electroscope. The number of lines of force between the leaf and the case is also less, since the induced positive charge on the leaf is now smaller, and hence the leaf falls.

This simple example demonstrates clearly the usefulness of the conception of lines of force to give a concrete picture of what is happening in the space between charges.

Faraday's ice-pail experiment.

We have stated that a line of electric force starts on, say, a positive charge, and ends on an equal negative charge.

Faraday based this important assumption on a famous experiment performed with an ice-pail, in which he proved that the total induced charge is always equal to the inducing charge.

Place a deep can on the disc of a gold-leaf electroscope. Charge a metal sphere, supported on an insulating ebonite handle, positively by induction with a rubbed ebonite rod. Lower this charged sphere into the can. It will induce a negative charge on the inside of the can and a positive charge on the outside and the electroscope (see Fig. 278(*a*)). The leaf diverges. Remove the sphere without touching the can. The leaf collapses, and conditions are the same as at the beginning.

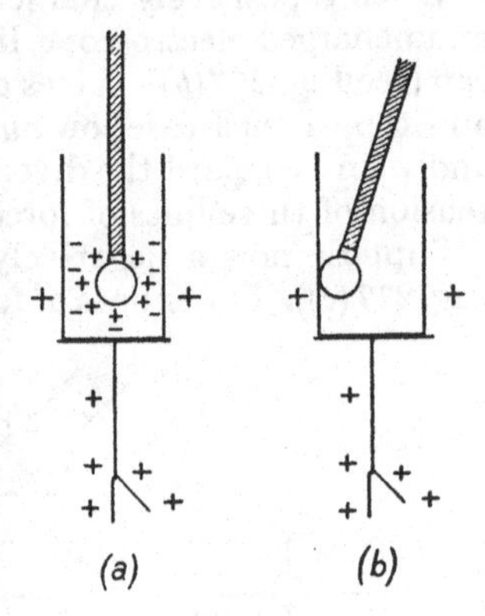

Fig. 278

Next introduce the sphere again and allow it to touch the bottom of the can (see Fig. 278(*b*)). No change takes place in the divergence of the leaf. Remove the sphere. *The leaf remains diverged to the same extent and on testing the sphere it will be found to be completely discharged.*

We therefore deduce that the induced negative charge on the inside of the can was exactly equal to the positive charge on the sphere and that they neutralised each other. The purpose of a deep can is that all the induced charge shall be on the can, i.e. that all lines of force from the sphere end on the can. When the sphere makes contact with the inside of the can we must suppose that all the lines of force contract and disappear.

Touching the inside of a can with a charged ball is clearly a good method of transferring the whole of the charge from the ball to the can.

Capacitance or capacity.

We have seen (p. 341) that if a large and a small can are connected to similar electroscopes and given equal charges, the leaves of the electroscopes diverge to different extents. The small can has a higher potential than the large can.

We say that the large can has a larger electrical capacitance than the small one. It will take more charge if its potential is raised to the same value as that of the smaller can, just as a wide can requires more water than a small can to bring the water level to the same height.

The **electrical capacitance** or **capacity** of a body is defined as **the quantity of charge required to raise its potential by one unit:**

$$C=\frac{Q}{V},$$

where C=capacity, Q=charge, V=rise in potential.

If a charge of 1 coulomb raises the potential of a conductor by 1 volt, the conductor is said to have a capacitance of 1 *farad.* This is a very large unit of capacitance and a smaller unit, the *microfarad,* is commonly used. The maximum capacitance of a variable wireless condenser is usually 0·0005 microfarad, and fixed wireless condensers of more than 2 or 3 microfarads are seldom required.

The condenser.

Charge a gold-leaf electroscope. Place the hand near to the disc. The leaf falls slightly. The potential, V, of the electroscope has therefore fallen slightly and, since its charge, Q, is unaltered, its capacitance has been increased. (If in the equation $C=\frac{Q}{V}$, V is decreased and Q kept constant, C will increase.)

A condenser is an arrangement for increasing the capacitance of a conductor in this way. The simplest form of condenser consists of two parallel plates.

Support a metal plate on an insulating paraffin-wax slab and connect it by means of a wire to an electroscope. Charge the plate positively by induction with a rubbed ebonite rod. The leaf of the electroscope diverges and gives a measure of the potential of the plate. Now gradually bring up a similar earthed plate to the insulated plate. The leaf of the electroscope will fall, showing that the potential of the insulated plate has fallen. Since $C=\frac{Q}{V}$ and the charge Q on the insulated plate is unaltered while the potential V is diminished, the capacitance C must have increased.

The drop in potential of the insulated plate when the earthed plate is brought near to it is due to the induced negative charge on the earthed plate (see Fig. 279). We know that the space surrounding a negative charge has a negative potential, and hence the positive potential of the insulated plate is lowered when the earthed plate is brought near.

Thus in a condenser there is "condensation" of charge on the insulated plate, in the sense that the potential is lowered by the presence of the earthed plate: hence the name.

The real job of a condenser, e.g. in radio circuits, is the storing of charge without unduly high potentials. It is an apparatus where electricity can accumulate and is essential for electric oscillations.

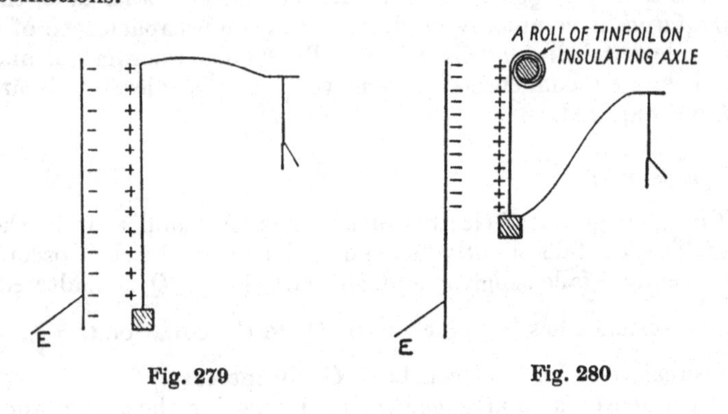

Fig. 279 Fig. 280

On what does the capacitance of a condenser depend?

The capacitance of a condenser depends on (1) the area of the plate, (2) the distance between the plates, (3) the insulator between the plates, called the *dielectric*.

(1) *Capacitance* $\propto$ *area of plates.*

Make a roller blind of tin-foil (preferably weighted at the bottom with lead) which rolls up on an insulating ebonite axle. Connect it by a wire to an electroscope. Charge it positively by induction with a rubbed ebonite rod. Bring near to it an earthed plate (see Fig. 280). When the area of the blind is reduced by rolling it up, the leaf of the electroscope diverges farther, showing that its potential has increased and its capacity, therefore, has diminished. When the area of the blind is increased by unrolling it, its potential diminishes, and hence its capacitance increases.

The capacitance of a variable wireless condenser (see Fig. 281) is altered by varying the area of one set of plates opposite to the other set—rather like rolling up the tin-foil blind.

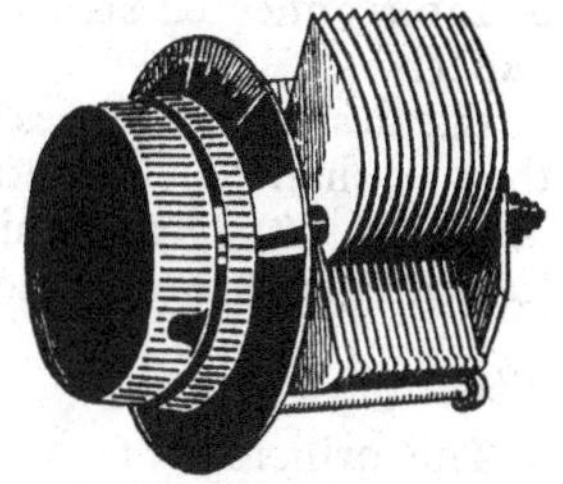

Fig. 281. Variable wireless condenser.

(2) *Capacitance* $\propto \dfrac{1}{\textit{distance between plates}}$.

Using the same apparatus as before, move the earthed plate towards the blind. The leaf of the electroscope tends to collapse, showing a decrease of potential and therefore an increase of capacitance.

(3) *Capacitance depends on the dielectric.*

Introduce a slab of paraffin wax between the plates. The leaf tends to collapse, showing a drop of potential and therefore an increase in capacitance. Repeat with other insulators, such as glass and ebonite.

It can be shown by accurate experiments that if the air between two condenser plates is completely replaced by mica, for example, the capacitance of the condenser is increased about six times. Mica is said to have a *specific inductive capacity* (S.I.C.) or *dielectric constant* (K) of 6.

Definition.

S.I.C. of a substance

$$= \frac{\text{Capacitance of a condenser with that substance as dielectric}}{\text{Capacitance of same condenser with air as dielectric}}.$$

Some Specific Inductive Capacities:

Substance	S.I.C.
Paraffin wax	2 approx.
Ebonite	3
Mica	6·5
Crown glass	7

Fixed capacitance wireless condensers are made of two sets of interconnected strips of tin-foil separated by mica or waxed paper.

The desirable features of a dielectric for use in a condenser are that it shall (1) be obtainable in thin sheets, (2) be a good insulator, (3) have a high S.I.C. These do not necessarily go together, as, for example, in the case of glass.

The Leyden jar.

The principle of the condenser was discovered in 1745 by Von Kleist, dean of the cathedral of Kanim, Pomerania, and also independently in 1746 by Pieter van Musschenbroek, professor of physics at Leyden University, Holland.

These men conceived the idea of storing up a charge of electricity in a bottle of water, which behaved as a condenser. The water inside the bottle acted as the insulated plate, the glass as the dielectric, and the hand holding the bottle as the earthed plate. Each man received an unexpected and tremendous shock on making contact with the water with the other hand. Von Kleist declared that the shock "stunned his arms and shoulders", while Musschenbroek vowed that he would not take another shock for the whole kingdom of France.

To-day condensers are made in the form of a glass jar with an insulated coating of tin-foil inside, connected to a knob for charging or discharging purposes, and an earthed coating of tin-foil outside (see Fig. 282). Such a jar is known as a Leyden jar and one of pint size has a capacitance of about 0·001 microfarad.

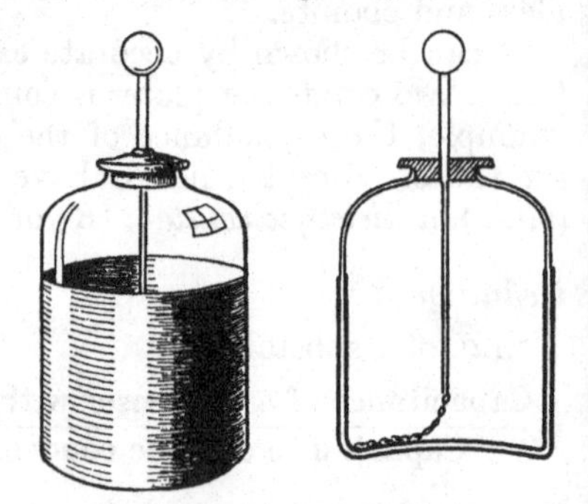

Fig. 282. The Leyden jar.

The fame of the Leyden jar spread far and wide and a shock from it was advertised by quacks as a cure for diseases. It is recorded that the Abbé Nollet, to amuse the King of France, gave a shock from a large Leyden jar to over a hundred austere monks linked up together.

In England, the Hon. Henry Cavendish (1731–1810), the brilliant but eccentric grandson of the Duke of Devonshire, compared the capacities of different Leyden jars, by charging them all up to the same potential and estimating their charges by the shock they gave him on discharge. He was using his body as a kind of galvanometer.

Cavendish made numerous important discoveries, including Ohm's law, but owing to his refusal to publish them, many were rediscovered years later. His papers were edited ultimately by Clerk Maxwell and published in the middle of the nineteenth century. It has been said that "as a physicist he was easily first in England, if not in Europe", and the world-famed physics laboratory at Cambridge University bears his name. He lived as a recluse and uttered very few words. He communicated with his servants by notes left on the hall table and any female servant coming into his sight was instantly dismissed. He met his nephew, whom he made his heir, for a few minutes once a year.

The identity of static and current electricity.

The static electricity produced by rubbing, or by induction with the electrophorus or Wimshurst machine, is identical in nature with the current electricity produced by an electric cell or dynamo. The apparent difference in their effects is due to the fact that static electricity is usually at a very high potential but small in quantity, whereas current electricity is usually generated at a low potential difference but in comparatively large quantity. This difference is strikingly revealed by the fact that the E.M.F. of a simple cell is 1 volt, whereas the P.D. required to produce a spark 1 cm. long between the knobs of a Wimshurst machine is of the order of 30,000 volts. The reason why the shock from a Wimshurst machine is not fatal is that the quantity of electricity it generates is so small: a shock from a dynamo or cells at a P.D. higher than 250 volts is extremely dangerous, since the quantity of electricity passing, in this case, depends almost solely on the resistance of the human body, and also the supply is continuous.

The electricity generated by a battery of several thousand cells will cause the leaf of an electroscope to diverge.

On the other hand, Benjamin Franklin, in 1751 magnetised

a sewing needle by the discharge from a Leyden jar. Slight electrolysis can be produced by a Wimshurst machine. If the wires connected to the knobs of the machine are placed on pole-finding paper (see p. 107), the characteristic coloration appears.

SUMMARY

The charge on a conductor resides on the outside surface and tends to collect in the more curved parts. The accumulation of charge on a point is so great that the charge leaks away in the form of an electric wind. Charges may be lost or gained by *the action of points*, *vide* Van de Graaff's generator and lightning conductors.

Potential is that condition of a body which determines the direction of the flow of electricity when the body is connected to the earth. It corresponds to water level or temperature. The potential of a body may be raised in two ways: (*a*) by giving it a positive charge, (*b*) by bringing a positively charged body near to it. The potential throughout a charged conductor is uniform.

The gold-leaf electroscope measures potential rather than charge.

The *electric field* is any space in which an electric influence can be detected and it may be represented by lines of electric force. A line of electric force is the path along which a free positive charge would tend to move. (For properties of lines of electric force see summary on p. 344.)

Faraday's ice-pail experiment proves that the induced charge equals the inducing charge and hence justifies the assumption that there are equal positive and negative charges at the ends of a line of force.

The *capacitance* (*or capacity*) of a conductor is the quantity of charge required to raise its potential by 1 unit:

$$C=\frac{Q}{V}.$$

The simple *condenser* consists of two plates: when one is insulated and charged and the other is earthed, the capacity of the first plate is increased as a result of the presence of the induced charge on the second plate.

The capacity of a condenser depends on (1) the area of the plates, (2) the distance apart of the plates, (3) the dielectric between the plates.

Specific Inductive Capacity (or dielectric constant, K)

$$= \frac{\text{Capacitance of a condenser with the substance as dielectric}}{\text{Capacitance of the same condenser with air as dielectric}}.$$

The *Leyden jar* is a form of condenser.

Static and current electricity are identical: but the former is usually small in quantity and at a high potential; the latter is comparatively large in quantity and at a low potential.

QUESTIONS

1. "Electrical charges reside only on the outer surface of a hollow conductor."

Describe fully any one experiment that you could make to verify the above statement.

Explain how you could charge two similar cans (*a*) with equal charges, (*b*) so that one has twice the charge of the other.

Two similar insulated conductors which will go inside the cans are provided, together with an electrophorus or Wimshurst machine.

(N.)

2. A hollow conical conductor *A* (see Fig. 283) is placed on an insulating stand and is charged with electricity. Samples of charge are taken from the various points on the outside and inside of *A* with a proof plane and transferred completely to an uncharged gold-leaf electroscope. Describe and explain the behaviour of the electroscope.

What differences would have been observed if a proof plane connected to the electroscope with a wire had been allowed to make contact with the various points on *A*? Give reasons. (O. and C.)

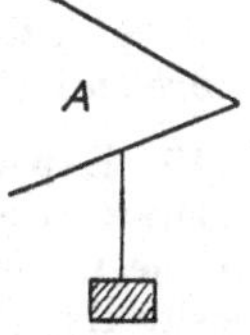

Fig. 283

3. Explain the following observations:

(*a*) If the palm of the hand is held near a sharp point which is attached to one of the knobs of a Wimshurst machine, a cold wind is felt when the machine is working, but no sparks pass into the hand.

(*b*) If the palm of the hand is held near to the cap of a charged electroscope the leaves collapse slightly, but resume their original position when the hand is removed. (O. and C.)

4. Explain:

When a charged glass rod is brought near a gold-leaf electroscope the leaves diverge, but fall together again when the rod is removed. If a needle is attached to the cap of the electroscope with its point upwards and the experiment is repeated, the leaves diverge when the charged rod is brought up, and remain divergent when the rod is removed. (O. and C.)

5. Describe the function of a lightning conductor. Why is a lightning conductor more necessary in the case of a brick chimney than of a steel smokestack of the same height?

6. Two insulated metal spheres, of unequal size, are both charged with positive electricity. When they are connected by a thin wire, what determines whether electricity will flow along the wire? Give a simple numerical illustration, if possible. (L.)

7. Describe and explain the changes in potential of a *positively charged* gold-leaf electroscope when a negatively charged ebonite rod is brought up so that the leaf drops and rises again.

8. Describe how a conductor may have (1) a positive charge, but zero potential, (2) two equal and opposite charges (i.e. no resultant charge), but a positive potential.

What happens in each case if the conductor is connected to the earth?

9. A rather insensitive gold-leaf electroscope is connected by a wire to the metal disc of an electrophorus. Describe and explain the indications of the electroscope, while the electrophorus is being used to charge a Leyden jar. (C.)

10. Explain what you understand by a line of electrostatic force.

Draw careful diagrams to show the distribution of the lines of force in the following cases: (*a*) a small insulated metal sphere charged positively, (*b*) when an insulated conductor is placed between the sphere and an earthed plate, (*c*) when the conductor is connected to the earthed plate by means of a metal wire. (L.)

11. Show by diagrams the distribution of electric lines of force in the following cases:

(*a*) In the neighbourhood of a small positively charged spherical conductor.

(*b*) In the same neighbourhood when a similar but uncharged insulated conductor is brought near.

(*c*) When the second conductor is earthed.

12. A metal ball hanging by a silk thread is (*a*) charged positively and brought in contact with the top of an electroscope, (*b*) similarly charged and brought in contact with the inside of a hollow insulated metal pot connected to the electroscope. What difference would you expect in the effect of each operation on the condition of the electroscope and of the metal ball? (L.)

13. What do you understand by electrostatic induction?

(*a*) An insulated, positively charged metal ball is lowered into a deep tin can standing on an insulating stand but is not allowed to touch the inside of the can.

(*b*) The can is now touched with the fingers and the ball withdrawn.

(*c*) The ball is again lowered into the can, allowed to touch the sides, and again withdrawn.

Explain carefully the nature and distribution of the charges on the can during each of the stages (*a*), (*b*) and (*c*). (O. and C.)

14. How may a charge equal to that on a metal sphere with an insulating handle be given to a hollow metal can so that (*a*) the sphere is completely discharged, (*b*) the charge on the sphere is unchanged?

15. If you were provided with a small insulated charged conductor, describe how you could, by means of it, communicate to a hollow metal vessel a charge many times greater than that of the conductor. (L.)

16. Explain how you would use an electroscope (1) to determine the sign of an electrostatic charge, (2) to compare the magnitudes of two charges, (3) to determine which of two insulated conductors was at the higher potential. (L.)

17. What is meant by the capacity of an electric condenser? On what factors does it depend and how do they affect it?

Describe how you would make a compact condenser of large capacity out of some tin-foil and waxed paper. (N.)

18. Two similar vertical insulated plates, P and Q, are placed parallel to each other and about an inch apart. Each is connected to the cap of an electroscope. State and explain the indications of the electroscopes (*a*) when a positive charge is given to P, and (*b*) when Q is afterwards earthed. (O.)

19. Carefully explain what is meant by a line of electric force.

Give diagrams of the lines of force in the field of a charged parallel-plate air condenser (*a*) when the uncharged plate is insulated, (*b*) when the uncharged plate is earthed.

How would you test the distribution of charge on the two sides of the charged plate in case (*b*) above, and what result would be obtained in such a test? (N.)

20. Describe the Leyden jar.

Why can it be carried about in the hand without losing its charge?

Why cannot it be charged appreciably when standing on a slab of paraffin wax?

21. A gold-leaf electroscope having a fixed plate and a single leaf is charged negatively so that the leaf stands at an angle of about 45° to the plate. Describe with full experimental details how you would produce (*a*) a small temporary increase, (*b*) a small permanent increase in divergence, (*c*) a small temporary decrease, (*d*) a small permanent decrease in the divergence of the leaf. (L.)

22. Explain what is meant by specific inductive capacity.

An insulated metal can is connected by a thin wire with an electroscope and then an insulated charged body is lowered into it without being allowed to touch the sides or the bottom. Describe what you expect the leaves of the electroscope to do, illustrating your description with diagrams. What changes would take place if the space between the charged body and the can were to be filled with a melted non-conductor, such as paraffin wax or resin? Explain why. (N.)

23. What are the most important points in which the electricity generated by a Wimshurst machine differs from that which you could obtain with a Daniell cell? Describe any simple experiments by which you could show the essential identity of the two. (C.)

24. Describe briefly three experiments; the first to show the production of electrostatic effects by means of voltaic cells, the second and third to show the production by means of a Wimshurst or other electrostatic machine of magnetic and true electrochemical effects (not effects due to the formation of ozone). (L.)

Chapter XIX

THE MODERN ERA

As the end of the nineteenth century approached, many physicists believed that their subject was almost rounded off and complete. They thought that the farthest limits of their explorations into the unknown werc in sight. The road, as far as they could see, had been broadly surveyed and the task they contemplated handing over to their successors was that of filling in the details.

There was, indeed, one set of phenomena, the discharge of electricity through rarefied gases, which defied explanation. Few scientific men suspected that the further investigation of those phenomena would reveal a new and vast tract of territory. Yet during the last forty years the discoveries in this field have followed so thick and fast that it is difficult for any but the specialist to keep up with them.

The discharge of electricity through rarefied gases.

It was found by Watson (1715–1787) that if the pressure of the air in a vessel is reduced, the possible length of the spark produced by a given P.D. becomes greater. Eventually the discharge is silent and beautifully coloured. The phenomenon was investigated by a number of men, such as Davy, Faraday and Plücker: but owing to their inability to frame a hypothesis or theory, their experiments were little more than observations, with no fruitful results. A skilful glass-blower, Geissler, became famous for the Geissler or "vacuum" tubes, containing traces of different gases which gave beautiful luminous discharges and were sold rather as interesting scientific toys.

We will describe the appearance of the discharge at various pressures. Suppose we have a long glass tube with two metal electrodes, at the ends, at a P.D. of several thousand volts. As the pressure of the air or gas in the tube is reduced by a pump, a silent discharge begins and a bright luminous glow, whose colour depends on the nature of the gas, fills the tube. At a pressure of the order of 1 mm. of mercury the bright glow in the tube begins to shrink, as it were, towards the anode. In Fig. 284, third line down, this bright glow, called the *positive column*,

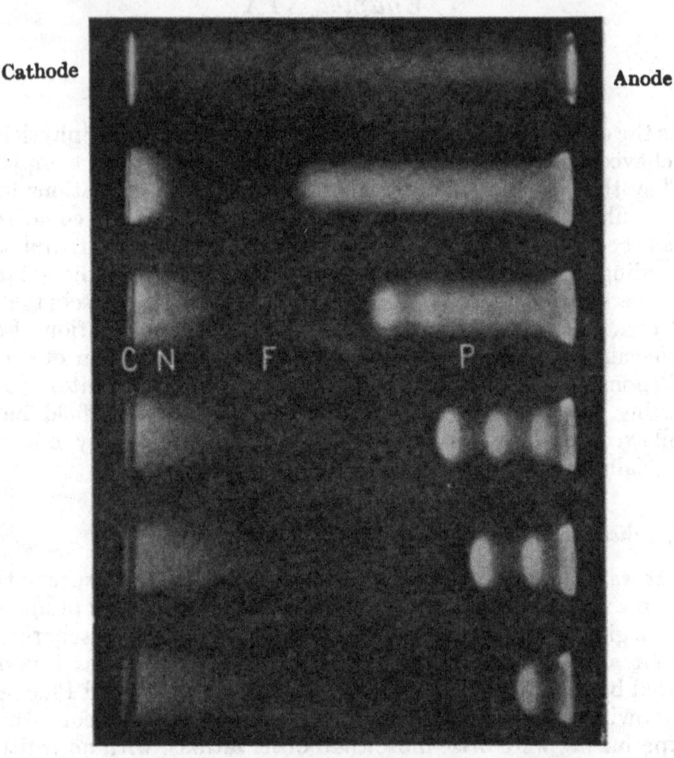

By courtesy of Messrs Williams and Wilkins Co., Baltimore, U.S.A., from Darrow's "Electrical Phenomena in Gases"

Fig. 284. Electrical discharges through a rarefied gas: the pressure is 1 mm. of mercury at the top and decreases to 0·1 mm. at the bottom. *C* is the Crookes dark space, *N* the negative column, *F* the Faraday dark space and *P* the positive column. Note how the discharge seems to grow away from the cathode and disappear at the anode as the pressure decreases, and also how the positive column becomes striated.

fills rather less than half the tube: behind it is a dark space called the *Faraday dark space*, then a bright glow called the *negative column*, another dark space called the *Crookes dark space*, and finally a narrow luminous glow round the cathode. As the pressure in the tube is further reduced the positive column shrinks more and more towards the anode and becomes striated while the Crookes dark space grows in size: eventually, at a pressure of about 0·02 mm. (much lower than that in the bottom line of Fig. 284), the positive column, Faraday dark space and negative column have disappeared, and the Crookes dark space fills the whole tube. The glowing gas has now disappeared, but the glass of the tube fluoresces green if soda glass, blue if lead glass, and so on. Finally, as a more perfect vacuum is approached, the discharge ceases completely.

Neon tubes.

The red neon and other coloured tubes which enliven Piccadilly Circus at night are discharge tubes in which the pressure is such that the positive column fills the tube. The tubes can be bent into letters or any fantastic shapes and the discharge still passes.

The red neon light is bright and will pierce slight fog or mist: hence it is used for beacons at aerodromes (see Fig. 285).

Other colours are obtained by using different gases or tinting the tube: argon gives a pale blue light; argon mixed with a little mercury vapour gives a brilliant blue light; helium gives a yellow-white light; while a green colour, not so brilliant as the others, is obtained by using yellow glass containing argon and mercury.

Cathode rays.

The first great discoveries were made by the examination of the Crookes dark space at such a pressure that it filled the discharge tube. Indeed, we shall have no more to say about the earlier stages of the discharge, which are capable of explanation but have proved to be of secondary importance.

Although the Crookes dark space is dark there must be something there since it causes the glass to fluoresce: furthermore, if other suitable minerals are placed inside the tube they too will fluoresce brilliantly. If an obstacle is placed in the discharge tube it will cast a shadow. In Fig. 286 the anode is at the side, and it is obvious from this experiment that something is being emitted

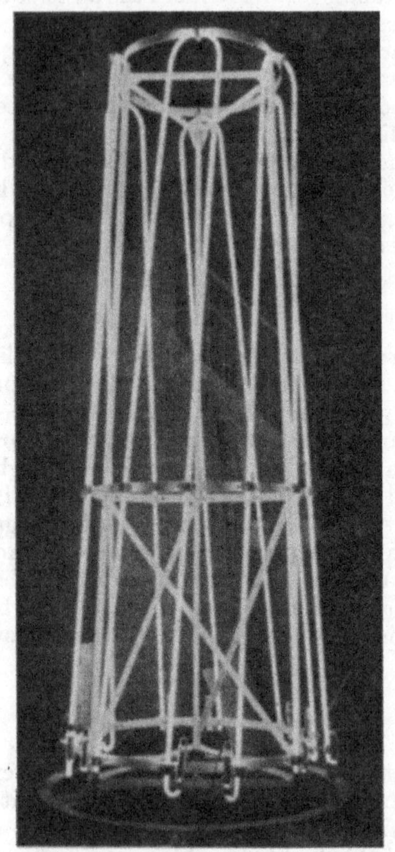

By courtesy of the General Electric Co. Ltd.

Fig. 285. A neon beacon at Croydon Airport. An electric discharge passes through neon gas contained in the tubes giving rise to a red light which has high powers of penetration in fog and is visible, under favourable conditions, at a distance of 50 miles. The beacon can be made to flash any Morse signal.

from the cathode which is stopped by the mica cross from reaching the end of the tube in the shadow.

What is this something? The German investigators thought it consisted of rays, a kind of invisible light, and Goldstein called them *cathode rays*, a name we have retained. Sir William Crookes hazarded the bold guess that it consisted of fast-moving particles, smaller than the atoms or molecules of the gas, "a fourth state of matter" or "radiant matter", as he called it. One of the experiments which prompted Crookes to make this assumption

B. F. Brown

Fig. 286. The shadow of a mica cross cast by cathode rays on the end of a discharge tube. The cathode is on the right and the anode is at the bottom of the vertical side tube.

was that shown in Fig. 287. A light mica paddle wheel on rails is made to revolve when the cathode rays strike it.

Again, and this is a very significant fact, cathode rays may be deflected by a magnet: their path is revealed by causing them to graze along a fluorescent screen (see Fig. 288). The direction in which they are deviated suggests that they may be negatively charged particles, which, by virtue of their motion, constitute an electric current.

In the year 1894, Sir J. J. Thomson, Cavendish professor at Cambridge University, began to investigate the phenomenon, and in a series of masterly researches proved that cathode rays consist of tiny, negatively charged particles which we now call

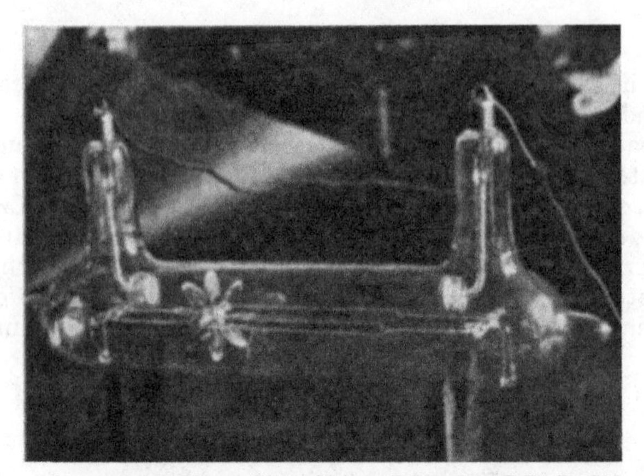

By courtesy of Messrs George Newnes Ltd. and the Tella Co. Ltd.

Fig. 287. A light mica paddle wheel which is made to run along glass rails under the impact of cathode rays. The two electrodes are the aluminium discs above the rails and either may be made the cathode. The wheel runs away from the cathode. The glass tube is, of course, highly exhausted (0·02 mm.).

By courtesy of Messrs George Newnes Ltd. and the Tella Co. Ltd.

Fig. 288. The deflection of cathode rays by the N. pole of a bar magnet. The rays are rendered visible by a fluorescent screen. The reader should apply Fleming's left-hand rule to this phenomenon. Remember that the positive direction of the field is away from the N. pole and that the cathode rays are equivalent to a current of positive electricity in the opposite direction.

electrons. In 1897, by deflecting the rays by means of a magnetic field, and also an electrostatic field, he was able to find their velocity and the ratio of the charge to the mass (e/m) of each particle. In 1898 he found the charge, e, of each particle by a new method and was able to calculate, knowing e/m, that the mass, m, of each particle was only about $\frac{1}{1840}$th of the mass of the lightest known atom, that of hydrogen. During the next few years, directing the most brilliant team of experimenters which has ever been gathered in one laboratory, he was able to gain a great deal of information about the electron, some of which we discussed in Chapter II.

There have been numerous other workers in this field. Townsend was the first to find the charge of the electron and Millikan obtained what was considered a very accurate value until it was checked recently by a new method utilising X-rays.

X-rays.

In the autumn of 1895, Wilhelm Konrad Röntgen, professor of physics at Würzburg, was working with a discharge tube when he found that some photographic plates in the same room, wrapped up, as usual, in black opaque paper, became fogged. After a few days he traced the cause of the fogging to the discharge tube. Some kind of radiation was coming from the discharge tube and it was sufficiently penetrating to pass through the wrapping round the plates. Röntgen called this mysterious radiation *X-rays*. He was able to show that the X-rays were produced when cathode rays hit the walls of the discharge tube, that they were not deflected by a magnet and that they were very penetrative but could be stopped by a comparatively small thickness of lead.

We now believe that they are waves in the aether of a nature similar to light or wireless waves, but of very short wave-length. Their wave-length varies between 10^{-7} and 10^{-9} cm.: the longer waves are the less penetrating and are called soft rays: the shorter ones are the more penetrating and are called hard rays.

X-rays may be used to radiograph any part of the body (see Fig. 289): since the bones or foreign bodies, such as a bullet, are denser than flesh, they reduce the intensity of the X-rays, and cast shadows on a fluorescent screen (usually of barium platinocyanide), or on a photographic plate. The digestive tract

By courtesy of Messrs George Newnes Ltd. and the Tella Co. Ltd.

Fig. 289. The shadow of a hand cast by X-rays on a fluorescent screen.

of a patient may be examined by giving him a meal of bread and milk containing 2 oz. of bismuth carbonate, which is comparatively opaque to X-rays. Flaws in steel up to 5 inches thick may be detected swiftly and reliably by X-rays and the method is coming into wide use.

Prolonged exposure to X-rays is dangerous. X-ray tubes are often enclosed in a lead box from which a beam may emerge only through a window. The early investigators suffered from a disease known as dermatitis and several lost fingers, arms, and even their lives, as a result.

Gas tubes.

X-rays are generated whenever high-speed electrons hit something.

If low-speed electrons strike a body, they knock some of the outer planetary electrons out of the atoms which they hit, and when these outer electrons return to their places visible light is emitted. Energy is required to knock them out of position and energy is radiated when they return. The light of a spark is emitted when millions of outer electrons, which have been wrenched out of atoms by the intense electric field or knocked out by fast-moving gaseous ions, return to their original positions.

Now high-speed electrons have sufficient energy to penetrate inside the electron swarm around the nucleus of an atom and dislodge some of the inner planetary electrons. X-rays are emitted when these inner electrons return to their positions.

The earliest type of X-ray tubes are known as gas tubes (see Fig. 290): they are simply highly evacuated discharge tubes, with a concave cathode to focus the cathode rays on to an anode or anticathode. The anticathode is simply put there to be hit by the electrons: it must be made of a metal with a high melting point, such as platinum or tungsten, since it gets very hot, and its surface is inclined so that the X-rays may be emitted sideways from the tube.

The penetrating power of the X-rays depends upon the speed of the electrons and therefore upon the voltage across the tube: the intensity of the X-rays depends on the number of electrons in the cathode rays, i.e. on the current through the tube.

A very high voltage is needed with the tube—up to 100,000

volts—and, in the early days, this was provided by an induction coil. To-day an A.C. transformer and rectifier are used.

The gas tube fluoresces an apple-green colour when in use: if the colour is blue this indicates that the "vacuum" is lower, and that the X-rays are therefore more intense but less penetrative.

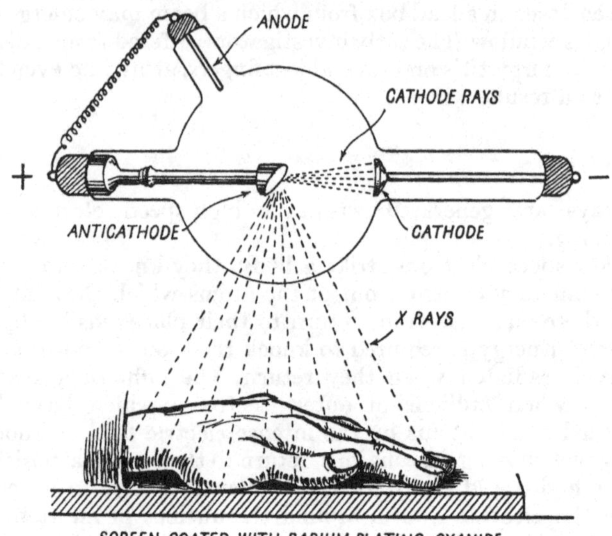

Fig. 290

The Coolidge tube.

The type of tube now used was invented by Coolidge about twenty-five years ago. It is exhausted so highly that no discharge would pass through it if used as a gas tube. The cathode, however, consists of a hot tungsten filament which emits electrons just as does the filament of a wireless valve. We may regard the free electrons in the filament as being "boiled out", and the rate of their emission may be controlled by means of the heating current through the filament. This is a great advantage.

The electrons are given a high speed by the P.D. maintained across the tube and strike an anticathode (see Fig. 291). As in the gas tube, X-rays are emitted when these high-speed electrons hit the anticathode.

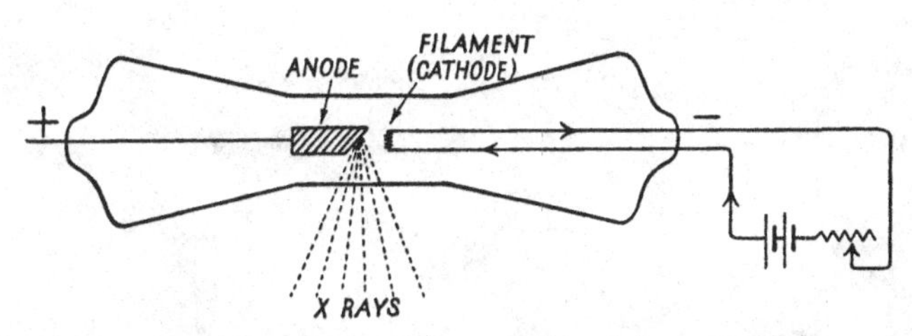

Fig. 291. Diagram showing the principle of a modern type of "Coolidge tube", known as a Metalix tube. The filament is here shown heated by means of a battery: in practice a transformer is used. A P.D. of 200 kV. is put across anode and cathode to give the electrons, emitted from the filament, a high speed.

The ionisation of gases.

When X-rays pass through air they render it conducting. They cause electrons to be ejected from the air molecules leaving the latter positively charged. The process is known as ionisation, and the charged molecules and electrons are called ions.

Air may also be ionised by the heat of flames, and by intense electric fields. The air in the neighbourhood of a charged point becomes ionised: if, say, the point is positively charged, the positive ions are repelled away as an electric wind, while the negative ions are attracted to, and neutralise the charge on, the point (see p. 330).

A charged body, however well insulated, gradually loses its charge in air. This is due to the slight ionisation of the air by the mysterious cosmic rays coming, probably, from beyond our world and which are being investigated by physicists at the present time.

Radioactivity.

X-rays produce fluorescence: a screen covered with barium platinocyanide, for example, fluoresces green in X-rays. Henri Becquerel set out to find whether fluorescent substances produce X-rays.

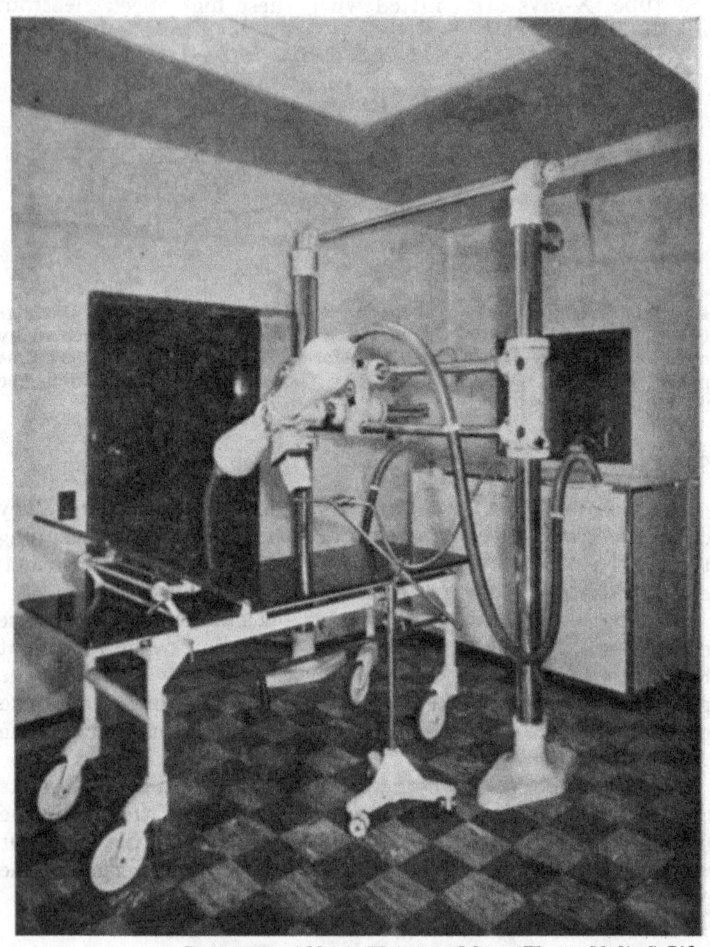

By courtesy of Messrs Watson and Sons (Electro-Medical) Ltd.

Fig. 292. A Metalix, X-ray, deep therapy installation. The X-rays are generated in the white, dumb-bell shaped, metal tube (compare with Fig. 291). The patient lies on the couch, and the gantry stand permits unlimited movement of the tube.

He examined pitchblende, a black shiny ore of the metal uranium, mined in Bohemia. He discovered in 1896 that this mineral affected a photographic plate although the latter was wrapped in opaque paper—as do X-rays. He found also that uranium salts would make air conducting. They were clearly giving off some kind of invisible radiation.

His experiments aroused the interest and enthusiasm of a professor in the School of Physics and Chemistry in Paris, Pierre Curie, and his wife. Together, the Curies began to investigate the phenomenon and found that salts of thorium behaved like those of uranium. They also succeeded in isolating a new active element, which they called *polonium*, in honour of Madame Curie's native land, Poland.

The Curies, as a result of their experiments, realised that there was another unknown and extremely active substance in pitchblende, present in minute quantities. They determined to extract it. Several sacks and eventually several tons of pitchblende were obtained from the mines at St Joachimsthal, Bohemia, and the arduous business of separation was carried on in an abandoned shed near the School of Physics, Paris. In December 1898 they announced that they had discovered *radium*.

Eight years later Pierre Curie was knocked down and killed by a car in a Paris street. Madame Curie continued the work and became head of the Radium Research Institute of Paris. In 1921 she went to the U.S.A. and received, as a gift from American women, 1 gm. of radium, of value about £20,000, extracted from 500 tons of Colorado carnotite ore.

The invisible rays from radium and similar substances affect the body. Becquerel carried a few milligrams of a radium salt in a glass tube in his waistcoat pocket for several days and it produced a sore on his body. The cells of cancer are destroyed by rays from radium, which is now used in the treatment of the disease.

The energy emitted by radium.

Some forty elements have been discovered which like radium spontaneously give out energy: they are said to be radioactive. All of them have a very high atomic weight, over 200; they are, in fact, the heaviest elements known.

The energy emitted by radium, in the course of a long period of time, is very large. One gramme of radium gives out 135 calories every hour. This will go on for thousands of years, the amount becoming slowly smaller as the years go by.

Physicists were deeply interested in the source of this energy. Where did it come from? Was the principle of the conservation of energy being violated? Madame Curie thought that there might be some very penetrating and hitherto undetected rays in the world, rather like hard X-rays, which were absorbed by very heavy atoms such as radium and that these were the source of the energy.

The rays from radium were carefully examined by the Curies, Rutherford and Villard. As a result of their researches it was found that there were three kinds of rays given off (see Fig. 293):

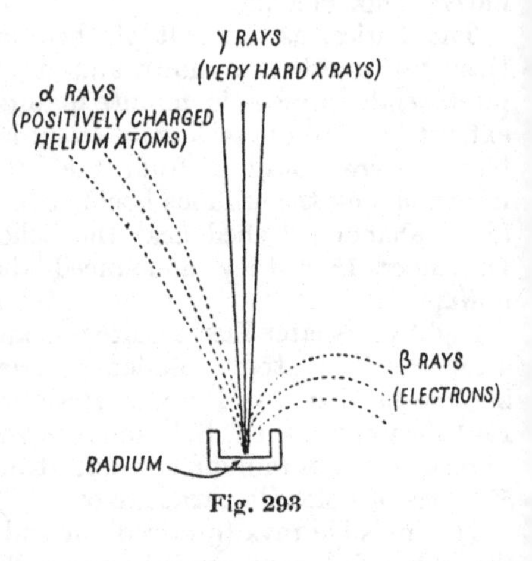

Fig. 293

1. *α-rays*, which could be deflected slightly by a very powerful magnet, and which proved to be *positively charged atoms of helium.*

2. *β-rays*, easily deflected by a magnet in a direction opposite to that of α-particles, more penetrative than α-particles, and which are *electrons.*

3. *γ-rays*, which will pass through a thickness of lead sufficient to stop α- and β-rays, which are undeflected by a magnet, and which are *like X-rays of very short wave-length.*

In 1900 Rutherford investigated a curious emanation from radium compounds now known as *radium emanation*, which he proved to be not the vapour of radium but a different substance. This heavy gas, when collected separately, gradually disappeared, being changed to a solid deposit.

The disintegration theory.

In 1903, Rutherford and Soddy presented their disintegration theory to the British Association. They suggested that the nuclei of radium atoms were disintegrating or exploding: they were being transformed into radium emanation, which in turn was being transformed into something else: meanwhile, high-speed α- and β-rays, and also γ-rays, were being emitted.

The theory was the subject of much discussion in which the great Lord Kelvin took the side of the opposition. The idea of atoms exploding was too revolutionary even for his flexible and original mind. He suggested, like Madame Curie, that the source of the energy must be sought outside the radium atom. In August 1906, just after the meeting of the British Association of that year, he initiated, by a strong letter, a dramatic controversy in *The Times* on the subject.

But Kelvin, who died in the following year, was wrong. The disintegration theory has now been established. Radium itself is formed as the result of the disintegration of uranium and, after passing through many transformations, finally becomes lead.

One gramme of radium is reduced to half a gramme through disintegration in 1730 years: in a further 1730 years it is again halved, and there is a quarter of a gramme left, and so on. This period of time is known as the *half value period.* Another way of putting the same thing is to say that 1 in every 1,000,000,000 radium atoms disintegrates each second.

The following is the beginning of the chain of substances formed when radium disintegrates, the half value period of each being placed below and the atomic weights in the circles. What we call radium is always a mixture of these products, which are continually being formed and breaking up themselves:

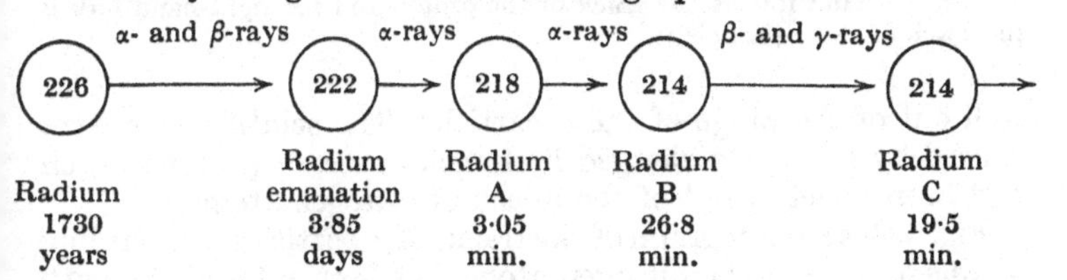

Radioactive transformations cannot be hastened or delayed: they are, apparently, uncontrollable.

Artificial disintegration.

In 1919 Rutherford attempted to disintegrate atoms of elements which are not radioactive. He bombarded nitrogen with the α-particles, moving at over 10,000 miles per sec., from radium. To smash a nitrogen atom a direct hit upon the nucleus is necessary. Rutherford achieved several random, direct hits which were registered by scintillations on a zinc-blende screen

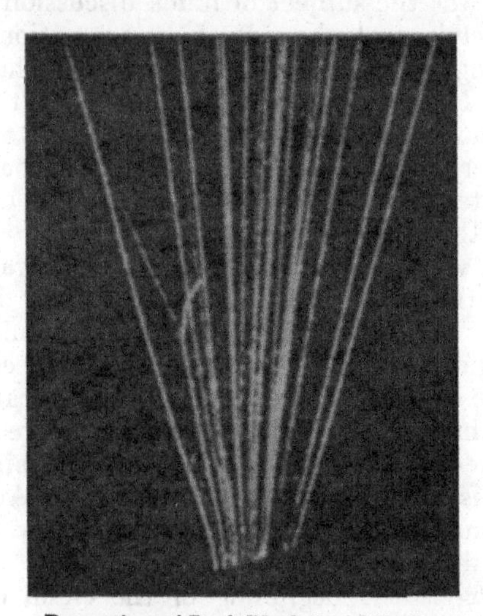

By courtesy of Prof. Blackett and the Royal Society

Fig. 294. The tracks of α-particles emitted from radium C and passing through nitrogen gas. The second track on the left is forked, showing that an α-particle has struck a nitrogen atom and knocked out a proton; the longer, left-hand fork is the track of the proton and the right-hand fork is the track of the α-particle.

well out of the range of the α-particles. The scintillations were caused by positively charged hydrogen atoms or protons which had been knocked out of the nuclei of nitrogen atoms.

Fig. 294 is a photograph of the tracks of α-particles bombarding nitrogen. One of the nitrogen atoms has been hit and the path of the proton can be seen. The tracks are made visible by the condensation of water vapour in an apparatus devised by C. T. R.

Wilson. Blackett counted 300,000 tracks of α-particles and only 8 were deflected; this was to be expected since, as we have said, a direct hit is essential.

Rutherford and Chadwick found that many elements could be transformed in this way. A particularly interesting case was that of beryllium. When bombarded with α-particles the beryllium atom was changed into carbon and there was ejected from its nucleus a new type of particle, with the mass of a proton but uncharged, which Chadwick called a neutron.

Cockcroft and Walton, in the Cavendish Laboratory, Cambridge, performed bombardment experiments with fast-moving protons which were given their high speed by making them fall through a P.D. of half a million volts. The Van de Graaff generator, which we described on p. 333, has been designed for the performance of similar experiments with much faster particles.

The dream of the old alchemists, the transformation of base metals into gold, is now an accomplished fact. Gold can be produced in minute quantities by bombarding mercury with α-particles, but the process is far more expensive than that of mining natural gold.

An immense amount of research in atomic physics is now proceeding throughout the world. The possibilities of the future are stirring and incalculable.

Wilson, Blackett [illegible] 400,000 tracks of [illegible] and only 8 were detected. It was to be expected since, as we have [illegible] [illegible] nucleus is [illegible].

Rutherford and Chadwick [illegible] [illegible] could be transformed in [illegible]. A particularly interesting case was that of beryllium. When bombarded with α-particles the beryllium nucleus was changed into carbon and there was ejected from its nucleus a new type of particle with the mass of a proton but no charge, which was named called a neutron.

Cockcroft and Walton [illegible] Cavendish laboratory, [illegible] [illegible] performed [illegible] [illegible] with [illegible] protons which are given a high speed by [illegible] [illegible] through [illegible] [illegible] volts. [illegible] [illegible] [illegible] [illegible] [illegible] has been designed for [illegible] [illegible] of [illegible] [illegible] [illegible] particles.

The dream of the old alchemists, the transformation of base metals into gold, [illegible] [illegible] [illegible] [illegible] [illegible] produced in minute quantities by bombarding mercury with α-particles, but the process is far more expensive than that of mining natural gold.

[illegible] [illegible] attempt [illegible] research in atomic physics [illegible] proceeding throughout the world. The possibilities of the future are amazing and incalculable.

ANSWERS TO QUESTIONS

CHAPTER III (page 50)

14. 1·9 ohms. **15.** 10 ohms, 20 volts.
16. 0·0596, 0·00542, 0·0542 amp. **17.** 0·54 amp., 3·7 ohms.
18. (*a*) $\frac{1}{2}$ amp., (*b*) 5 volts, (*c*) 12 volts.
19. (*a*) 4 ohms, (*b*) 50 volts, (*c*) 10 volts. **20.** 1·8 amps.
22. 5 cells. **23.** 0·41, 0·51 amp.; 0·28, 0·35 amp.
24. 1·5 volts. **25.** 5 : 3. **26.** 5 cells. **27.** 1·29 amps.

CHAPTER VI (page 98)

8. 0·00435, 0·00066 amp. **9.** 25 ohms.
10. (*a*) $\frac{5}{9}$ ohm in parallel, (*b*) 15 ohms in series. **11.** 0·15 ohm.
12. Series resistance of 19,900 ohms. **13.** 600 ohms.
14. (*a*) $\frac{1}{99}$ ohm in parallel, (*b*) 499 ohms in series.
15. (*a*) $4\frac{16}{21}$ ohms, (*b*) 0·00149 amp., (*c*) 0·000071 amp.
16. 99·9 ohms in series. **19.** 985 ohms in series.

CHAPTER VII (page 122)

11. 0·926 amp. **12.** 0·126 amp. Zero error.
13. 3·76 gm., 42 litres. **14.** 12 hr. 34 min.
15. 48·1 hr. **19.** +0·04 volt.

CHAPTER VIII (page 139)

9. 0·101 gm. **10.** 22·6 gm. **11.** 147,600 coulombs.
14. $17\frac{1}{2}$ ohms. **15.** $23\frac{1}{2}$ ohms. **16.** 100 hr.
18. 1·6 volts. **19.** 0·19 amp., 0·53 ohm.
20. 0·56 ohm. **21.** 0·47, 4·02 ohms.
22. (*a*) 2·59 ohms, (*b*) 0·59 ohm. **23.** (*a*) 1·5 amps., (*b*) 103·5 volts.
24. 1·1 volts, 2 ohms.

CHAPTER IX (page 156)

3. 5·0 microhm-cm.
4. 47·1 microhm-cm.
5. 11·4 microhm-cm.
7. 825 cm. **8.** 13·1 ohms. **9.** 0·816 : 1.
10. 0·413, 0·0459 ohm. **13.** 64·3 cm. from one end.
14. 71·4 cm. from one end.
16. 150 cm., 1 : 4. **17.** 2 volts.

19. (*a*) 2 ohms, (*b*) $\frac{1}{2}$ amp., (*c*) $\frac{1}{3}$ and $\frac{1}{6}$ amp., (*d*) $\frac{1}{3}$ volt, (*e*) zero, (*f*) no.
22. 4·25, 8·00 ohms. **23.** 1170° C.
24. 0·00875 volt per cm. **27.** 3·27 metres.
28. 0·4 amp. **30.** (*a*) 0·88 amp., (*b*) 37·9 cm.
31. (*a*) $\frac{1}{12}$ amp., (*b*) $\frac{5}{6}$ volt, (*c*) $41\frac{2}{3}$ cm. **32.** 92·3 cm.
33. 2·5 ohms. **34.** 0·206, 0·72 sq. cm.

CHAPTER X (page 176)

6. $201\frac{2}{3}$ ohms, 242 ohms, $109\frac{1}{11}$ watts. **9.** 2500 watts.
10. 55 ohms, 44 watts. **11.** 2·70 watts per c.p., £1. 5*s*. $2\frac{1}{2}$*d*.
12. 2860 cals., 7 min. 52 sec. **13.** 3·18 amps., 0·159 pence.
14. 71·4° C. **16.** (*a*) 857,000 cals., (*b*) 14·3 cals. per sec.
17. 120 ohms, 1800 joules, 240 ohms. **19.** $403\frac{1}{3}$ ohms.
20. 47·6, 119 cals. **21.** 8 amps., $31\frac{1}{4}$ ohms, 1*s*. 3*d*.
22. 23·5 yd. **23.** 34·4 ohms, 84 cals.
24. 252 ohms, 63 watts. **25.** (*a*) $3{\cdot}03 \times 10^6$ coulombs, (*b*) 1*s*. 5*d*.
26. 3*s*. 2*d*. **27.** 14 volts.

CHAPTER XII (page 223)

15. 78%. **16.** 0·51 H.P. **18.** 360 volts.
19. 99·5, 189·5 volts. **20.** 6·67 amps.
21. 96·5 volts. **22.** 4·5 amps.
23. 99 volts, 768 revs. per min. **24.** 840 revs. per min.

CHAPTER XIII (page 249)

2. $\frac{1}{8}$ ohm. **3.** 2·55 miles.
4. (*a*) 80 amps., (*b*) 96 kW., (*c*) 48,800 volts.
5. (*a*) 4 kW., (*b*) 0·16 kW.; 1*s*. $3\frac{1}{2}$*d*.
6. (*a*) 2222 volts, (*b*) 2·22 ohms.
8. 160,000 volts. **9.** 150 amps.

CHAPTER XV (page 296)

7. 2·2 oersted. **8.** 28 c.g.s. units. **9.** 20·25 c.g.s. units.
10. 4·5 cm. from weaker pole between the poles, 44·5 cm. outside the weaker pole. **11.** 49·7 c.g.s. units.
12. 0·14 dyne. **13.** 29° E. of N.
14. $62\frac{1}{2}$°. **16.** 0·15 oersted. **17.** 3·3 sec.
18. 0·51 oersted. **20.** 1 : 2·4. **21.** 34 c.g.s. units.
22. 12·6 cm., 10 cm. **23.** 381 c.g.s. units.
24. $\sqrt{8}$ and $\frac{1}{2}$ dyne parallel to first magnet.
25. 0·08 gauss in the direction *BD* produced.

26. 2·29 oersted at an angle of 29° 20′ with the line joining the nearest pole.

29. 27 dyne-cm.

30. $2\sqrt{2}$ sec.

31. 0·08 oersted, 19.

33. 1 : 1·56.

CHAPTER XVI (page 319)

15. 69°.

16. 0·29; 0·35, 0·62 oersted.

CHAPTER XVII (page 326)

1. 0·74 amp., 0·43.

2. 2, 3·46 amps.

6. 44°.

8. 0·2 gauss, N. 63° W.

10. 0·439.

11. 0·6 or 1·0 oersted.

INDEX

By courtesy of Messrs Siemens Bros. and Co. Ltd.

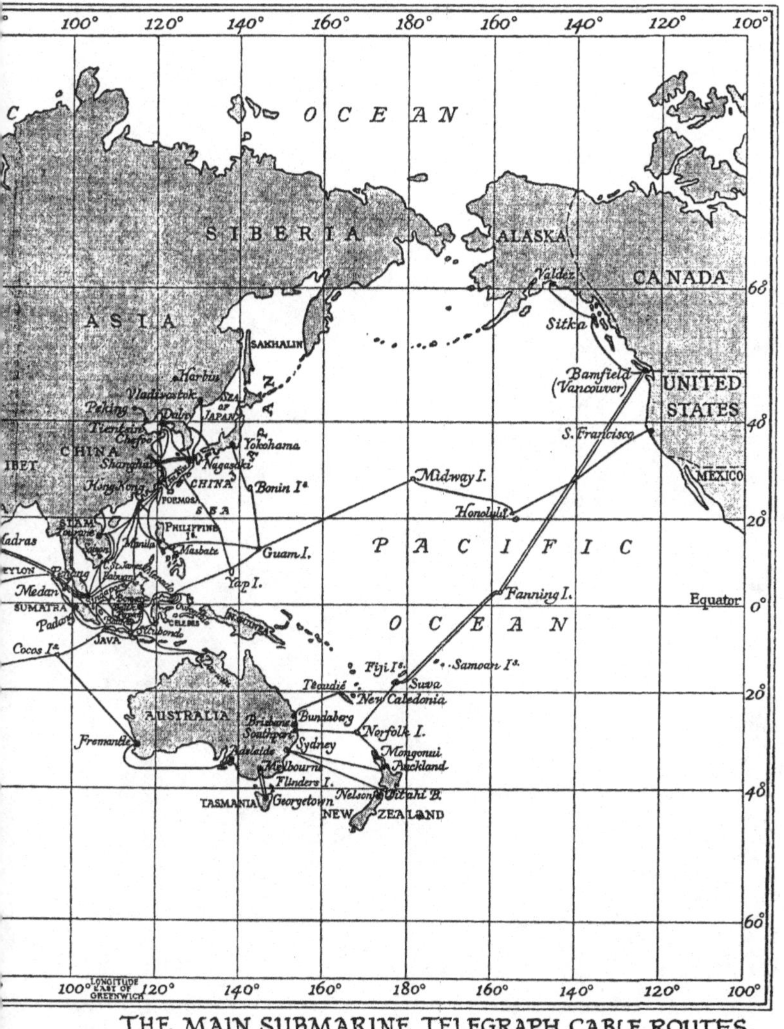

THE MAIN SUBMARINE TELEGRAPH CABLE ROUTES OF THE WORLD

Printed in the United States
By Bookmasters